MATHEMATICAL CONCEPTS IN CLINICAL SCIENCE

Steven Tiger, PA-C
Julienne K. Kirk, PharmD
Robert J. Solomon, MS, PA-C

Prentice Hall, Inc.
Upper Saddle River, NJ 07458

Library of Congress Cataloging-in-Publication Data

Tiger, Steven.
    Mathematical concepts in clinical science / Steven Tiger, Julienne K. Kirk, Robert J. Solomon.
        p. cm.
    Includes index.
    ISBN 0-13-011549-5
    1. Medicine—Mathematics.   2. Allied health personnel.   I. Kirk, Julienne K.
    II.   Solomon, Robert J. III. Title.

R853.M3 T54 1999
610′.1′51—dc21

99-049101

Publisher: *Julie Alexander*
Editor in Chief: *Cheryl Mehalik*
Acquisitions Editor: *Mark Cohen*
Marketing Manager: *Kristin Walton*
Production Manager: *Ilene Sanford*
Liaison: *Larry Hayden IV*
Managing Editor: *Patrick Walsh*
Full Service Production/Composition: *BookMasters, Inc.*
Printer/Binder: *Victor Graphics*
Cover Printer: *Victor Graphics*

© 2000 by Prentice-Hall, Inc.
Upper Saddle River, New Jersey 07458

Printed in the United States of America

10 9 8 7 6 5 4 3 2 1

ISBN 0-13-011549-5

Prentice-Hall International (UK) Limited, *London*
Prentice-Hall of Australia Pty. Limited, *Sydney*
Prentice-Hall Canada Inc., *Toronto*
Prentice-Hall Hispanoamericana, S. A., *Mexico*
Prentice-Hall of India Private Limited, *New Delhi*
Prentice-Hall Japan, Inc., *Tokyo*
Prentice-Hall (Singapore) Pte. Ltd.
Editora Prentice-Hall do Brasil, Ltda., *Rio de Janeiro*

Contents

Preface

Mathematics is an inevitable part of the study of clinical science, but most students will never have to deal with anything more complicated than basic algebra. More sophisticated mathematical methods may be required in certain advanced and specialized areas of study, but not in entry-level courses.

Yet some of the academically bright people who are preparing for careers in the clinical professions are phobic about mathematics. At the sight of numbers and formulas, they freeze up and recite the standard disclaimer, "But I'm *terrible* at math!"

Such fear is groundless and self-defeating. Being able to handle the mathematics of clinical science is crucial to mastering the basics—and students who have mastered the basics can learn more advanced material with greater ease. Certain mathematical concepts are routinely encountered in clinical science, and students cannot afford to be paralyzed by a needless fear of numbers.

We all have a natural curiosity about how things work. It is far more satisfying to understand why pressing the accelerator makes a car move than to regard the operation of a car as magic. We gain a sense of mastery when we finally grasp the workings of something that was initially mysterious. Students who can handle mathematics will gain that sense of mastery in the clinical sciences.

Let's take two examples: a formula for calculating the rate at which the blood is cleared of a substance called creatinine, and a graph showing the relationship between the amount of oxygen that is carried as a dissolved gas in the blood and the amount bound to hemoglobin in red blood cells. Some students tune right out when these concepts are explained, but others appreciate learning *why* the creatinine clearance formula works and *why* the oxygen-hemoglobin dissociation curve has its characteristic shape. Even students who don't find these and other mathematical applications intrinsically fascinating will feel more secure when they understand the concepts.

Mathematical Concepts in Clinical Science is a preparatory guide for students in medical and allied health programs, and applicants to those programs. It is a survey of clinical topics involving basic mathematical skills. The quiz at the end of each chapter is designed to test students' understanding of concepts, not their ability to memorize reams of information. The Appendix reviews basic mathematical tools and techniques, such as scientific notation, negative exponents, unit conversions, logarithms, and quadratic equations.

The goal is to give students—especially those with classic "math anxiety"—a running start in their coursework by helping them in understanding and handling the applied mathematics encountered in entry-level clinical science. To serve that goal, the underlying concepts are explained in some detail to reveal the logic behind the mathematics. At the same time, the discussion of basic science is kept on a practical level, which means that this book does *not* take the place of a comprehensive text in any aspect of clinical science.

We have tried to present this material in a way that is easy to read and perhaps even enjoyable. We take a spoonfeeding approach, assuming no prior knowledge except for the ability to handle elementary algebra and routine arithmetic operations with integers and fractions. Nevertheless, mastering these concepts still takes effort and concentration. Our hope is that students who put in the required effort will be rewarded with the sense of satisfaction and confidence that comes with clear understanding.

Steven Tiger, PA-C
Julienne K. Kirk, PharmD
Robert J. Solomon, MS, PA-C

Acknowledgments

As the primary author of *Mathematical Concepts in Clinical Science,* I want to express my appreciation for the help I received from several individuals.

Co-authors Julienne K. Kirk, Pharm D, and Robert J. Solomon, MS, PA-C, offered valuable suggestions at each stage of every chapter in the book. Dr. Kirk also contributed the chapter on pharmacokinetics (which I merely edited for stylistic consistency), and Mr. Solomon provided material on electrocardiography, for the chapter on circulatory function.

In addition, Charles Reed, MA, MPAS, PA-C, critiqued the chapter on diagnostic testing, and Robert Michielutte, PhD, patiently reviewed several drafts of the chapter on biostatistics. Others, too numerous to name individually, answered specific questions I had on a wide variety of topics.

With such assistance, I have tried to keep this book focused on its goal and accurate in its information. Any errors that may remain are solely my responsibility, and I invite readers to send comments and criticisms to the publisher.

I thank my wife Nancy and our son Joseph for their strong support and encouragement; and Mark Cohen, Senior Editor at Prentice Hall, for his patience, confidence, and enthusiasm throughout this project.

Finally, I want to acknowledge the students I have met and worked with over the years. Teaching is a two-way process. To teach is to learn—not only about the subject matter, but also about being an effective teacher. In an absolutely literal sense, this book represents what I have learned.

Steven Tiger, PA-C

Reviewers

Susan A. Allinger, J. D.
Pre-PA Student
Houston, TX

Ray Barlow II, CAN
BSN Student
University of Texas at Arlington
Arlington, TX

Nadja Vawryk Button, MS, EMT-P
Etters, PA

Deanna Currier
PA Student
Portland, OR

Karen S. Duncan
Medical Transcriptionist and
  Clinical Laboratory Science
  Student
Minot AFB, ND

Broheen Elias, PA-S
Yale University School of Medicine
Physician Associate Program
New Haven, CT

William Fenn, PA-C
Physician Assistant Department
Western Michigan University
Kalamazoo, MI

Scott French
PA Student
Nova Southeastern University
Davie, FL

Kristin Marie Groessl, MPH, PA-S
Physician Assistant Program
Western University of Health
  Sciences
Pomona, CA

Jennifer Huddleston
Physician Assistant Program
University of Southern California
  School of Family Medicine
Los Angeles, CA

Michael Olesen
Student of Epidemiology
University of Minnesota
Minneapolis, MN

Sheri Oswald
Physician Assistant Program
Wake Forest University School of
  Medicine
Winston-Salem, NC

Ross Querry, PhD
University of North Texas Health
  Science Center
Fort Worth, TX

Melissa Raue
Physician Associate Program
Yale University School of Medicine
New Haven, CT

Craig R. Sheerin, B.A.
Physician Associate Program
Yale School of Medicine
New Haven, CT

Sarah Shoaf, DDS, MEd, MS
Department of Dentistry
Winston-Salem University
Winston-Salem, NC

Tracy Thomason, PharmD
School of Pharmacy
University of North Carolina
Chapel Hill, NC

Nancy Warik
Physician Assistant Program
Cuyahoga Community College
Parma, OH

About the Authors

Steven Tiger, PA-C, is a professional medical writer and editor, formerly on the faculty of The Brooklyn Hospital Center Long Island University Physician Assistant Program, New York.

Julienne K. Kirk, PharmD, is Assistant Professor, Department of Family and Community Medicine, Wake Forest University Baptist Medical Center, Winston-Salem, North Carolina.

Robert J. Solomon, MS, PA-C, is Associate Professor and Academic Coordinator, Physician Assistant Program, The Community College of Baltimore County, Essex Campus, Baltimore, Maryland.

Six Practical Suggestions

To the student:

1. Depending on your field of clinical study, some of the topics covered in this book will be more relevant than others. Take what you want and leave the rest.

2. If certain mathematical operations are unfamiliar to you, consult the Appendix, which reviews basic mathematical tools and methods. Likewise, if certain complex units of measurement are unfamiliar, consult "Units Named for Scientists." In fact, it might be helpful to read through these sections *before* getting into the topic chapters.

3. In entry-level study, concentrate on understanding basic concepts and general rules. They do not always apply neatly in every clinical situation, but you should know the general rules before you study the variations and exceptions.

4. Follow each mathematical sequence in the text step by step; don't just look at the result. Likewise, try to work out the problems in each quiz; don't just look at the answers. If you can follow the math and do the problems, you understand the concepts. If you understand the concepts, you can handle the work.

5. Don't regard the numeric values presented in this book (for measurements, normal ranges, and so on) as universal standards. Other sources may cite slightly different values. Focus on grasping concepts, not on memorizing numbers.

6. Don't worry about not being a whiz at math—you don't have to be.

This symbol alerts readers to note the distinction between potentially confusing concepts.

chapter 1

Clinical Metrics

Chapter Outline

Metric Terms
Length
Volume
Weight
Temperature and Heat
SI Units

Mathematical Tools Used in This Chapter

Significant digits (Appendix A)
Scientific notation (Appendix B)
Negative exponents (Appendix C)
Dimensional units with negative exponents (Appendix D)
Unit conversions (Appendix E)

The metric system of weights and measures is based on factors of 10, which makes multiplication and division of metric units a simple matter of moving the decimal point some number of places to the right or left. The metric system is logical, consistent, and highly integrated; it is the basis of measurement in most parts of the world and in all fields of science. Certain metric units are used with great frequency in the clinical sciences.

METRIC TERMS

The *meter* is the basic metric unit of length; the *liter* is the unit of volume; the *gram* is the unit of weight; and the *centigrade degree* is the unit of temperature. These basic units may appear alone or with a prefix that defines the decimal factor (10, 100, 1000, etc.) by which the base-unit value is to be multiplied or divided.

Prefixes in the Metric System

Prefix	Symbol	Multiple of Base Unit
Tera-	T	one trillion ($10^{12} = 1{,}000{,}000{,}000{,}000$)
Giga-	G	one billion ($10^9 = 1{,}000{,}000{,}000$)
Mega-	M	one million ($10^6 = 1{,}000{,}000$)
Kilo-	k	one thousand (10^3)
Hecto-	h	one hundred (10^2)
Deka-	da	ten (10^1)

Prefix	Symbol	Fraction of Base Unit
Deci-	d	one-tenth ($10^{-1} = 0.1$)
Centi-	c	one-hundredth ($10^{-2} = 0.01$)
Milli-	m	one-thousandth ($10^{-3} = 0.001$)
Micro-*	μ, mc	one-millionth (10^{-6})
Nano-	n	one-billionth (10^{-9})
Pico-	p	one-trillionth (10^{-12})
Femto-	f	one-quadrillionth (10^{-15})

*In clinical usage, the prefix *micro-* is usually represented by the Greek letter mu (μ). Sometimes the abbreviation *mc-* is used; for example, mcg means micrograms.

Note how we form metric terms. For weight, the base unit is the gram (abbreviated g, not gm). And since the prefix *kilo-* means 1000 times the base unit, 1 kilogram (kg) equals 1000 grams:

$$1 \text{ kg} = 1000 \text{ g}$$

For volume, the base unit is the liter (l or L—the latter will be used in this book). And since *deci-* means $\frac{1}{10}$ of the base unit, 1 deciliter (dL) equals $\frac{1}{10}$ of a liter:

$$1 \text{ dL} = 0.1 \text{ L}$$

Theoretically, any prefix can be applied to any of the basic units of metric measurement. However, the prefixes designating the very largest and smallest multiples and fractions are less commonly seen, and certain theoretically possible constructions (such as "dekagram" and "centiliter") are not used at all.

Now let's look at each type of measurement—length, volume, weight, and temperature. We'll see which metric units are used for clinical purposes,

About Abbreviations

Do not write abbreviations in the plural: five grams is written 5 g, not 5 g's. Also, do not use periods in abbreviations.

their approximate equivalents in nonmetric units, and how to convert back and forth between metric and nonmetric units (a skill that becomes both automatic and irrelevant as one starts *thinking* in metric terms). Most important, we'll see how all of these metric units are interrelated and actually defined in terms of each other.

LENGTH

The basic metric unit of length, the meter, was originally defined as one-tenth of a millionth (10^{-7}) of the distance from the equator to either pole of the Earth along a meridian. Today, scientists use a more precise and reproducible standard based on the wavelength of certain types of light. For practical purposes, 1 meter (abbreviated m) is equal to 39.37 inches.

Since the prefix *centi-* means $\frac{1}{100}$ of the base unit, 1 centimeter (cm) is $\frac{1}{100}$ (0.01 or 10^{-2}) of 39.37 inches:

$$1 \text{ cm} = 0.01 \cdot 39.37 \text{ inches} = 0.3937 \text{ inch}$$

Therefore, taking the reciprocal of 0.3937 (that is, 1/0.3937):

$$1 \text{ inch} = 2.54 \text{ cm}$$

Height is recorded in meters or centimeters. Suppose a patient stands 5 feet 7 inches tall (67 inches). Since 1 inch equals 2.54 cm, the height in metric units is:

$$67 \text{ inch} \cdot \frac{2.54 \text{ cm}}{\text{inch}} \approx 170 \text{ cm} = 1.70 \text{ m}$$

Here we see the use of a *converting factor*—a ratio equal to 1, so that it can be used in either of its reciprocal forms, as needed. In this case, converting inches to centimeters, the converting factor is 2.54 cm · inch^{-1}. In the other direction, centimeters to inches, the converting factor would be inch · (2.54 cm)$^{-1}$.

Metric length units are used in recording findings on physical examination. For example, a clinician may note that the edge of the liver can be felt 2 cm below the right costal margin (Figure 1.1).

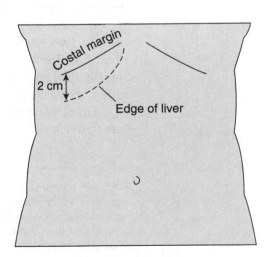

Figure 1.1 Physical examination measurement.

It is important to record all positive findings on physical examination, such as a surgical scar, 6 cm long, found 4 cm below left knee. Or a macule (a skin lesion that is flat but discernible by pigmentation) measuring 1.5 cm across. Or pupillary dilation to 8 millimeters (mm); since *milli-* means $\frac{1}{1000}$ of the base unit, $1 \text{ mm} = \frac{1}{1000}$ of a meter $= 0.001 \text{ m} = 0.1 \text{ cm}$ (Figure 1.2). Pathologists report the size of a tissue specimen in centimeters (Figure 1.3).

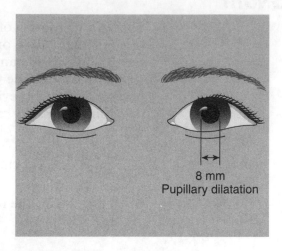

Figure 1.2 Physical examination measurement.

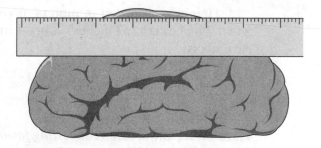

Figure 1.3 Pathology measurement.

Another common use of the length measure is *body surface area* (BSA). It is often important to know how "big" a patient is in order to calculate drug doses (especially for the highly toxic agents used in cancer chemotherapy). Height or weight alone is not adequate. Two men may both be 6 feet tall, but the one who weighs 150 pounds is smaller than the one who weighs 180 pounds. Two women may both weigh 120 pounds, but the one who stands 5 feet 5 inches is bigger than the one who stands 5 feet 2 inches. BSA combines height and weight into a single measurement that is a good guide to a patient's overall size.

To measure BSA, we use a *nomogram* (a chart that uses easily obtained data to derive information that would otherwise be difficult to calculate).[1] On

---

[1]BSA can be calculated as $\text{cm}^{0.725} \cdot \text{kg}^{0.435} \cdot 0.007184$. Using this formula requires a scientific calculator with a y^x function to deal with height and weight raised to decimal exponents. Using a nomogram is obviously much easier. For certain clinical applications—notably, the assessment of burn injuries—what counts is the *percentage* of BSA involved, rather than the actual measurement.

the BSA nomogram, mark the patient's height and weight on their respective scales and draw a straight line between the marks; the line intersects a third scale, which gives the corresponding BSA (Figure 1.4).

For an average adult, BSA is about 1.73 m². Note that BSA is usually measured in square meters rather than square centimeters. The nomogram gives only an approximation based on statistical averages, not a precise

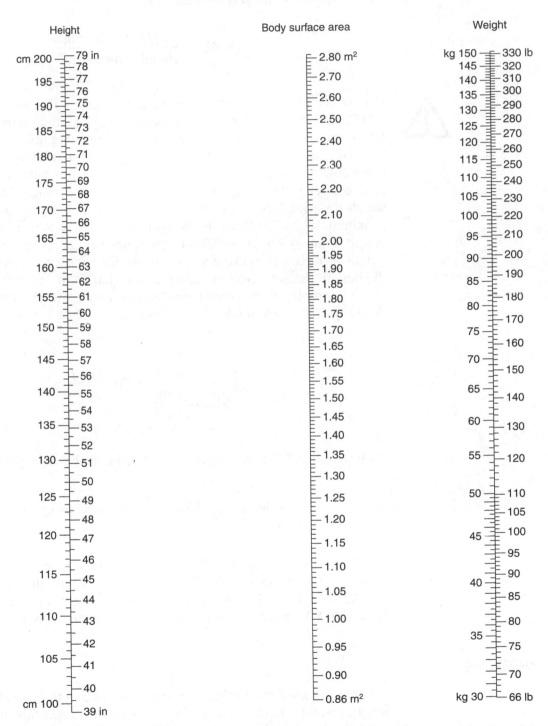

Figure 1.4 Nomogram for determination of body surface area from height and weight. Copyright 1970. Novartis. Reprinted with permission from the *Geigy Scientific Tables, 7th ed.,* edited by C. Lentner, M.D. All rights reserved.

measurement. Thus, with three significant digits, 1.73 m² reflects the limits of certainty, whereas 1.73 · (100 cm)² = 17,300 cm² would falsely suggest that BSA is precisely that value, to the nearest square centimeter.

Another useful measure of body size is *body mass index* (BMI). As with BSA, there are nomograms that provide the BMI for any given height and weight. However, BMI is easily calculated as the weight-in-kilograms divided by the square of the height-in-meters:

$$\text{BMI} = \frac{\text{weight-in-kilograms}}{(\text{height-in-meters})^2}$$

 Do not confuse the square of the height-in-meters (the denominator in the BMI formula) with BSA, which is reported in squared meters. They are different measures that both happen to utilize the same unit.

BMI can also be computed from weight-in-pounds and height-in-inches, using conversion factors for weight and height. For weight, 2.2 lb ≈ 1 kg (discussed later in this chapter); therefore, to get the weight-in-kilograms, the weight-in-pounds is divided by 2.2 or multiplied by the reciprocal of 2.2 (approximately 0.455). Thus, the weight-conversion factor is 0.455 kg/lb. For height, 1 inch = 2.54 cm = 0.0254 m; therefore, to get the height-in-meters, the height-in-inches is multiplied by 0.0254. But since we need the square of the height, the height-conversion factor is (0.0254 m/in)² = 0.000645 m²/in².

Combining these conversion factors for weight (the numerator of the BMI ratio) and height (the denominator):

$$\frac{0.455 \, \dfrac{\text{kg}}{\text{lb}}}{0.000645 \, \dfrac{\text{m}^2}{\text{in}^2}} \approx 705 \, \frac{\text{kg} \cdot \text{in}^2}{\text{lb} \cdot \text{m}^2}$$

Multiplying by 705 converts pounds and inches to the appropriate metric units:

$$\text{Value as } \frac{\text{lb}}{\text{in}^2} \cdot 705 \, \frac{\text{kg} \cdot \text{in}^2}{\text{lb} \cdot \text{m}^2} = \text{value as kg} \cdot \text{m}^{-2}$$

While BMI is calculated as kg/m², it is often reported as a pure number, with no dimensional units. For men and women, the normal range is 18.5–25.0; higher values suggest obesity; lower values suggest malnutrition. However, BMI must be interpreted on an individual basis.

VOLUME

Volume is a measure of three-dimensional space—the product obtained as length · width · depth or height. (*Width, depth,* and *height* are merely alternate terms for *length* in the two other mutually perpendicular spatial orientations.)

Picture a cube whose length along each edge is 10 cm (not quite 4 inches). The volume of this cube is:

How High Is BMI?

Two patients are to be evaluated. The first patient is a 60-year-old diabetic female who stands 5 feet 2 inches (about 157 cm) and weighs 170 pounds (about 77 kg). For the clinical assessment to be recorded in her chart, an evaluation more specific than "Patient is obese" would include calculation of her body mass index (BMI):

$$BMI = \frac{170 \text{ lb}}{(62 \text{ in})^2} \cdot 705 \approx \frac{170}{3844} \cdot 705 \approx 31.2$$

The second patient is a 26-year-old male body builder, who is being seen for a routine physical examination. He stands 185 cm = 1.85 m (about 73 inches) and weighs 107 kg (about 236 pounds). To calculate his BMI:

$$BMI = \frac{107 \text{ kg}}{(1.85 \text{ m})^2} \approx \frac{107}{3.42} \approx 31.3$$

BMI is almost the same in these two patients. Yet the first patient is obese while the second has a very low percentage of body fat. The high BMI in the body builder reflects his extreme muscular development (lean muscle tissue is relatively heavy because it has a far greater water content than adipose tissue). Otherwise, his weight would be excessive.

Lesson: Look at the patient as a whole person, not just as a set of data.

Metric Conversions: Length and Area

Use these approximations for conversion back and forth between metric and nonmetric units of length and area (the precise conversions are based on 1 inch = 2.54 centimeters):

$$1 \text{ cm} \approx 0.4 \text{ inch}$$

$$1 \text{ m} \approx 39 \text{ inches or } 3.3 \text{ ft}$$

$$1 \text{ ft} \approx 30 \text{ cm or } 0.3 \text{ m}$$

$$1 \text{ m}^2 \approx 1550 \text{ inch}^2 \text{ or } 10.8 \text{ ft}^2$$

$$1 \text{ ft}^2 \approx 930 \text{ cm}^2 \text{ or } 0.1 \text{ m}^2$$

$$10 \text{ cm} \cdot 10 \text{ cm} \cdot 10 \text{ cm} = 1000 \text{ cm}^3$$

This volume, 1000 cm³, is the basic metric unit of fluid measurement—the liter. Since 1 L = 1000 cm³ and 1 milliliter (mL) is 0.001 L, 1 mL is the volume of one of those thousand "cubic centimeters"; that is, 1 mL = 1 cm³ (Figure 1.5).

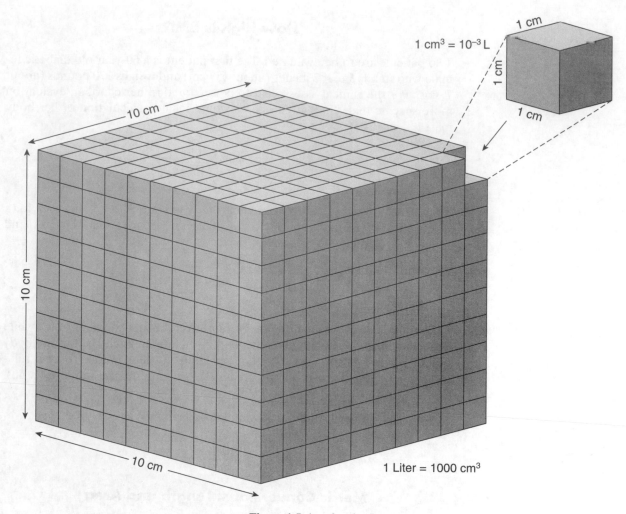

Figure 1.5 1 cm³ = 1 mL.

This unit of volume, cm³, is frequently encountered in clinical science and practice. In speaking, clinicians often use the phrase *cubic centimeters* or the initials *cc,* as in, "Get me fifty cc's of sodium bicarb, stat!" However, in writing, the proper term is cm³ or mL.

The liter itself is commonly used for measuring larger volumes of fluid. For example, the total amount of body water in an average adult is about 40 liters, and there may be 5 to 6 liters of blood in circulation. Also, the concentration of certain substances in the blood may be measured as the amount per liter (see Chapter 2).

There are no metric multiples of the liter in clinical science, but several metric fractions, down to the femtoliter, are used. The deciliter, by definition, equals one-tenth of a liter (0.1 L). And since 1 L = 1000 mL, 1 dL = 100 mL. The deciliter occupies a special place in medicine, as part of the expression for the concentration of a solution.

The milliliter is, as we saw, identical in volume to the cubic centimeter. Fluids added to or lost from the body can be measured in milliliters; and, like the deciliter and the liter, it is sometimes used as the unit of volume for expressing the concentration of a solution.

The microliter (μL) is a small cube measuring just 1 mm along each edge (Figure 1.6). That is, 1 μL = 1 mm³, just as 1 mL = 1 cm³. Is it clear why this is so? There are 10 millimeters to the centimeter. Therefore:

$$1 \text{ cm}^3 = 10 \text{ mm} \cdot 10 \text{ mm} \cdot 10 \text{ mm} = 1000 \text{ mm}^3$$

$$\frac{1000 \text{ mm}^3}{\text{cm}^3} \cdot \frac{1000 \text{ cm}^3}{\text{L}} = \frac{1{,}000{,}000 \text{ mm}^3}{\text{L}}$$

Since there are a million (10^6) cubic millimeters in a liter, 1 cubic millimeter represents one-millionth of a liter, or 1 microliter:

$$1 \text{ L} = 10^6 \text{ mm}^3$$

$$1 \text{ mm}^3 = 10^{-6} \text{ L} = 1 \, \mu\text{L}$$

The microliter is often used as the unit of volume in a complete blood count, a laboratory test that gives a tally of the cellular elements in the blood (as well as the concentration of hemoglobin in red blood cells). Here are some typical cell counts: erythrocytes (red cells), $5 \cdot 10^6/\mu$L; leukocytes (white cells), $6 \cdot 10^3/\mu$L; thrombocytes (platelets), $2 \cdot 10^5/\mu$L.

The femtoliter (fL) is the smallest unit of volume commonly used in the clinical sciences. Its cubic-length equivalent is the cubic micrometer (μm³), a microscopically tiny cube measuring 1 μm along each side (Figure 1.7).

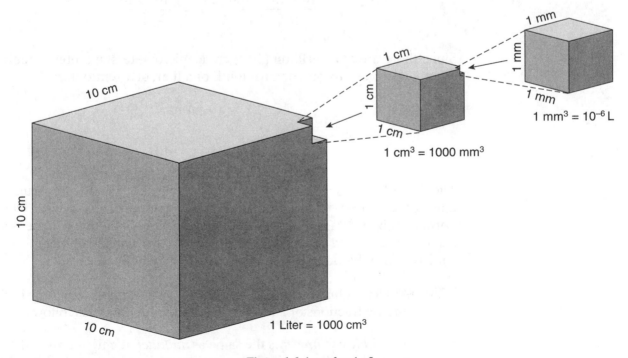

Figure 1.6 1 mm³ = 1 μL.

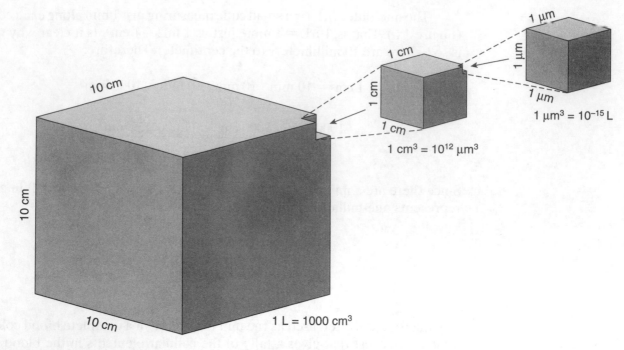

$1 \mu m^3 = 10^{-15} L$

$1 cm^3 = 10^{12} \mu m^3$

$1 L = 1000 cm^3$

Figure 1.7 $1 \mu m^3 = 1$ fL.

A centimeter is, by definition, 10^{-2} meter, and a micrometer is 10^{-6} meter; so there are 10^4 (that is, 10,000) micrometers to the centimeter. Therefore:

$$1 cm^3 = 10^4 \mu m \cdot 10^4 \mu m \cdot 10^4 \mu m = 10^{12} \mu m^3$$

$$\frac{10^{12} \mu m^3}{cm^3} \cdot \frac{1000 cm^3}{L} = \frac{10^{15} \mu m^3}{L}$$

Since there are a quadrillion (10^{15}) cubic micrometers in a liter, 1 cubic micrometer represents one-quadrillionth of a liter, or 1 femtoliter:

$$1 L = 10^{15} \mu m^3$$

$$1 \mu m^3 = 10^{-15} L = 1 fL$$

The femtoliter is sometimes used in reporting the "mean corpuscular volume" of red blood cells—that is, the average volume of a blood cell, which is normally about 90 fL. (To get some sense of how small this volume is, consider that red blood cells occupy less than half the volume of blood, and just 1 microliter of blood normally contains a few million red cells.)

REMEMBER: Volume may be reported in units of cubic-length or in liters or metric fractions of a liter. Equivalent volumes are as follows:

- *Cubic centimeter* is the same as *milliliter* ($1 cm^3 = 1 mL = 10^{-3} L$).
- *Cubic millimeter* is the same as *microliter* ($1 mm^3 = 1 \mu L = 10^{-6} L$).
- *Cubic micrometer* is the same as *femtoliter* ($1 \mu m^3 = 1 fL = 10^{-15} L$).

Metric Conversions: Volume

Approximate conversions between metric and nonmetric measures of fluid volume, based on 0.946 liter per quart, are as follows:

$$1 \text{ mL} \quad \approx 0.2 \text{ tsp or about } 12\text{–}15 \text{ drops*}$$

$$1 \text{ dL} \quad \approx 3.4 \text{ fl oz or } 0.1 \text{ quart}$$

$$1 \text{ L} \quad \approx 34 \text{ fl oz or } 1.06 \text{ quarts}$$

$$1 \text{ tsp} \quad \approx 5 \text{ mL}$$

$$1 \text{ tbsp} \approx 15 \text{ mL}$$

$$1 \text{ fl oz} \approx 30 \text{ mL}$$

$$1 \text{ quart} \approx 946 \text{ mL or } 9.5 \text{ dL or } 0.95 \text{ L}$$

*Drops are sometimes called *guttae* (abbreviated gtt), and a drop of water was a unit called a *minim* in the old apothecary system.

WEIGHT

The metric unit of weight or mass, the gram, is defined by the unit of fluid volume, the milliliter. By definition, 1 gram is the weight of 1 milliliter of pure water.[2] However, saying "One gram equals one cubic centimeter" is sloppy language, because it equates different types of measurement (weight and volume). Also, that relationship applies *only* to pure water—a milliliter of water containing a dissolved substance, or a milliliter of a different fluid, will weigh very slightly more or less than 1 gram.

Since 1 gram is the weight of 1 milliliter of pure water, 1 kilogram (that is, 1000 grams) is the weight of 1000 milliliters or 1 liter. The kilogram is used to measure body weight during physical examination. Another use is to determine the exact total daily dose of a drug that must be given in accordance with the patient's weight, again measured in kilograms. Dose-by-weight is the rule in pediatric patients; typically, it would be a certain number of milligrams (mg = 0.001 g) of drug per kilogram of body weight, per day:

$$\text{mg} \cdot \text{kg}^{-1} \cdot \text{day}^{-1}$$

In the clinical world, the kilogram is the only multiple of the gram in common use. However, every fraction, from the milligram down to the picogram,

---

[2]*Weight* reflects the gravitational acceleration on a given *mass*. However, in a constant gravitational field, weight is a measure of mass, and the terms are often used interchangeably. Thus, the gram, a physical unit of mass, is also understood as a unit of weight.

Metric Conversions: Weight

To convert between avoirdupois units and metric units, 1 pound (lb) = 0.45359237 kilogram; reciprocally, 1 kg ≈ 2.2 lb. Metric weights below a gram are too small to measure in avoirdupois units, so the only relevant relationships are as follows:

$$1 \text{ g} \approx 0.035 \text{ oz}$$

$$1 \text{ kg} \approx 35 \text{ oz or } 2.2 \text{ lb}$$

$$1 \text{ oz} \approx 28 \text{ g or } 0.03 \text{ kg}$$

$$1 \text{ lb} \approx 454 \text{ g or } 0.45 \text{ kg}$$

The old apothecary system used "troy units" (a system of weights for precious metals) plus a few of its own, just to make everyone's life miserable. A few of the old units may still be occasionally encountered:

$$1 \text{ grain} \approx 65 \text{ mg}$$

$$1 \text{ dram} \approx 60 \text{ grains or } 3.9 \text{ g or } \tfrac{1}{8} \text{ oz}$$

For example, a standard 325-mg tablet of aspirin is still sometimes labeled as "5 grains" (5 grains · 65 mg/grain = 325 mg).

is used in measuring the concentration of various substances in body fluids in terms of weight of substance per unit-volume of fluid: μg (microgram) = 10^{-6} g; ng (nanogram) = 10^{-9} g; pg (picogram) = 10^{-12} g.

Often, the lower the weight concentration, the smaller the weight unit used. Thus, plasma proteins, which are large, heavy molecules, are measured in grams per deciliter (g/dL); by comparison, glucose is measured in milligrams per deciliter (mg/dL).

TEMPERATURE AND HEAT

A centigrade scale is one that divides the temperature range between the freezing point and the boiling point of water—that is, the liquid phase of water—into 100 equal parts or *degrees.* On the *Celsius scale,* the freezing point of water is 0°C and the boiling point is 100°C. Normal body temperature on this scale is 37°C.

The *Fahrenheit scale* is not centigrade.[3] On this scale, the liquid-water temperature range is divided into 180 parts instead of 100; the freezing point

---

[3]These temperature scales are named for Swedish astronomer Anders Celsius (1701–1744) and German physicist Gabriel D. Fahrenheit (1686–1736). In temperature readings, C stands for Celsius, not centigrade. However, a Celsius degree may be described as a "centigrade degree" in terms of its *magnitude* ($\frac{1}{100}$ of the liquid-water temperature range). The Kelvin scale is the *absolute* centigrade scale, which means that absolute zero is set at 0 K. The kelvin unit of temperature is equal in magnitude to a centigrade degree, but kelvin temperatures are numerically 273.15 higher than the corresponding Celsius temperatures.

of water is 32°F and the boiling point is 212°F. Normal body temperature is 98.6°F.

The same temperature range is represented by 180 Fahrenheit degrees and 100 Celsius degrees; therefore, a degree on the Celsius scale is almost twice as large as a degree on the Fahrenheit scale:

$$1 \text{ Celsius degree} = \frac{9}{5} \text{ Fahrenheit degrees}$$

$$1 \text{ Fahrenheit degree} = \frac{5}{9} \text{ Celsius degree}$$

Also, the freezing point for water is set 32 degrees higher on the Fahrenheit scale than on the Celsius scale (Figure 1.8).

Conversions between Fahrenheit and Celsius temperatures must be made with some accuracy for clinical purposes. If we know the Fahrenheit temperature and we want to convert it to a Celsius reading, we must first subtract 32 degrees from the Fahrenheit temperature and then multiply the result by $\frac{5}{9}$:

$$C = (F - 32) \cdot \frac{5}{9}$$

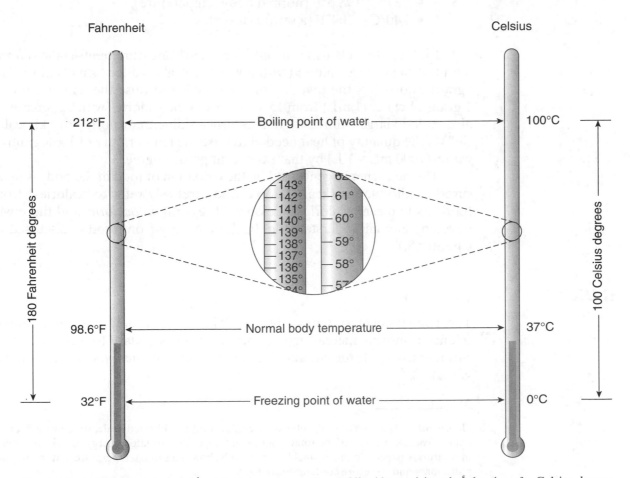

Figure 1.8 A Fahrenheit degree ($\frac{1}{180}$ of the temperature range of liquid water) is only $\frac{5}{9}$ the size of a Celsius degree ($\frac{1}{100}$ of that same range).

For example, to convert a normal body temperature of 98.6 on the Fahrenheit scale to the equivalent temperature on the Celsius scale:

$$(98.6 - 32) \cdot \frac{5}{9} = 66.6 \cdot \frac{5}{9} = 37°C$$

Going the other way—converting from Celsius to Fahrenheit—first multiply the Celsius temperature by $\frac{9}{5}$ and then add 32:

$$F = \left(\frac{9}{5} \cdot C\right) + 32$$

To convert a feverish temperature of 40 on the Celsius scale to the equivalent temperature on the Fahrenheit scale:

$$\left(\frac{9}{5} \cdot 40\right) + 32 = 72 + 32 = 104°F$$

REMEMBER: Two useful equivalent temperatures are as follows:

- 37°C = 98.6°F (normal body temperature).
- 40°C = 104°F (a serious fever).

Linking the Celsius scale and other metric measurements is the *calorie,* a unit of heat.[4] One calorie (with lowercase c; also called "small calorie" or "gram-calorie") is the quantity of heat needed to raise the temperature of 1 gram of water (1 mL) from 15°C to 16°C. One Calorie (with uppercase C; also called "large calorie," "great calorie," "kilogram-calorie" or "kilocalorie") is the quantity of heat needed to raise the temperature of 1 kilogram of water (1000 mL = 1 L) by that same centigrade degree.

The heat-energy generated by the oxidation of food in the body is measured in kilocalories (kcal). People who carelessly refer to "calories" from food really mean "Calories." Lesson: Always say *kilocalorie,* and there will never be any misunderstanding. (Caloric energy from food is discussed in Chapter 8.)

SI UNITS

The conventional system of metric units is still widely used, but clinical science is moving increasingly to the worldwide standard for scientific measurements—the International System (*Système International d'Units,* abbreviated SI).[5]

---

[4]*Temperature* is a measure of hotness or coldness on a calibrated scale, such as the Celsius scale. *Heat* is a measure of the total quantity of temperature-related energy in a given amount of matter. A paper clip and a steel bar may both show the same temperature, but the bar has more mass and therefore contains more heat.

 [5]Do not confuse the units in the International System with certain laboratory tests that are reported in "Units" or "International Units"; see Chapter 9.

Shortcut Temperature Conversions

In converting a patient's temperature from Fahrenheit to Celsius or vice versa, calculation by formula yields an exact result. As a quick alternative, the following shortcuts can be done mentally and yield reasonably accurate results.

Quick Conversion from Fahrenheit to Celsius:

- *If the Fahrenheit temperature is closer to 98.6 than to 104:* For each degree above or below 98.6, adjust the corresponding Celsius temperature (37) up or down by one-half degree.
- *If the Fahrenheit temperature is closer to 104 than to 98.6:* For each degree above or below 104, adjust the corresponding Celsius temperature (40) up or down by one-half degree.

For example, a patient's temperature is 102°F. This is *2 below 104,* so the adjustment is *down 1 from 40:*

$$40 - 1 = 39°C \text{ (compared to 38.9°C by formula)}$$

Another patient's temperature is 100.8°F. This is *2.2 above 98.6,* so the adjustment is *up 1.1 from 37:*

$$37 + 1.1 = 38.1°C \text{ (38.2°C by formula)}$$

Quick Conversion from Celsius to Fahrenhcit:

- *If the Celsius temperature is closer to 37 than to 40:* For each degree above or below 37, adjust the corresponding Fahrenheit temperature (98.6) up or down by two degrees.
- *If the Celsius temperature is closer to 40 than to 37:* For each degree above or below 40, adjust the corresponding Fahrenheit temperature (104) up or down by two degrees.

A shivering patient has a temperature of 35.8°C. This is *1.2 below 37,* so the adjustment is *down 2.4 from 98.6:*

$$98.6 - 2.4 = 96.2°F \text{ (96.4°F by formula)}$$

Another shivering patient's temperature is 40.8°C. This is *0.8 above 40,* so the adjustment is *up 1.6 from 104:*

$$104 + 1.6 = 105.6°F \text{ (105.4°F by formula)}$$

The small differences between the exact results by formula and the approximations by these shortcut methods are clinically insignificant in most circumstances. Temperature, blood pressure, heart rate, and respiratory rate—the "vital signs"—are subject to considerable variability owing to differences in clinicians' techniques in taking the readings, the equipment being used, and even the patient's emotional state.

Suppose we have a solution at a concentration of 35 μg/mL. We could vary the volume unit and express the same concentration as 3500 μg/dL or 35,000 μg/L. Likewise, if the concentration of another solution is 7 mg/dL, that information could also be expressed as 70 mg/L or 0.07 mg/mL.

To compare the concentrations of these two solutions from the original numbers, 35 μg/mL versus 7 mg/dL, we must convert one expression to the same units used in the other. Converting the second to the units of the first:

$$7\,\frac{mg}{dL} = 7000\,\frac{\mu g}{dL} = 70\,\mu g \cdot mL^{-1}$$

Other SI Units

The most frequent clinical use of SI units is in reporting the concentration of various substances in body fluids. However, other applications may also be encountered.

Heat and Energy

The SI unit for heat and energy is the *joule* (J). The conventional unit is the calorie (cal) or kilocalorie (kcal). Conversions are as follows:

$$1\,J \approx 0.24\,cal$$

$$1\,kilojoule\,(kJ) \approx 240\,cal = 0.24\,kcal$$

$$1\,cal \approx 4.2\,J$$

$$1\,kcal \approx 4200\,J = 4.2\,kJ$$

The joule is also a unit of electrical energy: 1 joule = 1 watt-second. Defibrillation devices deliver timed countershocks measured in joules.

Pressure

The SI unit for pressure is the *pascal* (Pa). The conventional unit is millimeters of mercury (mm Hg). Conversions are as follows:

$$1\,Pa^* = 0.0075\,mm\,Hg$$

$$1\,kilopascal\,(kPa^{**}) = 7.5\,mm\,Hg$$

$$1\,mm\,Hg \approx 133.3\,Pa = 0.1333\,kPa$$

*Not to be confused with P_a (P with subscript a), which, coincidentally, also relates to pressure. P_aO_2 and P_aCO_2 are abbreviations for the partial pressures of dissolved oxygen and dissolved carbon dioxide in arterial blood (see Chapter 4). An oxygen partial pressure of 90 mm Hg would be converted to SI units as follows:

$$P_aO_2 \approx 90\,mm\,Hg \cdot \frac{133.3\,Pa}{mm\,Hg} \approx 12,000\,Pa = 12\,kPa$$

**Not to be confused with pK_a, the negative logarithm of the dissociation constant for a weak acid (see Chapters 6 and 7).

So in terms of weight of dissolved substance, the concentration of the second solution is just twice that of the first—but this fact was not immediately obvious.

In different textbooks, journals, and laboratory reports, the same measurements may be reported in different units, which means that the numeric values may also be different. For example, the normal serum concentration of adrenocorticotropic hormone may be given as 15–70 pg/mL or as 1.5–7.0 ng/dL. The normal range for Vitamin A may be given as 0.15–0.6 μg/mL or as 150–600 ng/mL or as 15–60 μg/dL.

Such variability in the selection of units seems arbitrary and creates the potential for confusion and error. It would be simpler and safer if the same volume unit were used in all expressions of concentration—and in SI units, concentration is *always* measured as the amount per *liter*. Then all we have to look at is the amount itself, using any convenient metric units.

For example, suppose we compare two solutions whose concentrations are 3 ng/L and 600 pg/L. It is easy to see that 3 ng = 3000 pg, which is five times greater than 600 pg.

We can measure the concentration of a solution as either the *weight* of dissolved substance or the *number of particles* (atoms, ions, or molecules) of dissolved substance, per unit-volume of solution. In Chapter 2, we see why the number of particles is especially important, and how weight and particle count are related. In medicine, SI units are typically used to measure the *number of particles* of various dissolved substances *per liter* of fluid.

QUIZ

1. One patient stands 5 ft 7 in and weighs 190 lb. Convert these measurements into metric units. Another patient stands 1.8 m and weighs 73 kg. Convert to nonmetric units.

2. For each of the two patients in Question 1, determine body surface area (BSA) using the nomogram in Figure 1.4, and body mass index (BMI) using the formula for the metric height and weight, and then the formula for nonmetric units.

3. A man stands 5 ft 6 in; his BSA is 1.84 m². Use the nomogram in Figure 1.4 to estimate his weight, and then use that information to calculate his BMI by formula.

4. One patient's temperature is 38.6°C. Convert to Fahrenheit. Another patient's temperature is 103.2°F. Convert to Celsius. (Do these conversions both mentally and by formula.)

5. Convert: 36 mg = ____ g = ____ μg = ____ ng = ____ pg.

6. Convert: 64 mL = ____ L = ____ dL = ____ μL = ____ fL.

7. Convert: 48 mg/dL = ____ mg/L = ____ mg/mL =
   __ μg/L = __ μg/dL = __ μg/mL = __ μg/μL =
   __ ng/L = __ ng/dL = __ ng/mL = __ ng/μL =
   __ pg/L = __ pg/dL = __ pg/mL = __ pg/μL.

8. Solution A contains 35 mg of a substance dissolved in 80 mL of water. Solution B contains 0.84 g of the same substance dissolved in 2.1 L of water. Which solution is more concentrated?

9. A patient's hematology report shows 4,800,000 red blood cells per microliter ($4.8 \cdot 10^6/\mu L$). If the total blood volume is 5.2 liters, how many red cells are in the entire circulatory system?

10. The recommended dosage of a drug is $8.5 \text{ mg} \cdot \text{kg}^{-1} \cdot \text{day}^{-1}$. If this daily total is to be given in four equally divided doses, how large is each dose for a child who weighs 68 lb?

Answers and explanations at end of book.

Solutions and Body Fluids

Chapter Outline

Mathematical Tools Used in This Chapter

In general terms, a certain amount of *solute* dissolved in a certain amount of *solvent* forms a *solution* at a certain concentration. That concentration may be expressed as either the weight of solute per unit-volume of solution, or the number of particles (atoms, ions, or molecules) of solute per unit-volume of solution. To illustrate the difference: take 20 tablets, each weighing a half-gram, and distribute them evenly into five pillboxes. By weight, the "concentration per pillbox" is:

$$20 \text{ tablets} \cdot \frac{0.5 \text{ gram}}{\text{tablet}} = 10 \text{ grams}$$

$$\frac{10 \text{ grams}}{5 \text{ pillboxes}} = 2 \frac{\text{grams}}{\text{pillbox}}$$

By particle count (in this case, the particles are the tablets themselves), it is simply:

$$\frac{20 \text{ tablets}}{5 \text{ pillboxes}} = 4 \frac{\text{tablets}}{\text{pillbox}}$$

In the physiology of body fluids, the concept of concentration by weight or by particle count is almost as simple. Let's start with concentration by weight.

METRIC WEIGHT PER METRIC VOLUME

In simple arithmetic, percent means "amount per hundred." In referring to the concentration of a solution, percent means the amount of dissolved substance per 100 mL of fluid—that is, the metric weight of solute per deciliter of solution. An aqueous solution (a solution in which the solvent is water) containing N grams of solute per deciliter is called an "N-percent solution" of that substance.

This terminology applies only when the solvent is water and the weight of the solute is given as grams per deciliter of solution. For example, 5% glucose contains five grams of dissolved glucose per deciliter—5 g/dL. (Because another name for glucose is dextrose and the solution is in water, this formulation is often called "D-5-W.") Likewise, 50% dextrose-in-water contains 50 grams of glucose per deciliter (50 g/dL). And the concentration of an aqueous solution containing 70 grams of solute per liter (70 g/L) could also be expressed as 7 g/dL or 7%.[1]

The amount of solution present does not matter. We could have 10 L with 200 grams of a certain solute, or 1 mL with 20 milligrams of the same substance, or 3 dL with 6 grams—they are all 2% solutions, because 1 deciliter of any of these formulations in water contains 2 grams of the solute:

$$\frac{200 \text{ g}}{10 \text{ L}} = 20 \frac{\text{g}}{\text{L}} = 2 \text{ g} \cdot \text{dL}^{-1} = 2\%$$

$$20 \frac{\text{mg}}{\text{mL}} = 20 \frac{\text{g}}{\text{L}} = 2 \text{ g} \cdot \text{dL}^{-1} = 2\%$$

---

[1]A cultural reference point: Sherlock Holmes was supposedly addicted to a "seven-percent solution" of cocaine, according to his friend, Dr. John H. Watson.

$$\frac{6\,\text{g}}{3\,\text{dL}} = 2\,\text{g} \cdot \text{dL}^{-1} = 2\%$$

The deciliter is also used as the unit-volume in measuring the concentration ("level") of many substances found in serum.[2] A patient may have a serum glucose level of 90 mg/dL, an albumin level of 4 g/dL, and an iron level of 120 μg/dL.

Whereas the concentration of substances dissolved in water are often expressed as a "percent-solution," concentrations in serum or other physiologic fluids are sometimes expressed by a related phrase: "(metric weight)-percent," which again simply means the amount per deciliter. Thus, if a patient's serum glucose level is given as 90 mg%:

$$90\,\text{mg}\% = 90\,\frac{\text{mg}}{100\,\text{mL}} = 90\,\frac{\text{mg}}{\text{dL}} = 90\,\text{mg} \cdot \text{dL}^{-1}$$

Similarly, any of these expressions may be used for any physiologic substance whose concentration is given as metric weight per deciliter of body fluid.[3]

REMEMBER: These terms—"percent-solution" in water, and "(metric weight)-percent" in physiologic fluids—apply *only* when the concentration is given as metric weight of solute *per deciliter.*

For the sake of consistency, all concentrations could be converted to one standard volume unit, such as the liter (as in the SI unit system; see Chapter 1):

$$1.2\,\text{mg} \cdot \text{dL}^{-1} = 12\,\text{mg} \cdot \text{L}^{-1}$$

$$36\,\text{pg} \cdot \text{mL}^{-1} = 36{,}000\,\text{pg} \cdot \text{L}^{-1} = 36\,\text{ng} \cdot \text{L}^{-1}$$

$$200\,\text{ng} \cdot \mu\text{L}^{-1} = 200\ \text{million ng} \cdot \text{L}^{-1} = 200\,\text{mg} \cdot \text{L}^{-1}$$

And so on. However, varying units of volume are still commonly used, and clinicians must be able to convert any concentration given in one system of metric units to the corresponding value in another system.

THE NUMBER OF PARTICLES

For any element or compound, an amount equal to its atomic or molecular weight in grams (that is, its gram-atomic or gram-molecular weight) is a *mole.* Regardless of what the substance is, one mole (abbreviated mol) contains

---

[2]Whole blood minus the red cells, white cells, and platelets is plasma; and plasma minus certain proteins involved in the clotting process is serum.

[3]The concentration of substances in body fluids other than serum, such as the cerebrospinal fluid or joint-space fluid, may also be measured for clinical purposes.

about $6.02 \cdot 10^{23}$ atoms or molecules, plus or minus a few trillion; this number is called *Avogadro's Number*. Twice the gram-atomic or gram-molecular weight would be two moles, or about $1.20 \cdot 10^{24}$ atoms or molecules, and so on.

For example, ordinary carbon has an atomic weight of 12, so 12 grams of carbon is one mole or $6.02 \cdot 10^{23}$ atoms of carbon. Similarly, 24 grams is two moles, and 6 grams is a half-mole (about $3.01 \cdot 10^{23}$ atoms). Carbon dioxide (CO_2) is a compound whose molecular weight is $12 + (16 \cdot 2) = 44$. Therefore, 44 grams of CO_2 is a mole, 22 grams is a half-mole, and 66 grams is one and a half moles ($9.03 \cdot 10^{23}$ molecules).

REMEMBER: A mole of an element or compound is its atomic or molecular weight in grams. One mole contains approximately $6.02 \cdot 10^{23}$ particles (atoms or molecules). The number of moles in a given amount of substance reveals the number of particles present.

To find the number of moles in a given amount of substance, divide the weight-on-hand (in grams) by the gram-atomic or gram-molecular weight of that substance. For example, if we have 36 grams of carbon:

$$\frac{36 \text{ g}}{12 \text{ g} \cdot \text{mol}^{-1}} = 3 \text{ mol}$$

And if we have 33 grams of carbon dioxide (formula weight, 44):

$$\frac{33 \text{ g}}{44 \text{ g} \cdot \text{mol}^{-1}} = 0.75 \text{ mol}$$

Conversely, if the given information is that we have two-thirds of a mole of carbon and we want to know how many grams are present in that amount:

$$\frac{2}{3} \text{ mol} \cdot 12 \frac{\text{g}}{\text{mol}} = 8 \text{ g}$$

And if we have two and a half moles of carbon dioxide:

$$2.5 \text{ mol} \cdot 44 \frac{\text{g}}{\text{mol}} = 110 \text{ g}$$

Conversion between the metric weight of a given amount of substance and the number of metric-fractional moles represented by that weight is done with the same formulas that we just used—just keep the weight units straight.

With SI units, concentrations are expressed as the number of millimoles (or micromoles, nanomoles, etc.) of solute per liter of fluid. For example, a normal serum iron concentration is 120 μg/dL in conventional units of metric weight per deciliter. To convert this value to SI units, first find the weight per liter and then find the number of micromoles in that weight (again, the

Metric-Fractional Moles

The weight of most substances found in physiologic fluids is measured not in grams but in milligrams or smaller metric weight units. Because 1 mole is the atomic or molecular weight of a solute in grams

- 1 millimole (mmol) is the weight in milligrams, representing 10^{-3} mole or $6.02 \cdot 10^{20}$ atoms or molecules.
- 1 micromole (μmol) is the weight in micrograms, representing 10^{-6} mole or $6.02 \cdot 10^{17}$ atoms or molecules.
- 1 nanomole (nmol) is the weight in nanograms, representing 10^{-9} mole or $6.02 \cdot 10^{14}$ atoms or molecules.
- 1 picomole (pmol) is the weight in picograms, representing 10^{-12} mole or $6.02 \cdot 10^{11}$ atoms or molecules.

number of micromoles is the weight-on-hand in micrograms divided by the atomic weight, which for iron is about 56):

$$120\,\frac{\mu g}{dL} = 1200\,\mu g \cdot L^{-1}$$

$$\frac{1200\,\mu g \cdot L^{-1}}{56\,\mu g \cdot \mu mol^{-1}} \approx 21.4\,\mu mol \cdot L^{-1}$$

PARTICLES IN SOLUTION

When a given amount of a substance is completely dissolved, it means that all the individual atoms or molecules present are floating along as separate particles in solution. Then if we know how many moles of a substance are dissolved in a given quantity of fluid, we can express the concentration of the solution in terms of particle count (rather than weight) per unit-volume. Here are two chemical expressions for particle-count concentration:

Molarity is the number of moles of solute per liter of solution. Thus, a liter of solution containing 2 moles of glucose is a 2-molar solution; two liters of solution containing a total of 8 moles of glucose has 4 moles per liter, so it is a 4-molar solution. Molarity, then, gives the concentration as the number of moles of solute per unit-volume of solution.

Molality is the number of moles of solute per kilogram of pure solvent. Two moles of glucose dissolved in 1 kilogram of pure water is a 2-molal solution, and so on. Molality, then, gives the concentration as the number of moles of solute per unit-weight of pure solvent.

The molar expression tells us how much solute—but not how much pure solvent—is present in a given volume of solution. Molality provides a definite ratio between the number of solute molecules and the number of solvent molecules. For medical applications, the molal expression has greater theoretical relevance. However, at the low concentrations of solute in physiologic fluids, these two types of measurement are almost identical.

Now that we understand molality, we can tackle one of the most important concepts in the physiology of body fluids—*osmolality*.

OSMOSIS

Water molecules move freely (without any expenditure of energy) and randomly in either direction across a "semipermeable membrane," but the molecules of many dissolved substances cannot. The membrane lets water molecules pass through from one side to the other, but acts as a barrier to certain solutes—that is, it limits the *diffusion,* or passing through, of those solutes from one side to the other.

Osmosis is the free movement of water molecules across a semipermeable membrane toward the side where the particle-count concentration of solute is higher. Certain solutes will diffuse across such a membrane, just as water will pass through, to achieve equal concentrations in the fluids on either side of the membrane; but an *osmotically active* solute is a substance that cannot pass through the membrane.

Imagine a glass tank filled halfway with pure water, and divided into two compartments by a semipermeable membrane across the middle. Now suppose we add a decent amount of osmotically active solute to the water on one side of the membrane partition (Figure 2.1). The result is an *osmotic gradient*—a boundary between adjacent regions of differing solute concentration. Because the solute molecules cannot cross the membrane, all the solute stays on the side where it was added; therefore, the concentration is higher on that side. (The solute concentration remains zero on the other side, which still has nothing but pure water.)

Water molecules are always crossing the membrane, in both directions. When the solute concentration is the same on both sides of a semipermeable membrane, the rate of crossing is the same in both directions, so there is no net movement of water. *But when an osmotic gradient is present, the rate of crossing is greater going from the side at lower solute concentration to the side at higher concentration.* Some water molecules are still going in the other direction, but the *net* movement of water across the membrane is toward the side with higher concentration of solute. Thus, all references to water "moving" or "shifting" due to osmosis really mean *net* motion across the membrane (Figure 2.2).

The larger the osmotic gradient (that is, the greater the difference in solute particle-count concentration across the membrane), the greater the osmotic force pulling the water across the membrane toward the higher concentration. Formally, osmotic pressure is the amount of pressure, from some opposing force, that would be needed to stop the osmosis generated by a

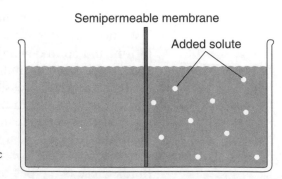

Figure 2.1 Creating an osmotic gradient.

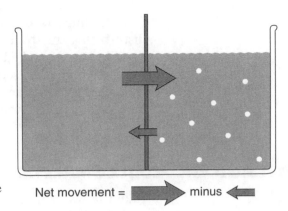

Figure 2.2 Net movement of water due to osmosis.

Net movement = ➡️ minus ⬅️

given osmotic gradient. But in casual usage, osmotic pressure is construed as the osmotic force itself.

The osmotic gradient diminishes as more and more water crosses the membrane to the side with the higher concentration of solute. (As the solute is "diluted," the particle-count concentration is reduced and thus brought closer to the lower concentration on the other side of the membrane.) Therefore, the osmotic force also falls, and the net movement of water gets slower and slower, but it will not stop as long as any osmotic gradient exists. The movement of water will go on until the gradient has been reduced to zero—the point of osmotic equilibrium, at which the concentration of osmotically active solute particles is once again equal on both sides of the membrane, and the net movement of water across the membrane is zero.

However, some other force may stop the process. In that partitioned tank of water, the net movement of water is toward the side where the solute was added. As a result, the water level rises on that side and falls on the other side. Indeed, viewed through the glass side of the tank, the water level forms a little "step" at the membrane-partition, going upward to the side with the solute.

No matter how much water crosses the membrane, the solute concentration on the rising side is higher than on the falling side, where it remains zero. Yet at a certain point, the net movement of water ceases. The weight of the water on the rising side is greater than on the falling side, and that weight pushes the water back toward the falling side. This weight-gradient increases as osmosis proceeds; at the same time, the osmotic gradient is diminishing. Eventually, when there is equilibrium between the gravitational force pushing in one direction and the osmotic force pulling in the other direction, the net movement of water across the semipermeable membrane will be zero (Figure 2.3).

Figure 2.3 On the right side of the tank, the presence of osmotically active solute has pulled water in from the left, creating a little "step" at the semipermeable membrane partition. But the extra weight of water on the right side pushes leftward, balancing out any further osmotic shift.

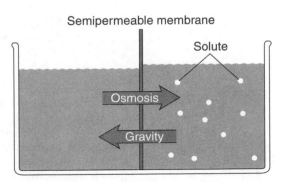

Semipermeable membrane

Solute

Osmosis

Gravity

Adding more water or solute to one or the other side of the tank will disturb this steady-state condition, and water will move again until a new equilibrium is reached between the osmotic and gravitational forces. But unless it is stopped by some other force (in this model, gravity), osmosis begins whenever there is any osmotic gradient across a semipermeable partition, and continues until osmotic equilibrium has been restored and the net movement of water is zero.

PHYSIOLOGIC FLUID COMPARTMENTS

Like the water in the partitioned tank, the body's water is divided into two main compartments separated by a semipermeable membrane. Water inside the cells (intracellular fluid, or ICF) is separated from water outside the cells (extracellular fluid, or ECF) by the semipermeable cell membranes.[4]

The total amount of body water in the average adult is about 40 liters, which includes 25 liters of ICF and 15 liters of ECF. This total body water accounts for approximately 60% of body weight in a normal adult.

The fluid within the circulatory system (the heart and blood vessels) is *intravascular* while everything outside the circulatory system is *extravascular.* Intravascular fluid includes both ICF (fluids inside the blood cells) and ECF (serum or plasma, the fluids surrounding the blood cells). Likewise, extravascular fluid includes both ICF (fluids inside other types of body cells) and ECF (interstitial fluids surrounding body cells; also called *lymph*).

Alternatively, we can say that intravascular ICF refers to the fluids inside blood cells, while extravascular ICF refers to the fluids inside other types of body cells. Similarly, intravascular ECF refers to serum or plasma, while extravascular ECF refers to interstitial fluid (Figure 2.4).

Osmosis is the basic mechanism that governs the movement of water in the body. The movement of water across the cell membranes—back and forth between the intracellular and extracellular compartments—is governed by the same osmotic forces that are at work in the partitioned tank. And in the body, gravity does not create any opposing weight-gradient. Therefore, *osmotic concentration must be the same in the ICF as in the ECF.* We may in-

Water Content in the Body

The percentage of body weight due to water is higher in lean individuals than in obese individuals, because there is much less water in adipose tissue than in muscle tissue. Age also plays a role—water content is significantly higher in infants (approaching 75%) than in older children and adults, and significantly lower in elderly patients (as low as 50%). Obviously, the notion that the human body is 99% water is nonsense.

---

[4]For this discussion, we can regard the cell membrane as impermeable to osmotic particles; later in this chapter, we will see that there actually is some leakage or diffusion of ions, which plays a crucial role in electrical events. Substances diffuse through the double lipid layer of the cell membrane at different rates—faster for small, non-ionized particles; slower for large molecules and ions. Passage directly through specialized gates or channels in the membrane is a separate phenomenon.

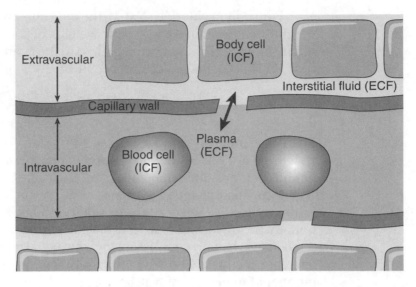

Figure 2.4 Fluid compartments.

troduce solute or water into the intracellular or extracellular compartment, but the resulting osmotic gradient will be quickly reduced to zero as water moves into or out of the cells to restore osmotic equilibrium.

Indeed, there is no such thing as a lasting osmotic gradient between the ICF and the ECF anywhere in the body except for the kidney, which relies on an elaborate mechanism to maintain an osmotic gradient (see Chapter 5).

FLUID OSMOLALITY

In chemistry, the number of atoms or molecules of solute in a solution is expressed as molarity or molality. In the body's fluid compartments, the particle-count concentration of osmotically active solutes is expressed as *osmolality*.

Here is the key point. *Osmolality depends only on the total combined particle-count concentration of all the various osmotically active substances in the body's fluids—and for all practical purposes, it makes no difference what those particles are.* Just add up the numbers of separate osmotically active particles in the ICF and in the ECF, and the total number of particles per liter will be the same in both compartments, even though the chemical constituents of the ICF and ECF are quite different. Consequently, the osmolality is the same in the ICF and ECF (Figure 2.5).

Figure 2.5 The chemistry is different but the total number of particles per liter is the same.

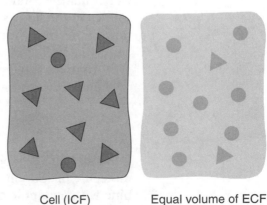

Cell (ICF)   Equal volume of ECF

For sodium chloride (NaCl), the molecular weight (also called "formula weight") is the sum of the atomic weights of sodium (23) and chlorine (35); so one mole of NaCl weighs 58 grams (to the nearest whole number). For glucose ($C_6H_{12}O_6$), the molecular weight is:

$$(12 \cdot 6) + (1 \cdot 12) + (16 \cdot 6) = 180$$

Then one mole of glucose weighs 180 grams—more than three times the weight of a mole of sodium chloride. Yet one mole of each of these compounds contains the same number of molecules.

Many compounds undergo ionization when placed into solution—that is, the molecules dissociate into separate ions. For different types of compounds, the dissociation process may be partial or complete. Sodium chloride dissociates virtually completely, yielding equal numbers of sodium ions (Na^+) and chloride ions (Cl^-). So there are twice as many separate particles as the number of molecules of NaCl placed in solution—and each separate particle, despite chemical differences, contributes independently and essentially equally to the overall osmotic pressure of the solution.

In contrast, glucose is a molecule that does not undergo ionic dissociation. Thus, a 1-molal solution of NaCl has twice the osmolality and exerts twice the osmotic influence of a 1-molal solution of glucose.

With a nondissociating solute, the osmolality is essentially equal to the molal concentration, and equimolal solutions of chemically different non-ionic substances will have the same osmolality and will exert the same osmotic effects. With an ionic solute, the osmolality is the equal to the molal concentration of the intact compound multiplied by the number of separate ions formed by the dissociation of each molecule in solution. Thus, a 1-molal solution of NaCl has an osmolality of 2 osmoles per liter (2 osmol/L)—two moles of osmotically active particles per liter.

REMEMBER: Osmosis and osmolality are functions of the particle-count concentration of a substance in solution, not the weight concentration. Osmotic pressure is directly proportional to the combined particle-count concentration of all the osmotically active solute present.

FROM MOLES TO EQUIVALENTS

If we know the metric weight concentrations of the various substances dissolved in the ECF, we can just convert to moles and add them up. The total will equal the total particle-count concentration in the ICF. But while many solutes in the body's fluids are measured in terms of metric weight per metric volume, some important substances are typically measured in units called "equivalents."

To understand equivalents, we must review some fundamental concepts in chemistry, starting with this simple idea:

$$Na^+ + Cl^- \rightarrow NaCl$$

Obviously, if one sodium ion combines with one chloride ion to form one molecule of sodium chloride, then it takes one mole of sodium ions to combine with one mole of chloride ions to form one mole of sodium chloride.

These ions combine one-to-one because they have the same valence-number (number of charges). Sodium is a *cation* (a positively charged ion) and chloride is an *anion* (a negatively charged ion), and both are *monovalent* (having a charge of magnitude 1). The compound formed by these ions is electrically neutral, because the charge on one sodium ion (+1) and the charge on one chloride ion (−1) cancel each other out.

Calcium is a *divalent* cation (an ion with a +2 charge); so it takes two chloride ions to combine with one calcium ion to form the electrically neutral compound calcium chloride. Therefore, two moles of chloride ions combine with one mole of calcium ions to form one mole of calcium chloride:

$$Ca^{+2} + 2Cl^{-} \rightarrow CaCl_2$$

Combining ratios may also be expressed in terms of *equivalents* (eq). Technically, 1 eq is the amount of substance that can combine with or replace 1 mole of hydrogen ions. For example, 1 gram of hydrogen (the approximate weight of 1 mole) will react with 35 grams of chlorine (1 mole) to form 36 grams (1 mole) of hydrogen chloride (HCl; also called hydrochloric acid). Because both chlorine and hydrogen are monovalent, their combining ratio is one-to-one.

Therefore, 1 equivalent of chlorine—the amount that will combine with 1 mole of hydrogen—is 1 mole. But as we saw earlier, calcium is a divalent ion, and the combining ratio of calcium to chlorine is one-to-two. So it takes only a half-mole of calcium to combine with 1 mole (that is, 1 eq) of chlorine in forming calcium chloride. Therefore, 1 equivalent of calcium is half of a mole.

Similarly, the *trivalent* nitrogen anion combines with three hydrogen ions to produce a toxic compound called ammonia:

$$N^{-3} + 3H^{+} \rightarrow NH_3$$

So it takes one-third of a mole of nitrogen to combine with 1 mole of hydrogen, and 1 equivalent of nitrogen is one-third of a mole.

In short, 1 equivalent is the amount of ion that will balance out the electric charge in 1 mole of oppositely charged monovalent ions; indeed, equivalent means "equal valence." *And the valence number gives the number of equivalents per mole.* That is, 1 mole of monovalent chlorine is 1 equivalent; 1 mole of divalent calcium is 2 equivalents; and 1 mole of trivalent nitrogen is 3 equivalents.

We can also define equivalents in terms of metric weight. The atomic weight of calcium is 40; therefore, 1 mole of calcium is 40 grams. However, 1 equivalent of calcium is a half-mole, so 1 *equivalent weight* of calcium is 20 grams.

 Don't get mixed up! To find the metric weight of 1 equivalent of an ionic substance, divide its gram-atomic weight by its valence (which, again, is equal to the number of equivalents per mole). For calcium:

$$\frac{40 \text{ g} \cdot \text{mol}^{-1}}{2 \text{ eq} \cdot \text{mol}^{-1}} = 20 \text{ g} \cdot \text{eq}^{-1}$$

Moles and Equivalents

For *monovalent* ions, such as sodium (Na^+) or chloride (Cl^-):

1 mol = 1 eq
1 molar weight (or gram-atomic weight) = 1 equivalent weight

For *divalent* ions, such as calcium (Ca^{+2}) or oxygen (O^{-2}):

1 mol = 2 eq; 1 eq = 0.5 mol
1 molar weight = 2 equivalent weights
1 equivalent weight = 0.5 molar weight

For *trivalent* ions, such as aluminum (Al^{+3}) or nitrogen (N^{-3}):

1 mol = 3 eq; 1 eq = 0.33 mol
1 molar weight = 3 equivalent weights
1 equivalent weight = 0.33 molar weight

Conversely, to find out how many equivalents are present in a given amount of a substance, multiply the number of moles by the valence. For 80 grams of calcium:

$$\frac{80 \text{ g}}{40 \text{ g} \cdot \text{mol}^{-1}} = 2 \text{ mol}$$

$$2 \text{ mol} \cdot 2 \frac{\text{eq}}{\text{mol}} = 4 \text{ eq}$$

Or just divide the weight-on-hand by the known weight of 1 equivalent:

$$\frac{80 \text{ g}}{20 \text{ g} \cdot \text{eq}^{-1}} = 4 \text{ eq}$$

Because most solutes in body fluids are measured in milligrams and millimoles rather than grams and moles, we use the unit *milliequivalent* (1 meq = 10^{-3} eq). For a monovalent ion, 1 meq = 1 mmol; for a divalent ion, 1 meq = 0.5 mmol; and so on.

Now let's count particles in the body fluids.

MEASURING OSMOLALITY

As we have already seen, serum is part of the ECF compartment. If we can measure the osmolality of serum, we will also know the osmolality of the ICF, because the osmotic concentrations in both fluid compartments must be the same—osmosis will see to that.

The total concentration of all the solutes in physiologic fluids such as serum is low enough to be measured in millimoles rather than moles per liter; thus, the standard unit of measurement for serum osmolality is milliosmoles per liter (mosmol/L; 1 mosmol $= 10^{-3}$ osmol).

In the laboratory, serum osmolality can be determined with considerable precision by a measurement called *freezing-point depression.*[5] However, we can get a fairly good estimate of serum osmolality using readily available data from a standard serum chemistry profile.

In the serum chemistry report, the ionic substances (sodium, potassium, chloride, bicarbonate) are measured in milliequivalents per liter—and because they are all monovalent, the number of milliequivalents equals the number of millimoles (and in SI units, these substances are reported in millimoles per liter).

The concentrations of these ionic substances (called *electrolytes,* because they carry electric charge) are already given as particle counts per liter. Sodium accounts for the huge majority of cations in the ECF, and we know that there must be an equal number of anions (mostly chloride, Cl^-, and bicarbonate, HCO_3^-) to balance the total electric charge represented by that amount of cation. Therefore, doubling the sodium concentration gives a fair approximation of the particle-count total for the ionic substances in the ECF. Using the middle of the normal range for sodium:

$$2 \cdot Na^+ = 2 \cdot 140 = 280 \, \frac{meq}{L} = 280 \, mmol \cdot L^{-1}$$

Standard Serum Chemistry Values

Sodium (Na^+)	135–145 mEq $\cdot$ L^{-1}
Potassium (K^+)	3.5–5.0 mEq $\cdot$ L^{-1}
Chloride (Cl^-)	100–106 mEq $\cdot$ L^{-1}
Bicarbonate (HCO_3^-)	21–28 mEq $\cdot$ L^{-1}
Glucose	70–110 mg $\cdot$ dL^{-1}
Blood urea nitrogen (BUN)	8–25 mg $\cdot$ dL^{-1}

In this six-channel profile of the major ionic and nonionic substances in serum, note that the two cations (sodium and potassium) add up to more than the total for the two anions (chloride and bicarbonate). The reason is because many other substances are present in the serum besides these six. So a difference of about 8–15 meq/L is normal; however, a much larger "anion gap" signifies a serious metabolic disturbance.

---

[5]The temperature at which a solution freezes or boils is a *colligative* property—one that varies with its particle-count concentration rather than its chemical nature. The freezing point of 1 kg of pure water is lowered (from 0°C) by 1.86°C for each mole of nondissociating solute that is present. For example, if an aqueous solution freezes at a temperature of −4.65°C, the freezing-point of water has been lowered by 4.65 degrees. Without knowing the chemical nature of the solute, we can determine the molal concentration: $4.65/(1.86 \cdot mol^{-1}) = 2.5$ mol (per kg of water).

With the nonionic substances, we first convert from deciliters to liters, and then from metric weights to particle counts (again, using mid-normal readings). For glucose, with a molecular weight of 180:

$$90 \frac{mg}{dL} = 900 \; mg \cdot L^{-1}$$

$$\frac{900 \; mg \cdot L^{-1}}{180 \; mg \cdot mmol^{-1}} = 5 \; mmol \cdot L^{-1}$$

Note that by converting to units of mmol/L, we have gone from conventional metric units to SI units. For BUN, the weight[6] is 28:

$$15 \frac{mg}{dL} = 150 \; mg \cdot L^{-1}$$

$$\frac{150 \; mg \cdot L^{-1}}{28 \; mg \cdot mmol^{-1}} \approx 5.4 \; mmol \cdot L^{-1}$$

For both glucose and BUN, we first multiplied the weight-per-deciliter by 10 to get the weight-per-liter, and then divided by the milligram-molecular weight. In one step, we can just divide the weight-per-deciliter by one-tenth of the milligram-molecular weight. For glucose, one-tenth of 180 is 18:

$$\frac{90 \; mg \cdot dL^{-1}}{18} = 5 \; mmol \cdot L^{-1}$$

Likewise, for BUN, one-tenth of 28 is 2.8; but to make life easier, we can round it off to 3 (after all, the formula is only an estimate anyway):

$$\frac{15 \; mg \cdot dL^{-1}}{3} = 5 \; mmol \cdot L^{-1}$$

Now we can put all the serum chemistry data together into a formula that will often provide a fair estimate of fluid osmolality:

$$\text{Osmolality} \approx 2(Na^{+}) + \frac{\text{Glucose}}{18} + \frac{\text{BUN}}{3}$$

(Alternatively, we can multiply glucose by 0.056, the reciprocal of 18; and multiply BUN by 0.36, the reciprocal of 2.8.) Using the same numbers as before:

---

[6]The molecular weight of urea is 60, but what contributes to ECF osmolality is the nitrogen content of urea—that is, blood urea nitrogen (BUN). The weight of BUN is 28, which is the weight of the two nitrogen atoms in the urea molecule.

Chapter 2 / Solutions and Body Fluids

$$\text{Osmolality} \approx 2(140) + \frac{90}{18} + \frac{15}{3} = 280 + 5 + 5 = 290 \text{ mmol} \cdot \text{L}^{-1}$$

Because they are all osmotically active particles, we can also give this reading as milliosmoles per liter (mosmol/L). And indeed, the normal osmolality in the serum or other body fluids is about 280–300 mosmol/L. In a common variation of this formula, potassium (with a normal ECF concentration of about 4 meq/L) is included along with sodium as another cation:

$$2(140 + 4) + \frac{90}{18} + \frac{15}{3} = 298 \text{ mosmol} \cdot \text{L}^{-1}$$

REMEMBER: Electolytes and nonelectrolytes both contribute to osmolality, which is the total number of separate osmotically active particles in solution, per unit volume. Physiologic osmolality is measured in units of *milliosmoles per liter.*

For a nonelectrolytic substance such as glucose, 1 millimole (1 formula weight, in milligrams) represents 1 milliosmole. However, with electrolytes, the total number of particles in solution depends on the extent of dissociation that occurs. Assuming complete dissociation, 1 millimole of sodium chloride yields 2 millimoles of separate particles (1 millimole each of sodium ions and chloride ions). Likewise, 1 millimole of calcium chloride ($CaCl_2$) yields 3 millimoles of separate particles (1 millimole of calcium ions plus 2 millimoles of chloride ions). The osmolality of a solution depends on the sum total of all the separate osmotically active particles that are present, regardless of any chemical differences between them, as we see in the "Osmotic Soup Recipe."

REMEMBER: If the osmolality in the ECF is normal, so is the osmolality in the ICF; if the ECF osmolality were higher or lower than normal, the ICF osmolality would show the same abnormality.

Osmolality *must* be the same in both fluid compartments, because water *will* move between compartments until osmotic equilibrium is achieved; nothing can stop it. The fact that the chemical composition of the ICF is different from that of the ECF is not relevant to the osmolality in each compartment. (As it happens, the ratio of sodium to potassium is reversed in the ICF, where potassium is the main cation and sodium is found at much lower levels.)

Nonosmotic solutes can pass through the cell membrane; they contribute to the overall solute concentration in both compartments, but not to the osmotic force in either compartment. Urea, for example, *partially* crosses cell membranes, adding to the overall solute concentration in both the ECF and ICF (which is why the osmolality formula includes BUN rather than urea itself).

Now let's continue with sodium, the main determinant of osmolality in the extracellular compartment—the site of most clinical measurements.

Osmotic Soup Recipe

Suppose we mix the following all together in one container:

3 deciliters of a 2-millimolal solution of NaCl

6 deciliters of a 4-millimolal solution of $CaCl_2$

12 deciliters of a 3-millimolal solution of glucose

Assuming complete dissociation of the ionic compounds, what is the osmolality of the resulting soup?

We'll work with each ingredient separately and then add them up. The 2-millimolal solution of NaCl contains 2 mmol/L (for convenience, we'll use the millimolar expression, which is virtually identical in concentration), and we have 0.3 L; and each millimole of NaCl yields 2 milliosmoles:

$$2 \frac{mmol}{L} \cdot 0.3 \, L = 0.6 \, mmol$$

$$0.6 \, mmol \cdot \frac{2 \, mosmol}{mmol} = 1.2 \, mosmol$$

The 4-millimolal solution $CaCl_2$ contains 4 mmol/L, and we have 0.6 L; and each millimole of $CaCl_2$ yields 3 milliosmoles:

$$4 \frac{mmol}{L} \cdot 0.6 \, L = 2.4 \, mmol$$

$$2.4 \, mmol \cdot \frac{3 \, mosmol}{mmol} = 7.2 \, mosmol$$

The 3-millimolal glucose contains 3 mmol/L, and we have 1.2 L; and because glucose is nonionic and nondissociating, the number of millimoles equals the number of milliosmoles:

$$3 \frac{mmol}{L} \cdot 1.2 \, L = 3.6 \, mmol = 3.6 \, mosmol$$

The totals are $1.2 + 7.2 + 3.6 = 12$ milliosmoles in $0.3 + 0.6 + 1.2 = 2.1$ liters. Then the osmolality of the soup is:

$$\frac{12 \, mosmol}{2.1 \, L} \approx 5.7 \, mosmol \cdot L^{-1}$$

THE DOMINANCE OF SODIUM

Sodium cations (and their accompanying chloride or bicarbonate or other anions) normally account for more than 90% of the osmotically active particles in the ECF. That is, sodium *alone* produces almost half of the osmotic pressure in the ECF, and osmolality is often a direct reflection of sodium levels: hypernatremia (too much sodium in the blood) is usually associated with hyperosmolality, while hyponatremia (too little sodium) is associated with hypo-osmolality.

For example, if a patient's serum sodium concentration is 130 meq/L (just 10 meq/L below mid-normal), osmolality will be slightly low even if glucose and BUN are at normal levels. Using the osmolality formula:

$$2(130) + \frac{90}{18} + \frac{15}{3} = 270 \text{ mosmol} \cdot \text{L}^{-1}$$

Now suppose that the sodium concentration is normal but the concentrations of glucose and BUN (and everything else) are zero. Ignoring the fact that such a state is incompatible with life, note that the osmolality will still be borderline normal—again, showing the dominance of sodium in ECF osmolality:

$$2(140) + \frac{0}{18} + \frac{0}{3} = 280 \text{ mosmol} \cdot \text{L}^{-1}$$

In abnormal conditions, osmolality may be elevated by some other substance.[7] If a diabetic patient's serum glucose level rises tenfold, to 900 mg/dL, glucose would contribute 50 mosmol/L (900/18) to the total, instead of the typical 5 mosmol/L. Likewise, an alcoholic patient may have a large amount of ethanol in the blood.[8] Because the standard serum chemistry profile does not include blood alcohol levels, the calculated osmolality by formula would be falsely low; however, measurement by freezing-point depression would reveal the hyperosmolar state.

Because sodium is the main determinant of ECF osmolality, saline (sodium chloride in water) is often used as a replacement fluid. A 0.9% solution is casually called "normal saline," but the proper name is "physiologic saline." By definition, 0.9% NaCl (approximate formula weight, $23 + 35 = 58$) contains 0.9 gram of NaCl per deciliter of solution. First, we convert this weight per deciliter to particle count per liter (that is, SI units):

$$0.9 \frac{\text{g}}{\text{dL}} = 9 \frac{\text{g}}{\text{L}} = 9000 \text{ mg} \cdot \text{L}^{-1}$$

$$\frac{9000 \text{ mg} \cdot \text{L}^{-1}}{58 \text{ mg} \cdot \text{mmol}^{-1}} \approx 155 \text{ mmol} \cdot \text{L}^{-1}$$

Then, because the dissociation of NaCl is virtually total:

$$155 \text{ mmol NaCl} \rightarrow 155 \text{ mmol Na}^+ + 155 \text{ mmol Cl}^-$$

---

[7]In abnormal conditions in which a substance such as glucose is extremely elevated, the laboratory measurement of sodium concentration may be falsely depressed; that is, serum osmolality may be high although the total *amount* of sodium in the body is normal and the sodium *concentration* appears lower than normal.

[8]Ethanol (consumable grain alcohol) is often referred to as "ETOH" in medical chart notations; the −OH signifies part of the molecular structure of an alcohol, and the ET− signifies that the alcohol is ethanol (in contrast to methanol or "wood alcohol," which is poisonous).

So we have a total of 310 milliosmoles (310 millimoles of separate, osmotically active particles) per liter—close enough to the normal osmolality of body fluids.[9] Similarly, half-normal saline is a 0.45% solution, delivering about 150 mosmol/L, and 1.8% saline would be twice-normal, delivering over 600 mosmol/L.

Now if sodium accounts for so much of the osmolality of the ECF, what about the ICF, which has far less sodium? Won't osmolality be lower there? *No*—osmolality *must* be the same in both fluid compartments. Potassium is the dominant cation in the ICF, as sodium is in the ECF; but the total number of particles per liter is the same in both compartments, because water will move in either direction across the cell membrane until osmotic equilibrium is achieved.[10]

FLUID MOVEMENT BETWEEN COMPARTMENTS

A standard biology course experiment involves looking through a microscope at red blood cells placed into solutions of differing osmotic concentration. In a medium with an osmotic concentration equal to that of the fluid inside the cell, nothing happens—the cell does not change in size, because the rate of water moving into and out from the cell is equal in both directions, so there is no net movement of water across the cell membrane (Figure 2.6).

In a medium with an osmotic concentration higher than that of the fluid inside the cell, the cells shrink (a process called *crenation*). Why? Because of osmosis—the net movement of water is outward. Water is drawn from inside the cells, through the semipermeable cell membranes, to the outside medium, where the particle-count concentration is higher. As water moves out of the cells, the intracellular solute is contained in a smaller volume of water, so the ICF osmolality goes up; and this water coming out of the cell dilutes the surrounding medium, so the concentration of solute in the medium goes down.

Thus, the osmotic gradient—the difference in particle-count concentration between the ICF and the surrounding medium—is reduced. But as long as any gradient still exists, water keeps leaving the cells and the cells keep

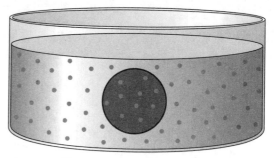

Figure 2.6 Isotonic environment.　　　No net movement of water

---

 [9]Be careful. *Normality* is another measure of chemical concentration (equivalents of solute per liter, just as molarity is moles per liter). "Normal saline" just means saline at a concentration close to the normal osmolality of body fluids.

[10]This chemical segregation—ECF high in sodium and low in potassium, ICF high in potassium and low in sodium—is established and maintained by specialized cell-membrane structures that pump sodium outward and potassium inward.

shrinking, until the gradient is zero and osmotic equilibrium is achieved (Figure 2.7).

When cells are placed into a medium with an osmotic concentration lower than that of the fluid inside the cell, they swell up. Why? Because of osmosis—the net movement of water is inward. Water from the surrounding fluid is drawn through the semipermeable cell membranes to the interior of the cells, where the particle-count concentration is higher. As water leaves the surrounding medium and enters the cells, the solute concentration in the medium goes up and the intracellular osmolality goes down.

Thus, the osmotic gradient is reduced. But as long as any gradient is present, water keeps entering the cells and the cells keep expanding, until the gradient is zero (Figure 2.8).

Water moves
out of cell...

... resulting in
crenation

Figure 2.7 Hypertonic medium.

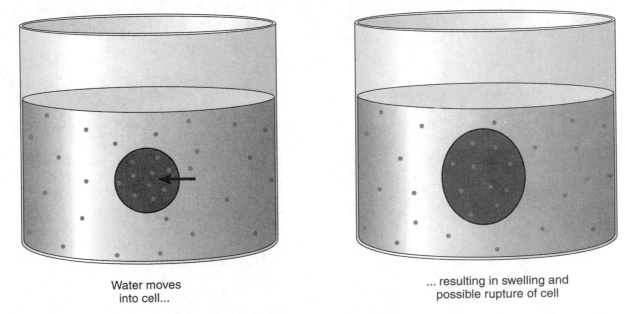

Water moves
into cell...

... resulting in swelling and
possible rupture of cell

Figure 2.8 Hypotonic medium.

In fact, when cells are placed into pure water, they burst apart—for no matter how much water enters the cells, intracellular osmolality remains higher than the zero concentration of the surrounding medium. Because osmotic equilibrium can never be achieved, water just keeps coming in until the cell membranes, stretched past their physical limits, are literally ripped apart.

"The Infusion Question" illustrates how the concentrations and volumes of the ICF and ECF can be adjusted and calculated by infusing solutions that are hypo- or hyperosmotic.

The Infusion Question

A generally healthy man is admitted into the hospital for a routine procedure. His weight is 70 kg; total body water is 40 L (25 L in the ICF, 15 L in the ECF); osmolality is 300 mosmol/L.

At teaching rounds, an instructor poses this hypothetical question to the students: "Suppose we were to infuse this patient with two liters of half-normal saline. What would happen to his body fluids in terms of volume and osmolality?"

To answer this question, start by figuring out how much solute is in the patient's body fluids before the infusion. Each of the 25 liters of ICF and 15 liters of ECF contains 300 milliosmoles of solute. Therefore:

$$ICF: \quad 25 \text{ L} \cdot \frac{300 \text{ mosmol}}{\text{L}} = 7500 \text{ mosmol}$$

$$ECF: \quad 15 \text{ L} \cdot \frac{300 \text{ mosmol}}{\text{L}} = 4500 \text{ mosmol}$$

The total for the body is 12,000 milliosmoles in 40 liters of water. Half-normal saline contains around 150 mosmol/L, so infusing 2 liters of this formulation delivers an additional 300 milliosmoles of solute. For the whole body, the solute load becomes $12,000 + 300 = 12,300$ milliosmoles, in $40 + 2 = 42$ liters of water. So the new osmolality is:

$$\frac{12,300 \text{ mosmol}}{42 \text{ L}} \approx 293 \text{ mosmol} \cdot \text{L}^{-1}$$

While the solute stays where it is, the water will end up wherever osmosis pulls it. All of the added solute stays in the ECF, because it cannot cross the cell membrane. So now the ECF has $4500 + 300 = 4800$ milliosmoles of solute, while the ICF still has just its original 7500 milliosmoles. And since we know that the new osmolality is 293 mosmol/L in both compartments, the new volume distribution is as follows:

$$ECF: \quad \frac{4800 \text{ mosmol}}{293 \text{ mosmol} \cdot \text{L}^{-1}} \approx 16.4 \text{ L}$$

$$ICF: \quad \frac{7500 \text{ mosmol}}{293 \text{ mosmol} \cdot \text{L}^{-1}} \approx 25.6 \text{ L}$$

Tonicity vs. Osmolality

The osmotic concentration of a medium is often described in terms of tonicity. *Tonicity* is defined by the movement of water: an isotonic medium is one in which cells neither crenate nor enlarge; a hypertonic medium is one in which cells crenate; and a hypotonic medium is one in which cells expand and may even burst.

The reason for using these terms rather than iso-, hyper-, or hypo-osmotic is that an isosmotic medium (fluid at an osmotic concentration equal to that of the ICF) is not necessarily isotonic. That can happen if the medium also contains some nonosmotic substance, such as urea. The osmotic concentration in the medium may be the same as inside the cell; but because urea can cross the cell membrane, it will enter the cell following a chemical gradient (the cell contains little or no urea).

As a result, the overall solute load is raised in the ICF and lowered in the medium. This movement of urea *creates* an osmotic gradient that pulls water into the cell, causing the cell to expand, which defines the medium as *hypotonic,* even though it was *isosmotic.*

Here are three more questions as follow-up to the case. *"If the saline infusion enters the plasma, which is part of the extracellular compartment, why do we calculate the new osmolality on the basis of the whole body?"* Answer: Osmolality *must* be the same *everywhere* in the body, which means intravascular and extravascular, intracellular and extracellular; therefore, osmolality for the whole body is the same as osmolality for any part of the body.

"At the end, why do we divide the solute load in each compartment by the new osmolality?" Answer: We want the final volume in each compartment. Here is a simple analogy. Suppose 24 marbles are distributed into a certain number of little bags, each holding three marbles. Obviously, eight bags would be needed—but if we didn't instantly realize that $3 \cdot 8 = 24$, how would we know that? What we are really doing is dividing the total marble count by the known number of marbles per bag to find the number of bags that would account for that number of marbles:

$$\frac{24 \text{ marbles}}{3 \text{ marbles} \cdot \text{bag}^{-1}} = 8 \text{ bags}$$

Now instead of bags, each holding a certain number of marbles, think about liters of fluid, each containing a certain amount of solute. In each fluid compartment, we divide the total milliosmole load by the number of milliosmoles per liter—that is, the osmotic concentration—to find the number of liters that would account for that amount of solute.

"Why does the extracellular compartment expand by less than the amount of fluid that was infused?" Answer: When the hypotonic solution is introduced into the ECF, ECF osmolality is momentarily reduced, creating an osmotic gradient with the ICF—the ECF medium becomes hypotonic in relation to the ICF. Water immediately begins moving from the lower-osmolality ECF, across the cell membranes, to the higher-osmolality ICF, until osmotic equilibrium is restored; therefore, the ICF also expands. Of

the 2 liters of fluid infused, 1.4 liters stay in the ECF while osmosis pulls 0.6 liter into the ICF.

MEASURING THE VOLUME OF THE FLUID COMPARTMENTS

So far, we have assumed standard ICF and ECF volumes. To find out the actual volumes in an individual patient, we use the same kind of formula that we used in calculating the final ECF and ICF volumes at the end of "The Infusion Question."

In the dilution technique, a known quantity of traceable radioactive water is injected. Because water distributes itself freely throughout the body, the concentration of this traceable water will be the same everywhere in the body. After allowing time for full dispersal throughout the body, the serum concentration (as measured by the level of radioactivity per unit volume) is determined.

Now we can figure out what volume of fluid at that concentration would account for the total amount injected—how much fluid it would take to dilute the amount injected to its measured concentration:

$$\text{Volume} = \frac{\text{amount injected}}{\text{concentration}}$$

Note that this formula is analogous to the one we used to find the volume of the ECF and ICF compartments in "The Infusion Question." Using traceable water, the calculation gives us the volume of total body water.

Is the concept clear? Suppose we want to find out how much water is in a large tank. Stir in a measured amount of a traceable substance until it dissolves and disperses evenly throughout the water. Now take a sample of the water and measure the concentration. To find the quantity of water in the tank, divide the total amount of substance introduced by its concentration in the water. If we stir in 20 grams and the concentration is 5 grams per liter:

$$\frac{20 \text{ g}}{5 \text{ g} \cdot \text{L}^{-1}} = 4 \text{ L}$$

Obviously, it takes 4 liters of water to account for 20 grams of substance if each liter contains 5 grams. Just be sure that the units are consistent. If the final concentration is 5 mg/L:

$$20 \text{ g} = 20,000 \text{ mg}$$

$$\frac{20,000 \text{ mg}}{5 \text{ mg} \cdot \text{L}^{-1}} = 4000 \text{ L}$$

It takes 4000 liters of water to account for 20 grams of substance if each liter contains only 5 milligrams.

The same formula is used to measure the extracellular compartment, with traceable sodium injected instead of traceable water. Radioactive sodium, like ordinary sodium, is osmotically active—it stays within the ECF. Once it is evenly dispersed throughout the ECF, the amount injected divided by the measured concentration in the serum (representing the ECF) reveals the ECF volume in which that amount was distributed. And of course, once we know the volume of total body water and the volume of the extracellular compartment, the intracellular volume is simply the total volume minus the ECF volume.

Neither Extracellular nor Intracellular

The ECF and ICF compartments seem to account for all the water in the body. However, in various regions, water is sequestered from the rest of the body. Collectively, these regions constitute a "third space." The biggest third-space region is the gastrointestinal tract, which at any time may hold a significant volume of fluid. An abnormal shift of body water into the third space implies a corresponding loss of fluid from the ECF and ICF, which can be a serious clinical problem if the loss is large.

FLUID MOVEMENT AT THE CAPILLARY LEVEL

So far, we have been looking at fluid movement at the level of the cell membrane, where osmosis is the driving force that pulls water back and forth between the intracellular and extracellular compartments. Different forces direct fluid into and out of the capillaries. First, let's briefly review the pertinent anatomy.

Blood has both ICF (fluid inside the "formed elements"—the red cells, white cells, and platelets) and ECF (plasma, or, without the clotting proteins, serum). Arteries carry blood away from the heart, branching into smaller and smaller vessels, culminating in the arterioles and metarterioles.

The next level is the true capillaries, which are like tiny tubes with pores through which various molecules can pass. The pores are *not* large enough to allow the formed elements to escape, so blood cells and platelets stay within the circulatory system. However, many substances carried in the serum can pass through the capillary pores to enter the interstitial fluids surrounding the body's cells; substances in the interstitial fluid can pass through the pores to enter the serum. Past the capillary level, venules join together to become the veins that return blood to the heart.

Thus, it is at the capillary level that substances are exchanged between the body cells and the bloodstream (the bloodstream brings oxygen and nutrients to the cells and carries away carbon dioxide and other waste products from the cells). The capillary is the boundary between the serum and the interstitial fluid just as the cell membrane is the boundary between the ICF and

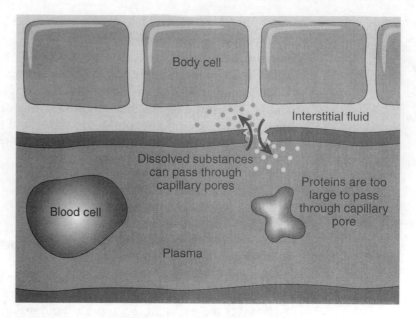

Figure 2.9 The capillary pore.

ECF. Serum and interstitial fluid are continuous with each other across the capillary pore (Figure 2.9). Chemically, these two fluids are similar, except for the fact that proteins tend to stay in the intravascular space—and that fact has important consequences.

THE STARLING HYPOTHESIS

Osmosis is not effective at the capillary level, because capillary pores are big enough to let molecules like sodium ions pass through. Two other forces, working in opposite directions, control the movement of fluid in and out of the capillaries.

Hydrostatic pressure pushes a filtrate of serum out of the capillaries, into the interstitial spaces around the body cells. ECF outside the capillary is called interstitial fluid. Blood is always pressing outward against the walls of all the blood vessels; the capillaries just happen to be the only blood vessels with pores that will let the fluid out.

The force that pulls ECF back into the capillaries may be compared to osmosis, in that water is drawn across a boundary toward an area of higher concentration of something that cannot pass through that boundary. In this case, the boundary is the capillary pore rather than the cell membrane; and the "something" that cannot cross the boundary is plasma proteins rather than osmotic solutes. Proteins, with molecular weights in the thousands, are vastly larger than the ions and molecules that exert osmotic force across the cell membrane. In short, most protein molecules are simply too big to fit through the capillary pores. Albumin is the most plentiful of the proteins in the plasma, and the curled-up proteins called globulins account for most of the remainder.

Plasma proteins are *colloids*—heavy molecules that are too large to pass through the capillary pores (see the sidebar, "Crystalloids vs. Colloids"). Because proteins normally stay inside the capillaries, colloid density is much higher in the intravascular than the extravascular space; this is the main difference between

Crystalloids vs. Colloids

A system comprising a substance in a medium can take several forms, depending on the particle size and miscibility of the substance:

- *Solution:* A homogeneous ("one-phase") system in which a solute is completely dissolved and cannot be separated by physical means.
- *Crystalloid:* A system of crystalline particles that can pass through a semipermeable barrier.
- *Colloid:* A system of heavy, finely dispersed particles (typically larger than solutes) that cannot pass through a semipermeable barrier.
- *Suspension:* A system of intact, individually discernible particles (typically larger than colloids) that are dispersed but not dissolved; a cloudy suspension of particles that are not individually discernible but are large enough to scatter light is sometimes called a *colloidal.*

The key clinical distinction is between crystalloids and colloids. Technically, saline is a crystalloid; although the ions are osmotically active at the cell membrane, they can diffuse through, very slowly; that leakage is offset by the action of membrane pumps that push sodium out and pull potassium in. More significantly, crystalloids pass through capillary pores while colloids do not. Since crystalloids produce no osmotic effects at the capillary pore, they increase total body water but are not especially effective in restoring intravascular volume. Colloids do produce osmotic effects at the capillary pore, so they expand intravascular volume without a large increase in total body water (intravenous administration of colloid raises colloid osmotic pressure, which draws excess interstitial fluid back into the blood vessels).

serum and interstitial fluid. Water is drawn inward through capillary pores toward the higher density of colloid within the capillary; this effect is called *colloid osmotic pressure* (COP). COP, working at the capillary pore, is analogous to the osmotic force created by electrolytes, glucose, and BUN at the cell membrane.

In short, hydrostatic pressure pushes fluid out of the capillaries, while COP pulls it back in. This mechanism is called the *Starling hypothesis*[11] (Figure 2.10).

Hydrostatic pressure is highest at the arteriolar end of the capillary. As blood moves toward the venous end, fluid filters out and the hydrostatic pressure drops (see Chapter 3). With less pressure pushing outward, COP can pull more fluid in. Overall, however, hydrostatic pressure pushes more fluid out than COP pulls in. Obviously, that "lost" fluid, along with the small amount of protein that leaks through the capillary barrier, must be returned to the circulatory system.[12]

---

[11]Ernest H. Starling (1866–1927) was a British physiologist. The Starling Hypothesis is well established, but it is still called a hypothesis to avoid confusion with Starling's Law of the Heart; see Chapter 3.

[12]From every part of the body, excess interstitial fluids and proteins that have leaked out from the capillaries are returned to the systemic circulation by a network of *lymphatic channels.* These channels collect and empty into the subclavian veins.

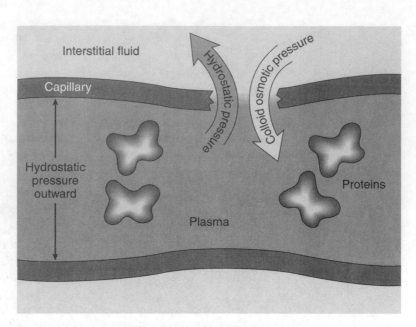

Figure 2.10 The Starling Hypothesis.

MEASURING PLASMA VOLUME AND TOTAL BLOOD VOLUME

We measure the volume of total body water with traceable water, and the volume of the ECF compartment with traceable sodium. Likewise, we can measure the intravascular volume with a traceable substance that combines with plasma proteins after intravenous administration. Because the proteins remain inside the blood vessels, the tracer is distributed throughout the plasma; but it does not leave the blood vessels or enter the blood cells.

Once again, the plasma concentration of the tracer is measured, and the amount injected divided by the concentration gives the volume of plasma in which the amount injected must have been distributed.

With that information, the total volume of blood is easily calculated. The hematocrit is simply the percentage of the blood volume occupied by the blood cells; the normal figure is about 40%, which means that the plasma occupies about 60% of the total blood volume. We can express the hematocrit as a decimal and subtract it from 1 to find the plasma fraction:

$$1 - 0.40 = 0.60$$

Knowing the actual plasma volume and the percentage of the total blood volume represented by the plasma, the total blood volume equals the plasma volume divided by the plasma fraction. For example, if the tracer measurement shows a plasma volume of 3.3 liters, and the hematocrit is 37% (which means that the plasma fraction is 0.63), then the total blood volume would be:

$$\frac{3.3 \, \text{L}}{0.63} \approx 5.2 \, \text{L}$$

IONS AND TRANSMEMBRANE ELECTRICAL POTENTIAL

Apart from their contribution to osmolality, electrolytes create *voltage* (electrical potential) across the cell membrane. Voltage results from a *charge gradient*—the distribution of electrical charge is not equal inside and outside the cell.

In terms of osmotic effects, we have looked at sodium and other ions as if they were totally nondiffusible. In fact, the cell membrane does permit slow diffusion or leakage of ions, at a rate directly proportional to the ion's permeability through the membrane, and to the magnitude of the concentration gradient. For any given ion, a greater permeability coefficient and a larger difference between intracellular and extracellular concentrations means faster and more extensive diffusion (per unit-area of cell membrane surface). Diffusion of sodium and potassium along their individual concentration gradients is offset by the action of the sodium-potassium pump in the cell membrane.

At equilibrium, the overall distribution of ions on either side of the cell membrane creates a charge gradient—*the interior of the cell is relatively negative to the exterior.* One of the mechanisms that contributes to this gradient is the cell membrane ion pump. The exchange of sodium and potassium is *not* one-for-one; three sodium ions are pushed out for every two potassium ions pulled in, which means that more positive charge moves outward than inward.

Another factor that affects ionic distribution and transmembrane voltage is the presence of proteins in the plasma. Here is how it works. The protein molecule contains a four-atom group called *carboxyl* ($-COOH$); in plasma, this group can reversibly dissociate to release a hydrogen ion:[13]

$$-COOH \rightleftarrows -COO^- + H^+$$

Proteins obviously cannot cross the cell membrane. In the dissociated state, with the hydrogen ion split off, the remainder of this extracellular protein molecule carries a net negative charge, like an anion.

At the start, the total amounts of positive and negative charge on either side of the cell membrane are equal. Outside the cells, negative charge is produced by dissociated protein and electrolytic anion; but inside, the same negative charge is produced solely by the anion. Therefore, the amount of electrolytic anion inside the cell must be greater than the amount outside, which means that there is a concentration gradient that favors diffusion of anion outward. As anion diffuses out of the cell, cation follows. As a result, more positive charge is found outside the cell than inside. (Note that these outward shifts of ion alter the osmotic balance, causing some water to move out from the cells.)

This phenomenon is called the *Donnan effect* (Figure 2.11) or *Gibbs-Donnan equilibrium.*[14] In the simplest model (one monovalent cation and one monovalent anion), the equilibrium is expressed mathematically as follows:

---

[13]This mechanism is one of the main buffer systems in the body; see Chapter 6.

[14]J. Willard Gibbs (1839–1903) was an American mathematician and physicist; Frederick Donnan (1870–1956) was an English chemist.

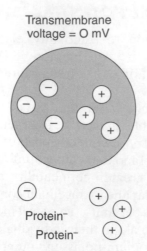

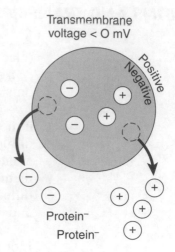

Figure 2.11 Donnan effect. On left, the anion concentration is higher inside cell. On right, anion leaves cell and cation follows, creating voltage.

$$\frac{[\text{Cation}]_{\text{ECF}}}{[\text{Cation}]_{\text{ICF}}} = \frac{[\text{Anion}]_{\text{ICF}}}{[\text{Anion}]_{\text{ECF}}}$$

$$[\text{Cation}]_{\text{ECF}} \cdot [\text{Anion}]_{\text{ECF}} = [\text{Cation}]_{\text{ICF}} \cdot [\text{Anion}]_{\text{ICF}}$$

(Square brackets denote concentrations.) At equilibrium, the concentration ratios for each ion are the same, and the cation-times-anion concentration products are the same inside and outside the cells.

The net result of all these factors—chemical concentration gradients, permeability coefficients, ionic valences, the action of the cell membrane ion pump, and the Donnan effect—is an equilibrium distribution of ions such that an electrical gradient is established across the cell membrane. *This charge gradient constitutes the transmembrane voltage.*

At equilibrium, the intracellular-extracellular distribution of each type of ion is affected by the electrical gradient as well as by the concentration gradient; for different ions, the two gradients may work in the same direction or in opposite directions. Sodium and potassium are both cations, and they both tend to move inward along the electrical gradient (toward the relatively negative interior of cells). However, while sodium tends to move inward along its concentration gradient, potassium tends to move outward. So the gradients work in the same direction for sodium, but in opposite directions for potassium.

For a given ion, there is a certain transmembrane electrical potential at which the charge gradient would exactly counterbalance the effect of the concentration gradient. This equilibrium potential (E) is defined by the *Nernst equation:*[15]

$$E = \frac{\text{gas constant} \cdot \text{absolute temperature}}{\text{faraday constant} \cdot \text{valence}} \cdot \ln\left(\frac{[\text{ion}]_{\text{ECF}}}{[\text{ion}]_{\text{ICF}}}\right)$$

---

[15]Walther Nernst (1864–1941) was a German physicist.

The equation looks complex, but the meaning becomes clearer when we study its units. Let's start with the gas constant, which relates energy in *joules* to the absolute (Kelvin scale) temperature:

$$\text{Gas constant} = 8.314 \text{ joules} \cdot \text{mole}^{-1} \cdot \text{kelvin}^{-1}$$

The energy in a gaseous system would be 8.314 joules per mole of gas present, per kelvin (the unit of temperature on the absolute scale). Here, mole^{-1} applies to the *faraday* (F), which is the total amount of electric charge *per mole* of electrons (charge is measured in *coulombs;* F = 96,500 coul). The numbers of moles or equivalents of ion outside and inside the cell are represented in the concentration ratio at the end of the equation. (The concentration ratio may vary in different types of tissue; thus, the equilibrium potential for a given type of ion may vary in different sites.)

Multiplying the faraday by the valence gives the magnitude of the charge per mole for any type of ion. For a monovalent ion, the charge per mole is F; for a divalent ion, 2F; for a trivalent ion, 3F. The algebraic sign of the valence determines whether the charge is positive or negative.

Now let's restate the Nernst equation as a proportionality of units:

$$E \propto \frac{(\text{joules} \cdot \text{mole}^{-1} \cdot \text{kelvin}^{-1}) \cdot \text{kelvin}}{(\text{coulombs} \cdot \text{mole}^{-1})}$$

Valence and the natural logarithm of the concentration ratio are not shown, because they are pure (dimensionless) numbers. The units reduce as follows:

$$\frac{(\text{joules} \cdot \text{mole}^{-1} \cdot \text{kelvin}^{-1}) \cdot \text{kelvin}}{\text{coulombs} \cdot \text{mole}^{-1}} = \frac{\text{joules}}{\text{coulomb}}$$

Because the Nernst equation is a measure of electrical potential, we must confirm that joules per coulomb is an expression for voltage. Electrical current (flow of charge) is measured in *amperes;* flow at a rate of 1 coulomb per second is a current of 1 amp:

$$\text{amp} = \text{coul} \cdot \text{sec}^{-1}$$

Power (energy expended or work done per unit-time) is measured in *watts;* energy expended at a rate of 1 joule per second is 1 watt:

$$\text{watt} = J \cdot \text{sec}^{-1}$$

But another definition of a watt is an amp · volt; therefore, 1 volt = 1 watt · amp^{-1}. Substituting the preceding expressions for watts and amps in this expression for volts:

$$\text{volt} = \frac{\text{watt}}{\text{amp}} = \frac{J \cdot \text{sec}^{-1}}{\text{coul} \cdot \text{sec}^{-1}} = J \cdot \text{coul}^{-1}$$

Now that the meaning of the Nernst equation is a little clearer, let's simplify its form. Still leaving the concentration ratio and its natural logarithm aside for the moment, we plug in the numeric values for the gas constant and the faraday, and set the absolute temperature at 310 K (normal body temperature, equivalent to 37°C):

$$E = \frac{(8.314 \text{ J}) \cdot 310}{(96{,}500 \text{ coul}) \cdot \text{valence}} \approx \frac{0.0267 \text{ J} \cdot \text{coul}^{-1}}{\text{valence}} = \frac{0.0267 \text{ V}}{\text{valence}}$$

Note that the units are already reduced to joules per coulomb, or volts; again, the valence is a pure number, but it certainly must be included in actual computations.

Because the value in volts is so small, we can convert to millivolts ($mV = 10^{-3}$ V). Thus, 0.0267 V = 26.7 mV; and as long as we remember that the constant 26.7 yields an answer in millivolts, we can even omit that unit so as to make the equation as clear and simple as possible. At this point, we have:

$$E = \frac{26.7}{\text{valence}} \cdot \ln\left(\frac{[\text{ion}_{ECF}]}{[\text{ion}_{ICF}]}\right)$$

Finally, we replace the natural logarithm with the common (base 10) logarithm. There is a constant ratio between $\ln x$ and $\log x$, defined by the fact that $\ln 10 \approx 2.303$ and $\log 10 = 1$. For any value $x > 0$, $\ln x \approx 2.303 \log x$. Therefore:

$$E = \frac{26.7}{\text{valence}} \cdot 2.303 \log\left(\frac{[\text{ion}_{ECF}]}{[\text{ion}_{ICF}]}\right) \approx \frac{61.5}{\text{valence}} \cdot \log\left(\frac{[\text{ion}_{ECF}]}{[\text{ion}_{ICF}]}\right)$$

This is the more familiar form of the Nernst equation. In clinical applications, the main ions are monovalent; with a valence of 1 in the denominator, the *magnitude* of the equilibrium potential (in millivolts) for a given ion is 61.5 times the common logarithm of the ECF/ICF concentration ratio. The algebraic sign of the valence determines whether that equilibrium potential is positive or negative. (Positive versus negative voltage indicates the direction of the charge gradient across the cell membrane; positive voltage means that the interior of the cell is relatively positive to the exterior; negative voltage means that the interior is relatively negative to the exterior.)

Let's see how it works, using standard values for extracellular and intracellular ionic concentrations. For sodium:

$$E = \frac{61.5}{+1} \cdot \log\left(\frac{140 \text{ meq} \cdot \text{L}^{-1}}{10 \text{ meq} \cdot \text{L}^{-1}}\right)$$

$$= 61.5 \cdot \log 14 \approx 61.5 \cdot 1.146 \approx 70 \text{ mV}$$

For potassium:

$$E = \frac{61.5}{+1} \cdot \log\left(\frac{4 \text{ meq} \cdot \text{L}^{-1}}{140 \text{ meq} \cdot \text{L}^{-1}}\right)$$

$$\approx 61.5 \cdot \log 0.029 \approx 61.5 \cdot (-1.544) \approx -95 \text{ mV}$$

Finally, just to see the effect of a negative valence, we'll compute the equilibrium potential for chloride, a monovalent anion:

$$E = \frac{61.5}{-1} \cdot \log\left(\frac{100 \text{ meq} \cdot \text{L}^{-1}}{6 \text{ meq} \cdot \text{L}^{-1}}\right)$$

$$\approx -61.5 \cdot \log 16.7 \approx -61.5 \cdot 1.223 \approx -75 \text{ mV}$$

Because the electrical and chemical concentration gradients work in opposite directions for potassium, the inward and outward forces are roughly balanced. Consequently, the equilibrium potential for potassium (-95 mV) is not far from the actual transmembrane voltage (typically about -70 mV).[16]

However, the equilibrium potential for sodium is far from the actual transmembrane voltage. Since both gradients drive sodium inward, it would take a complete reversal of the direction of the charge gradient (into the positive range, so that sodium would be pushed outward) to counterbalance the inward push of the concentration gradient. Indeed, the only reason that sodium does not flood into the cells is that its membrane permeability is extremely low; the cell membrane ion pump efficiently pushes out the small amount of sodium that leaks inward.

It is crucial to understand that the Nernst equation does not express the actual transmembrane voltage; it expresses the voltage that would, in theory, exactly counterbalance the effect of the concentration gradient for a given ion.

The actual transmembrane voltage is a function of *all* the intracellular and extracellular ions present in the intracellular and extracellular fluids. Each ion's contribution to transmembrane voltage is represented by the product of its concentration and its membrane permeability coefficient; there are separate products for each ion's presence inside and outside the cell. The voltage (V) thus produced can be measured directly or predicted by the *Goldman equation:*[17]

---

[16]If the equilibrium potential for an ion is exactly equal to the actual transmembrane voltage, it means that the electrical and concentration gradients are exactly balanced. With potassium, these values are close but not equal, so there must be another factor at work, keeping the system stable; that other factor is the cell membrane ion pump. The electrical gradient pulling inward is weaker than the concentration gradient pushing outward, but the inward-pumping action of the membrane makes up the difference.

[17]American physiologist David E. Goldman was born in 1911.

$$V = \frac{\text{gas constant} \cdot \text{absolute temperature}}{\text{faraday constant}} \cdot \ln\left(\frac{A}{B}\right)$$

Except for the absence of valence, the format is identical to that of the Nernst equation; indeed, it could be simplified to $V = 61.5 \cdot \log(A/B)$, with the obtained value in millivolts. However, instead of taking the logarithm of the ECF/ICF concentration ratio for a single ion, we take the logarithm of a ratio designated here as A/B, where A is the sum of the concentration-permeability products for extracellular cations and intracellular anions, and B is the sum of the products for intracellular cations and extracellular anions.

In the actual equation, A and B are impressively long strings of numbers and units; but because the units in A and B are identical, the *ratio* A/B reduces to a pure number. Therefore, the Goldman equation, like the Nernst equation, reduces to units of joules per coulomb.

QUIZ

1. In the hypothetical case presented in "The Infusion Question," the patient was left with 42 liters of total body water (25.6 L in the ICF, 16.4 L in the ECF) and an osmolality of 293 mosmol/L. Now suppose 1.5 liters of twice-normal saline are administered in an attempt to reverse the osmotic disturbance. Compute the patient's final osmolality and the final volumes in the ECF and ICF. (Ignore the effects of urinary output of salt and water.)

2. What is the osmolality of a 1.2% solution of $MgCl_2$? (Magnesium is a divalent cation, atomic weight 24; chloride is a monovalent anion, weight 35. Assume complete ionic dissociation of the compound in solution.)

3. What is the osmolality of 5% glucose? (Formula weight is 180.)

4. What is the *molar concentration* of a solution produced by dissolving a total of 12.6 grams of sodium bicarbonate ($NaHCO_3$) into 4 liters of water? With complete dissociation of the compound into ions of sodium (Na^+) and bicarbonate (HCO_3^-), what is the *osmolality* of the solution (assume molality is approximately equal to molarity)? Use the following atomic weights:

   Sodium (Na), 23
   Hydrogen (H), 1
   Carbon (C), 12
   Oxygen (O), 16

5. How many equivalent weights are in 100 grams of calcium (atomic weight 40, valence +2)? How many equivalents of calcium ions does it take to combine with two moles of chloride ions to form calcium chloride ($CaCl_2$)? If the serum concentration of nonionized calcium is 9.0 mg/dL, convert to SI units (mmol/L).

6. If plasma volume = 3.3 L and total blood volume = 5.2 L, hematocrit = ____%.

7. Given *identical* total volumes of a 2-molar and a 2-molal formulation of the same substance, which formulation is (slightly) more concentrated?

8. Copper (atomic weight 63.5) is a trace element in the human body. If a patient's plasma level of copper is given in SI units as 20 μmol/L, what is the equivalent value in conventional weight-units (μg/dL)? Compute the conversion factor to go directly from conventional units to SI units.

9. What is the approximate serum osmolality of a patient with the following laboratory values, which are given in SI units?

> Sodium, 136 mmol/L
> Potassium, 3.2 mmol/L
> Chloride, 100 mmol/L
> Bicarbonate, 21 mmol/L
> Glucose, 6.4 mmol/L
> BUN, 9.5 mmol/L

What is the approximate serum osmolality of a patient with the following laboratory values, which are given in conventional units?

> Sodium, 142 meq/L
> Potassium, 4.4 meq/L
> Chloride, 104 meq/L
> Bicarbonate, 18 meq/L
> Glucose, 108 mg/dL
> BUN, 22 mg/dL

10. What is the equilibrium potential for magnesium ion (Mg^{2+}) if the ECF concentration is 2.5 meq/L and the ICF concentration is ten times greater?

Answers and explanations at end of book.

Circulatory Function

Chapter Outline

Overview
The Cardiac Cycle
Murmurs
Pressure Changes During the Cardiac Cycle
Blood Pressure
Measurement of Pressure
Capillary Pressure
Cardiac Output
Stroke Volume and Ejection Fraction
Preload and Starling's Law
Afterload
Contractility and Heart Rate
Voltage and the Action Potential
Automaticity and Conduction
The Electrocardiogram
The ECG Leads
Cardiac Vector
Voltage Magnitude
Electrical Axis
Components of the ECG Tracing

Mathematical Tools Used in This chapter

Negative exponents (Appendix C)
Dimensional units with negative exponents (Appendix D)
Unit conversions (Appendix E)
Order of magnitude (Appendix G)
Vectors (Appendix I)

Circulatory function is best explained in terms of hemodynamics—the pumping action of the heart pushing blood through a system of vessels. This chapter deals with the *movement* of blood within the circulatory system. The *composition* of blood (as pertains to gas transport) is examined in Chapter 4.

OVERVIEW

Let's start with an overview of cardiovascular anatomy and physiology. *Arteries* carry blood away from the heart; they have a substantial layer of smooth muscle, which helps to regulate blood pressure. *Veins* carry blood toward the heart; veins in the limbs have valves, which prevent backflow. *Capillaries* are microscopic vessels linking the smallest arteries to the smallest veins; their structure is porous and permeable, which allows fluid and dissolved substances to move into and out from the vascular system (by the mechanisms described in the Starling Hypothesis; see Chapter 2).

The left and right sides of the heart are divided by a *septum,* creating two circulation systems that operate simultaneously. Aerated blood pumped from the left side of the heart is carried by the systemic arteries to every part of the body.[1] At the systemic capillaries, oxygen and nutrients in the blood are delivered to the cells while carbon dioxide and other waste products from the cells are taken away. Now unaerated, the blood returns to the right side of the heart via the systemic veins.

At the same time, unaerated blood pumped from the right side of the heart is carried by the pulmonary arteries to the lungs. At the pulmonary capillaries, inhaled oxygen enters the blood and carbon dioxide leaves the blood. Now aerated, the blood returns to the left side of the heart via the pulmonary veins.

On both sides of the heart, veins empty into the upper chambers, the *atria.* The pulmonary veins deliver aerated blood to the left atrium; the systemic veins deliver unaerated blood to the right atrium. The blood then moves down from the atria to the lower chambers, the *ventricles.* The left ventricle pumps aerated blood out to the body via the aorta, which is the beginning of the systemic arterial system; the right ventricle pumps unaerated blood out to the lungs via the pulmonary arteries (Figure 3.1).

The forward course of blood through the arteries depends on the pumping action of the ventricles. However, that impetus dissipates as blood goes through smaller arteries to reach the systemic capillaries. Past the capillaries, the course of blood through the veins depends largely on routine body movements that press in on the veins and push the blood along; since venous valves prevent *retrograde* (backward) flow, the flow is always *antegrade* (forward) toward the heart.

Because the direct pumping action operates only in arteries, blood spurts rhythmically from a cut artery but trickles steadily from a cut vein. Pressure is generally higher in the arterial system than in the venous system, and arteries must be structurally stronger than veins of comparable diameter, to cope with the higher pressure.

---

[1]The term *aerated* refers to systemic arterial blood, which is fully primed with oxygen from the lungs—that is, oxygenated blood. *Unaerated* refers to systemic venous blood, which has delivered much of its oxygen to the body cells. Venous blood is *not* totally deoxygenated; it still carries a significant amount of oxygen, as well as the carbon dioxide it has picked up from the cells.

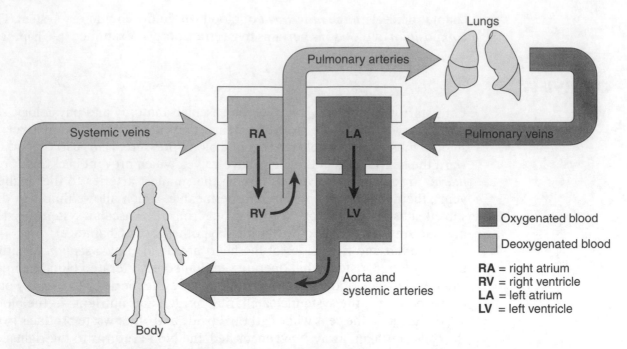

Figure 3.1 Schematic of blood flow through right and left sides of heart.

REMEMBER: The left side of the heart receives aerated blood from the lungs via the pulmonary veins and pumps it out to the body via the systemic arteries. Simultaneously, the right side of the heart receives unaerated blood from the body via the systemic veins and pumps it out to the lungs via the pulmonary arteries.

The heart has four valves, two on each side—an atrioventricular[2] (AV) valve and an outflow valve. The left-side AV valve is the *mitral or biscuspid valve;* the right-side AV valve is the *tricuspid valve.* The left-side outflow valve is the *aortic valve;* the right-side outflow valve is the *pulmonic valve* (Figure 3.2).

The cardiac valves serve the same function as venous valves—they prevent backflow. Valve action is mechanical; they are pushed open and closed by the hydrostatic pressure of still blood or the hydrokinetic force of blood in motion.

THE CARDIAC CYCLE

The mathematical expression of circulatory function begins with the mechanical events of the *cardiac cycle* (myocardial contraction and relaxation) and the electrical events that trigger them (depolarization and repolarization).

The cardiac cycle consists of a ventricular contraction (*systole*) and the relaxation period leading to the next contraction (*diastole*).[3] In terms of

---

 [2]*Atrioventricular* is abbreviated AV; the AV valves lie between the atria and the ventricles. *Arteriovenous* is abbreviated A-V; an A-V shunt allows blood to pass directly from an artery to a vein. Don't confuse these abbreviations.

[3]Systole and diastole refer to ventricular events. "Atrial systole" (the atrial contraction) actually occurs as the final event in diastole.

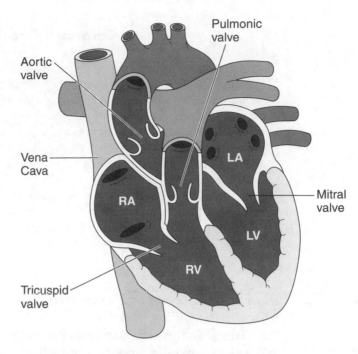

Figure 3.2 Cardiac valves.

time, systole represents about one-third of the cardiac cycle; diastole, about two-thirds.

At the start of ventricular contraction, the AV valves are open and the outflow valves are closed. The contraction pushes the blood forward along the outflow tracts and also backward toward the atria; the hydrokinetic force of the retrograde movement catches the undersides of AV valve cusps and pushes them shut, thus preventing backflow into the atria. This event produces the *first heart sound* (S_1), which marks the end of diastole and the beginning of systole.

With all four valves closed, the blood in the ventricles cannot go anywhere; and because a liquid cannot be compressed, the ventricular contraction at this point is *isometric,* which means that the hydrostatic pressure in the ventricles is rising as myocardial tension increases. A split second later, the outflow valves are pushed open from below as the fast-rising pressure in the ventricles surpasses the pressure of the blood resting on top of these valves. With the opening of the outflow valves, blood is ejected from the ventricles into the arteries. *In short, systole begins with a split second of isometric contraction before the start of the ejection phase.*

At the end of ventricular contraction, the AV valves are still closed and the outflow valves are still open. With the driving force spent, blood starts falling backward in the aorta and pulmonary artery; this retrograde movement of blood fills the cusps of the outflow valves and pushes them shut, thus preventing backflow into the ventricles. This event produces the *second heart sound* (S_2), which marks the end of systole and the beginning of diastole.

Again, with all four valves closed, blood cannot move but intraventricular pressure falls as myocardial tension decreases. A split second later, the AV valves are pushed open from above as the falling pressure in the ventricles slips below the slowly rising pressure of blood entering the atria. With the opening of the AV valves, blood descends into the ventricles. *In*

Coronary Circulation

The heart's own blood supply is provided by the *coronary arteries*. These vessels arise from openings in the wall of the aorta, behind the aortic valve cusps. When the aortic valve is open (during the ejection phase of systole), the apertures are blocked by the valve cusps. When the aortic valve closes (at the beginning of distole), the apertures are uncovered, and aerated blood above the valve flows in. After delivering oxygen and nutrients and picking up carbon dioxide and other waste products, this blood returns to the right atrium via the coronary veins.

Narrowing or blockage of the coronary arteries restricts oxygen delivery to areas of the heart, causing chest pain (angina pectoris), tissue damage, or death of heart muscle tissue (myocardial infarction).

short, diastole begins with a split second of isometric relaxation before the start of ventricular filling.

Blood in the atria descends passively into the ventricles as soon as the AV valves open—a gravitational effect. An additional amount is pushed down into the ventricle when the atria contract, just before systole. (Later in this chapter, under "Preload and Starling's Law," we'll see the purpose of the atrial contraction.)

REMEMBER: Closure of the AV valves (at S_1) marks the end of diastole and the beginning of systole; a split second later, the outflow valves open and the ejection phase begins. Closure of the outflow valves (at S_2) marks the end of systole and the beginning of diastole; a split second later, the AV valves open and ventricular filling begins.

MURMURS

Cardiac murmurs are noises indicating turbulent blood flow. One cause is valvular *insufficiency* (also called *incompetence* or *regurgitation*)—the valve is partially open when it should be closed tightly, allowing blood to move backward. Another cause is *stenosis*—the valve is partially obstructed when it should be wide open, interfering with the forward flow of blood (Figure 3.3).

AV valve insufficiency causes a murmur from backflow into the atria, starting when these valves *should* close tightly—at the very beginning of systole (that is, at S_1). Likewise, outflow valve insufficiency causes a murmur from backflow into the ventricles, starting at the very beginning of diastole (at S_2).

Outflow valve stenosis causes a murmur from impeded ejection, starting when these valves *should* open completely—at the beginning of the ejection phase of systole (that is, a split second after S_1). Likewise, AV valve stenosis causes a murmur from impeded ventricular filling, starting at the beginning of the ventricular-filling phase of diastole (a split second after S_2).

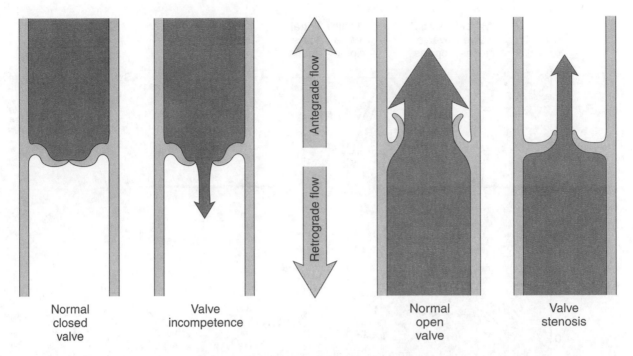

Figure 3.3 Two types of valvular abnormality.

PRESSURE CHANGES DURING THE CARDIAC CYCLE

Everything that occurs during the cardiac cycle—valvular openings and closings, and the movement of blood through the heart—depends on the pressure at each instant, at each anatomic location (Figure 3.4). Note that when a cardiac valve is open, the structures on either side are continuous with each other, and so the pressures on either side are essentially equilibrated (unless the valve is severely stenotic). Pressures on either side of a closed cardiac valve diverge widely (unless the valve is severely insufficient).

Follow the left-sided events shown in Figure 3.4. During the split-second period of isometric contraction immediately following the closure of the mitral valve at S_1, the left ventricular pressure rises rapidly toward the steady aortic pressure. At the point where the rising left ventricular pressure surpasses the aortic pressure, the aortic valve is pushed open and the ejection phase begins. The equilibrated left ventricular and aortic pressures rise to a peak and then fall.

During the split-second period of isometric relaxation immediately following the closure of the aortic valve at S_2, the left ventricular pressure falls rapidly while the aortic pressure (after a bounce from the sudden closure of the aortic valve) is once again steady. At the point where the falling left ventricular pressure slips below the left atrial pressure, the mitral valve is pushed open and ventricular filling begins. The equilibrated left ventricular and left atrial pressures rise slowly until the atrium contracts, just before the next systole.

The same relationships apply on the right side of the heart. With the pulmonic valve open during the ejection phase of systole, there is equilibration between right ventricular pressure and pulmonary artery pressure; and with

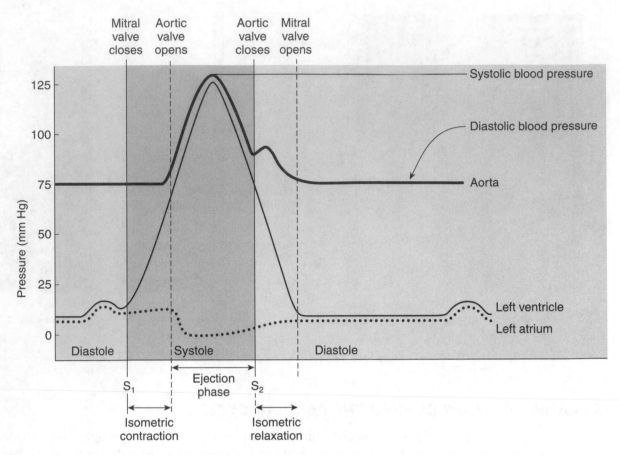

Figure 3.4 Pressure during the cardiac cycle. (Timing is not to scale on the horizontal axis.)

the tricuspid valve open during the ventricular-filling phase of diastole, there is equilibration between right ventricular pressure and right atrial pressure.

BLOOD PRESSURE

Pressures in the left ventricle and aorta are notably higher than pressures in the right ventricle and pulmonary artery. In a healthy adult, the *systolic blood pressure* (the upper of the two numbers reported, corresponding to the peak pressure in the left ventricle and aorta during the ejection phase; see Figure 3.4) is typically in the range of 120 ± 20 mm Hg. In contrast, the peak pressure in the right ventricle and pulmonary artery during systole is typically below 25 mm Hg.

Likewise, the *diastolic blood pressure* (the lower number, corresponding to the resting pressure of blood in the aorta between ventricular contractions) is typically in the range of 75 ± 15 mm Hg, whereas the resting pressure in the pulmonary artery during diastole is around 10 mm Hg.

But what is the source of the pressure within the arterial system? Two main sources are *tension in arterial smooth muscle*, pressing inward upon the blood, and the *ventricular contraction*, driving the blood forward.

Mean Arterial Pressure

Because diastole occupies twice as much time as systole, *mean arterial pressure* (MAP; the average pressure throughout the cardiac cycle) is estimated as one-third the sum of the systolic pressure plus twice the diastolic pressure. For an individual with a blood pressure of 130/80:

$$MAP = \frac{systolic + (2 \cdot diastolic)}{3} = \frac{130 + (2 \cdot 80)}{3} = \frac{290}{3} \approx 97 \text{ mm Hg}$$

Another version of the MAP formula is diastolic pressure plus one-third of *pulse pressure* (the difference between systolic and diastolic pressure):

$$Diastolic + \frac{systolic - diastolic}{3} = 80 + \frac{130 - 80}{3} = 80 + \frac{50}{3} \approx 97 \text{ mm Hg}$$

Either version may be encountered; they are mathematically identical and interchangeable rearrangements of each other.

Let's look at two physical laws that are pertinent to blood pressure. *Pascal's law*[4] states that pressure applied *anywhere* against a contained liquid is distributed uniformly and in every direction *everywhere* throughout the liquid (a consequence of the fact that a liquid cannot be compressed); the liquid therefore presses outward against every part of the containing vessel. Thus, the pulsating pressure from the heart is distributed throughout the arterial blood.

However, pressure is *not* uniform at every point within the arterial system, because the system is dynamic—blood is moving forward from the aorta through ever-narrowing arteries approaching the capillaries.

Laplace's law[5] relates *transmural pressure* (the difference between the pressures on either side of the wall of a hollow structure) to the tension in the wall and the interior dimensions of the structure. For a long cylindrical structure such as a blood vessel:

$$Pressure = \frac{tension \text{ in wall}}{radius \text{ of lumen}}$$

Tension in the wall of an artery is regulated by the tone of its smooth muscle layer. If a distensible vessel is neither shrinking nor swelling, the inward pressure from wall tension and the outward pressure from the blood must be at equilibrium. Also, the radius itself can be adjusted by changes in smooth muscle tone.[6]

---

[4]Blaise Pascal is acknowledged in "Units Named for Scientists."

[5]Pierre S. de Laplace (1749–1827) was a French mathematician.

[6]With tension measured in newtons per meter and radius as a fraction of a meter, the quotient gives pressure as newtons per square meter ($N \cdot m^{-2}$), demonstrating that pressure is force per unit-area.

Typical pressure (mm Hg)	$\frac{125}{75}$	$\frac{120}{70}$	$\frac{60}{50}$	20		10	5

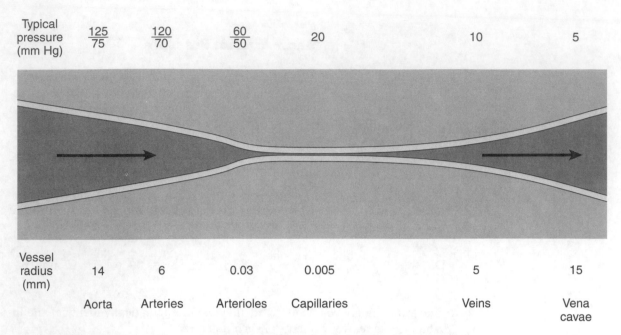

Vessel radius (mm)	14	6	0.03	0.005		5	15
	Aorta	Arteries	Arterioles	Capillaries		Veins	Vena cavae

Figure 3.5 Pressure declines with diminishing arterial radius.

While transmural pressure *is* numerically equal to the ratio of tension to radius, it would be wrong to infer that the ratio of two independent variables (tension and radius) determines a dependent variable (pressure). *In fact, all three elements vary simultaneously and directly with each other*—pressure and wall tension are both highest in the widest arteries, and lowest in the narrowest arteries (Figure 3.5).

Note that there is little drop in pressure as blood goes through the larger arteries (which is why blood pressure can be measured from the brachial artery in the arm). However, the rate of drop-off increases at the arterioles. Also note that the pulse pressure (systolic minus diastolic pressure) shrinks to near zero by the time blood reaches the capillaries. Within the capillaries and veins, there is no significant pulsatile rise and fall in pressure corresponding to systole and diastole.[7]

What the Laplace equation shows is that in the smallest arteries, where the transmural pressure is lowest, very little inward pressure from wall tension is required to achieve equilibrium with the outward pressure of the blood.

The decrease in blood pressure going from larger to smaller arteries can be compared to the voltage drop that occurs in an electric circuit where the current goes through a source of *resistance* (voltage and current are analogous to pushing pressure and blood flow, respectively). Narrowing in blood vessels creates added resistance (as discussed later in this chapter). In simple terms, pushing through resistance uses up some of the pressure-related energy of blood.

As blood pressure *decreases* going from larger to smaller arteries, the total cross-sectional area of all the blood vessels at each level *increases*. The

---

[7]"Jugular venous pulsations" are *retrograde reflections* of cardiac pulsations associated with contraction and filling, transmitted backward into the veins; they do *not* represent a force driving the return of blood to the heart.

Chapter 3 / Circulatory Function

cross-sectional area of a vessel can be computed with the familiar formula for the area of a circle. If the radius (r) of the aortic lumen is 14 mm = 1.4 cm:

$$\text{Aortic cross-sectional area} = \pi r^2 \approx 3.14 \cdot (1.4 \text{ cm})^2 \approx 6.2 \text{ cm}^2$$

The aorta is the largest artery. In contrast, each individual capillary is microscopic in radius and cross-sectional area. Because of the vast number of capillaries, however, their *combined* cross-sectional area is several hundred to a thousand times greater than that of the aorta—an increase approaching three orders of magnitude.

The *velocity* of blood moving forward through the arteries is a function of the pressure driving it and the total cross-sectional area of the vessels it is traversing.[8] In general terms, velocity varies directly with pressure and inversely with total cross-sectional area:

$$\text{Velocity} \propto \frac{\text{pressure}}{\text{total cross-sectional area}}$$

Thus, blood velocity is highest at the aorta, where the pressure is highest and the cross-sectional area is smallest. Velocity decreases to a minimum at the capillaries, where the pressure is lowest and the total cross-sectional area is greatest. (Only about 5% of the total volume of blood in the body is traversing the capillaries at any given moment, but the slow velocity of blood in the capillaries allows time for effective delivery of oxygen and nutrients to the cells and take-up of waste products from the cells.)

Pressure is relatively constant within the venous sytem, but total cross-sectional area decreases substantially going from the venules to the largest veins. From the proportionality above, the result is predictable. If the denominator of the ratio is smaller, the quotient is larger—that is, velocity actually *increases* slightly as blood approaches the right atrium via the systemic veins.

REMEMBER: Ventricular and arterial pressures are higher on the left side than on the right side. Arterial pressures are higher than venous pressures. Blood pressure and blood velocity are both highest in the largest arteries, decreasing to low values in the capillaries and veins.

The Laplace equation shows that transmural pressure varies directly with wall tension. Recall that wall tension is controlled by vascular smooth muscle tone, and that smooth muscle is substantial in arteries but minimal in veins. Thus, wall tension is lower in veins than in arteries, and the outward pressure of blood causes more distention in veins than in arteries. Veins therefore serve as *capacitance vessels* (accommodating increased volume with little rise in pressure); arteries are *resistance vessels* (responding to increased volume with increased pressure).

---

[8]Don't confuse the *velocity* of blood in the arteries (distance over time) with the *flow rate* (volume over time).

Renal Regulation of Blood Pressure

The action of the kidneys in regulating arterial pressure depends on *perfusion pressure* and a substance called *renin*, which is stored in certain pressure-sensitive cells within the kidney.

Perfusion pressure within any localized region of the circulatory system is simply the difference between the pressures at the arterial and venous ends of the region. Renal perfusion pressure is the pressure inside the renal arteries minus the pressure inside the renal veins.

If arterial pressure approaching the kidneys is low, renal perfusion pressure will be low. In response, the pressure-sensitive cells release renin into the bloodstream. Renin reacts with a plasma protein component called *angiotensinogen*, splitting off a molecule called *angiotensin I*. As blood passes through the lungs, angiotensin I is converted to a more potent form, angiotensin II, by the action of *angiotensin converting enzyme* (ACE). The direct effect of angiotensin II is a widespread increase in vascular smooth muscle tone; and as wall tension is increased, arterial blood pressure rises.

Renin also causes the adrenal glands to secrete a substance called *aldosterone*, which causes the kidney to retain sodium and water. This action also raises the blood pressure systemically (in simple terms, pressure is higher when a larger volume of liquid is contained within a finite system of vessels).

High blood pressure is often treated with *ACE inhibitors* (drugs that inhibit the action of the angiotensin converting enzyme in the lungs) and *diuretics* (drugs that reduce circulating volume by limiting the reabsorption of sodium and water, thus causing increased urination).

If renal perfusion pressure rises, the activity of the entire renin-angiotensin-aldosterone system declines, allowing blood pressure to fall.

MEASUREMENT OF PRESSURE

Blood pressure as measured by arm cuff reflects aortic and left ventricular pressure during systole, and aortic pressure during diastole. The measurement of other pressures—right ventricular and pulmonary artery pressures, systemic venous pressure (blood returning to the right atrium), and pulmonary venous pressure (blood returning to the left atrium)—is more difficult.

On the right side, *central venous pressure* is measured directly, by inserting a catheter into the right atrium via a large peripheral vein. Measurement of right ventricular and pulmonary artery pressures during systole and diastole requires catheterization *through* the right atrium, *past* the tricuspid valve, *into* the right ventricle, and then past the pulmonic valve into the pulmonary artery.

On the left side, the pulmonary veins are not accessible for catheter insertion, so measurements are made indirectly. A catheter is inserted all the way through the right side of the heart and into the pulmonary arteries, until the branches are too narrow to permit further advancement. A balloon around the tip of the wedged-in catheter is inflated, blocking the flow of pulmonary arterial blood past that point in that vessel. With no flow distally, the manometer at the catheter tip records the *wedge pressure*—back pressure from the pulmonary capillaries, which reflects the pressure in the pulmonary veins and the left atrium (Figure 3.6).

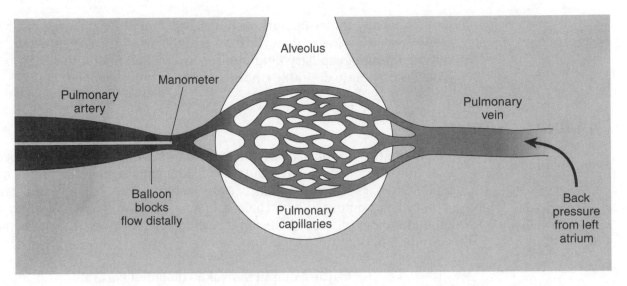

Figure 3.6 Wedge pressure.

Arterial pressure is reported as millimeters of mercury (mm Hg), but venous pressure is often reported as centimeters of water (cm H$_2$O). The specific gravity of mercury is about 13.6 (that is, 13.6 times greater than the specific gravity of water, which is 1.000, by definition). Then since the conventional unit with water is centimeters, and 1 cm = 10 mm:

$$1 \text{ mm Hg} \approx 1.36 \text{ cm H}_2\text{O}$$

$$1 \text{ cm H}_2\text{O} \approx 0.735 \text{ mm Hg}$$

CAPILLARY PRESSURE

Precapillary sphincters control the entry of blood into the systemic capillaries; the hydrostatic pressure of blood entering the capillaries is held at around 35 mm Hg. This outward pressure is partially counterbalanced by the inward pull of *colloid osmotic pressure* (COP) from plasma proteins.[9] COP is about 25 mm Hg; therefore, the net pressure near the arterial end of a systemic capillary is 35 − 25 = 10 mm Hg (pushing outward, since hydrostatic pressure exceeds COP).

As fluid is pushed out through the porous capillary wall, the hydrostatic pressure declines to about 16 mm Hg, while COP remains constant. Therefore, the net pressure near the venous end is 25 − 16 = 9 mm Hg (pulling inward, since COP now exceeds hydrostatic pressure). The small amount of fluid lost into the interstitial spaces is returned to circulation via the lymphatic channels (see Chapter 2).

---

[9]These pressures are the factors that control fluid movement at the capillary level (the Starling Hypothesis). Hydrostatic pressures as shown include a small adjustment (1–2 mm Hg) for the effect of interstitial fluid; if this effect is much larger (as in tissue edema), the outward push may be reduced to the point where delivery of oxygen and nutrients to the cells is seriously compromised.

The same forces operate within the pulmonary capillaries, but hydrostatic pressure is lower than COP even at the arterial end. Thus, there is no outward push of capillary fluid into the alveoli and interstitium of the lung, which is obviously desirable (the alternative is pulmonary edema).

CARDIAC OUTPUT

The ultimate criterion of heart function is *cardiac output* (volume pumped per unit-time). In "heart failure," output is inadequate to meet the body's needs, and blood backs up in the veins because output is lower than venous return.

Cardiac output reflects the combined effects of several complex, interrelated factors (discussed later in this chapter). However, the mathematical definition of output is simple:

$$\text{Cardiac output} = \text{stroke volume} \cdot \text{heart rate}$$

Stroke volume is the quantity of blood ejected into the arteries at each ventricular contraction. *Heart rate* is the number of contractions per minute. The product of these measurements is the total volume of blood pumped out in 1 minute. If left ventricular stroke volume is 80 mL and the heart rate is 72 beats per minute, then left-side cardiac output is computed as follows:

$$80\,\frac{\text{mL}}{\text{beat}} \cdot 72\,\frac{\text{beats}}{\text{min}} = 5760\,\frac{\text{mL}}{\text{min}} \approx 5.8\,\text{L} \cdot \text{min}^{-1}$$

The right side of the heart pumps out the same volume per minute.

Left-Side and Right-Side Output

For blood to circulate normally, the volume of aerated blood moving through the left side of the heart must be equal to the volume of unaerated blood moving through the right side. If these volumes are different, blood will accumulate behind the side that is pumping out the lesser volume, as we see in *congestive heart failure*.

In right-side failure, more blood approaches the right atrium than the right ventricle can pump out, so blood backs up in the systemic veins. If the backup reaches all the way to the systemic capillaries, the increased hydrostatic pressure pushes excess fluid into the interstitial spaces, causing *peripheral edema*.

In left-side failure, more blood approaches the left atrium than the left ventricle can pump out, so blood backs up in the pulmonary veins. If the backup reaches all the way to the pulmonary capillaries, the increased hydrostatic pressure pushes excess fluid into the pulmonary interstitium and the alveolar spaces, causing *pulmonary edema*.

Although output should be equal on both sides of the heart, the total volume of blood within the systemic circulation is many times greater than the total volume within the pulmonary circulation—the systemic loop is larger than the pulmonary loop, but flow rates are equal in both loops.

One approach to measuring cardiac output employs a basic physiologic concept called the *Fick principle*. As a substance is extracted from arterial blood, its concentration in venous blood is correspondingly decreased. The product of the arteriovenous (A-V) difference and the blood flow rate gives the rate at which the substance—oxygen—is being extracted and consumed:

$$\text{Rate of consumption} = \text{A-V difference} \cdot \text{blood flow rate}$$

If a certain amount of oxygen is consumed from each liter of arterial blood in 1 minute, then multiplying the A-V difference (amount removed per liter) by the flow rate (liters per minute) yields the total amount consumed per minute.

Here is the tie-in to cardiac output. The rate at which blood flows through the arteries is the rate at which it is pumped from the heart—that is, flow rate *is* cardiac output. Therefore, we can rewrite the Fick equation as follows:

$$\text{Cardiac output} = \frac{\text{rate of extraction}}{\text{A-V difference}}$$

Oxygen consumption is obtained as the ventilatory volume per minute times the difference in the percentage of oxygen in inspired versus expired air; see Chapter 8. The A-V difference is obtained from direct measurements of the oxygen content in arterial versus venous blood.

If arterial oxygen content is 20 mL/dL and venous content is 15 mL/dL, the A-V oxygen difference is 20 mL/dL − 15 mL/dL = 5 mL/dL. Then if oxygen consumption is measured at 240 mL/min:

$$\text{Cardiac output} = \frac{240 \text{ mL} \cdot \text{min}^{-1}}{5 \text{ mL} \cdot \text{dL}^{-1}} = 48 \text{ dL} \cdot \text{min}^{-1} = 4.8 \text{ L} \cdot \text{min}^{-1}$$

Note that mL cancels out of the fraction, leaving cardiac output in units of dL/min (which we then convert to L/min).[10]

The value obtained in the preceding calculation—4.8 liters per minute—might be normal or abnormal. Output varies considerably among healthy children and adults, men and women, tall people and short people. To adjust for differences in body size, the *cardiac index* is frequently used in the clinical setting. Cardiac index is simply the cardiac output divided by the body surface area (BSA; see Chapter 1). If cardiac output is 5 L/min and BSA is 1.60 m², the cardiac index is computed as follows:

$$\text{Cardiac index} = \frac{\text{cardiac output}}{\text{BSA}} = \frac{5 \text{ L} \cdot \text{min}^{-1}}{1.6 \text{ m}^2} \approx 3.1 \text{ L} \cdot \text{min}^{-1} \cdot \text{m}^{-2}$$

---

[10]The Fick method (named for German physician Adolf Fick, 1829–1901) can also be used to measure flow within an organ, by measuring the A-V difference from the vessels entering and leaving that organ. In current practice, cardiac output is computed from the rate of decline in the arterial concentration of an injected tracer substance, or the rate of temperature change in blood after injection of a measured bolus of cold saline; these methods are more mathematically sophisticated.

REMEMBER: *Cardiac output* is stroke volume times heart rate, reported as liters per minute. *Cardiac index* is cardiac output divided by body surface area, reported as liters per minute, per square meter.

Cardiac output as the product of stroke volume and heart rate is intuitively logical. Another expression of output is somewhat more complex and less obvious:

$$\text{Cardiac output} = \frac{\text{mean arterial pressure}}{\text{peripheral resistance}}$$

Let's analyze the dimensional units. Although blood pressure is measured clinically as mm Hg, a scientific definition of pressure is *force per unit-area*. Resistance is the product of *force, time,* and *the negative fifth power of length*. With time measured in seconds, length in centimeters, and area in square centimeters:

$$\text{Cardiac output} \propto \frac{\text{force} \cdot \text{cm}^{-2}}{\text{force} \cdot \text{sec} \cdot \text{cm}^{-5}} = \frac{\text{cm}^3}{\text{sec}}$$

(Without numeric factors, the relationship of units is shown as a proportionality rather than an equation.) We are left with cubic centimeters per second—that is, volume per unit-time, which *is* flow rate or output. Although force simply cancels out of the ratio, it is worth noting that with length in centimeters and mass in grams, force is measured in *dynes* (dyne = g · cm · sec^{-2}). In SI units, with length in meters and mass in kilograms, force is measured in *newtons* (N = kg · m · sec^{-2}; 1 N = 10^5 dynes). Therefore, pressure (force per unit-area) is measured as N · m^{-2}; 1 N · m^{-2} = 1 pascal (Pa); 1 Pa = 0.0075 mm Hg; 1 mm Hg ≈ 133.3 Pa.

STROKE VOLUME AND EJECTION FRACTION

On either side of the heart, the total volume of blood in the ventricle at the beginning of systole is called the *end-diastolic volume*—the sum of the amount of blood remaining after the last ventricular contraction, the amount that descended passively through the open AV valve, and the amount pushed in by the atrial contraction. Self-evidently, stroke volume (the amount ejected from the ventricle) is some fraction of the end-diastolic volume (the amount that was in the ventricle). Expressed as a percentage, that fraction is called the *ejection fraction*:

$$\text{Ejection fraction (\%)} = \frac{\text{stroke volume}}{\text{end-diastolic volume}} \cdot 100$$

For example, if the end-diastolic volume is 90 mL and stroke volume is 72 mL:

$$\text{Ejection fraction} = \frac{72 \text{ mL}}{90 \text{ mL}} \cdot 100 = 80\%$$

A high ejection fraction indicates good pump efficiency (stroke volume is close to end-diastolic volume, leaving little blood in the ventricle after contraction). A low ejection fraction indicates poor pump efficiency (stroke volume is well below end-diastolic volume, leaving considerable blood in the ventricle after contraction).

Adequate cardiac output depends on efficient pump action. Malfunctioning cardiac valves and weakness of the ventricular contraction can interfere with pump efficiency. Valvular insufficiency allows backflow, while stenosis impairs forward flow; a weak contraction means a feeble ejection of blood into the arteries. Such conditions reduce stroke volume and the ejection fraction.

Cardiac output is determined by the combined effects of four interrelated factors—*preload, afterload, contractility,* and *heart rate.* In the following sections, we'll see how these determinants affect each other and cardiac output.

PRELOAD AND STARLING'S LAW

Preload refers to the stretching of the ventricular myocardial tissue before contraction. The greater the end-diastolic volume, the more the myocardium is stretched to contain that volume—and up to a certain physiologic limit, the more the myocardium is stretched, the more forcefully the ventricle contracts.[11]

This phenomenon is the essence of *Starling's Law of the Heart.* The relationship between preload and the efficiency of the ventricular contraction is shown in the *Frank-Starling curve*[12] (Figure 3.7). The left side of the curve illustrates normal function. As myocardial stretching increases, the efficiency of contraction increases; or as end-diastolic volume increases, stroke volume increases. The very purpose of the atrial contraction is to maximize end-diastolic volume, thereby increasing preload and improving the efficiency of

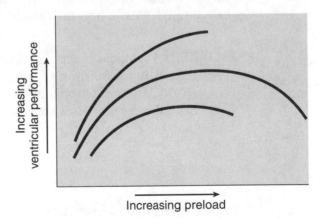

Figure 3.7 The Frank-Starling curve. The different lines indicate variability in myocardial contractility.

---

[11]Likening the more forceful contraction of stretched myocardium to the more forceful snap of a stretched rubber band is a questionable but inevitable simile.

[12]Named for Ernest H. Starling (see Chapter 2) and German physiologist Otto Frank (1865–1944).

Filling Pressure

With the AV valves open during the ventricular-filling phase of diastole, ventricular pressure is normally equilibrated with atrial pressure. This pressure is called the *filling pressure*.

Filling pressure in the ventricle represents potential resistance to the atrial contraction, which is a relatively weak event. Normally, filling pressure is low, so there is little resistance. If filling pressure is high, the atrial contraction may be ineffective in terms of pushing blood down into the ventricle; then excess blood is left in the atria, impeding the entry of venous blood returning to the heart.

In heart failure, a large volume of blood is left in the ventricle at the end of systole. Thus, filling pressure is higher, venous return is impaired, and blood backs up into the veins. In right-side failure, backup into the systemic veins causes elevation in central venous pressure; in left-side failure, backup into the pulmonary veins causes elevation in wedge pressure.

---

the ventricular contraction. As a result, stroke volume—which can only be as large as some fraction of the end-diastolic volume—increases.

The right side of the curve illustrates progressive dysfunction. As preload rises past a certain point, efficiency flattens out and then starts falling. In this condition, the heart is overly stretched and dilated, and further increases in preload will result in *decreased* ventricular efficiency and cardiac output.

Changes in preload occur as part of a homeostatic mechanism. Normally, when venous return to the atria increases, end-diastolic ventricular volume is also greater. In turn, preload rises, the ventricular contraction is more forceful, and stroke volume is increased. In this way, the heart accommodates the larger volume of blood entering via the veins—by pumping that same larger volume forward, into the arteries.

REMEMBER: Preload reflects myocardial stretching to contain the end-diastolic volume. Starling's Law of the Heart states that the efficiency of ventricular contraction improves—that is, stroke volume increases—as preload increases. However, past a certain point, further increases in preload will *reduce* pump efficiency.

AFTERLOAD

Afterload refers to all forms of resistance to the ejection of blood into the arterial system and to the forward flow of blood within the arteries; that is, the total resistance confronting the ventricle during systole. Other things being equal, an increase in afterload causes a decrease in cardiac output.

One of the major components of afterload is the arterial blood pressure. Recall that cardiac output can be expressed as the quotient of mean arterial pressure divided by total peripheral resistance; rearrangement yields:

$$\text{Peripheral resistance} = \frac{\text{arterial pressure}}{\text{cardiac output}}$$

The point is that peripheral resistance varies directly with blood pressure. And because afterload is the totality of resistance to flow, high blood pressure places the burden of additional afterload on the heart.

Two noteworthy biophysical variables that contribute to afterload are *blood viscosity* and *arterial radius.* (A highly viscous fluid is one that is thick and slow-flowing.) In general terms, resistance varies directly with blood viscosity and inversely with vessel radius:

$$\text{Resistance} \propto \frac{\text{viscosity}}{\text{radius}} = \text{viscosity} \cdot \text{radius}^{-1}$$

That is, resistance is higher when blood is more viscous and the vascular radius is smaller, and lower when blood is less viscous and the vascular radius is larger.

Now let's see how these variables affect blood flow. Flow within a vessel varies inversely with resistance:

$$\text{Flow} \propto \frac{1}{\text{resistance}}$$

That is, flow increases with lower resistance, and decreases with higher resistance. Since we have related resistance to viscosity and radius:

$$\text{Flow} \propto \frac{1}{\text{viscosity} \cdot \text{radius}^{-1}} = \frac{\text{radius}}{\text{viscosity}}$$

Flow increases with less viscous blood in wider vessels, and decreases with more viscous blood in narrower vessels.

Viscosity is largely a function of the hematocrit—the fraction of the whole-blood volume represented by erythrocytes and other cellular elements. When the hematocrit is high (as in a condition called *polycythemia*), viscosity is increased. When the hematocrit is low (as in various anemias), viscosity is reduced. However, it takes a relatively large change in hematocrit to affect resistance and blood flow.

In contrast, small changes in vessel radius (mediated by adjustments in arterial smooth muscle tone, mainly at the arteriolar level) yield surprisingly large effects. The reason is mathematical—resistance and flow vary with the *fourth power* of the radius. For a vessel of radius *r*:

$$\text{Resistance} \propto r^{-4}$$

$$\text{Flow} \propto r^4$$

These proportionalities reveal the effect of any change in the radius of a blood vessel. A 10% expansion in the width of a vessel (to 1.1*r*) results in a 33% *decrease* in resistance and a 50% *increase* in blood flow:

$$\text{New resistance} = \frac{1}{1.1^4} \approx \frac{1}{1.5} \approx 0.67 \cdot \text{original resistance}$$

$$\text{New flow} = 1.1^4 \approx 1.5 \cdot \text{original flow}$$

Radius, Pressure, and Resistance

The relationship between vessel radius and arterial pressure may seem self-contradictory. Transmural pressure *falls* as the arteries branch and narrow; yet increased arteriolar smooth muscle tone constricts the vessels and *raises* the blood pressure.

The seeming contradiction arises out of confusion between the local transmural pressure *at the arteriolar level,* and the systolic and diastolic pressures *as measured at the brachial artery.*

Vessel narrowing (aside from obstructive lesions) is due to normal anatomic branching plus the constricting effect of arteriolar smooth muscle tension. Arteriolar narrowing creates *peripheral resistance.* While the pressure within the narrowing vessel drops, the pressure measured at the brachial artery varies directly with the peripheral resistance.

With higher resistance, more driving pressure is needed to push the blood through, which is reflected in a higher systolic pressure. And with higher resistance distally, more blood is left proximally (above the aortic valve) at the end of the ejection phase, which is reflected in a higher diastolic pressure.

Conversely, a mere 5% contraction in width (to $0.95r$) results in a 23% *increase* in resistance and a 19% *decrease* in blood flow:

$$\text{New resistance} = \frac{1}{0.95^4} \approx \frac{1}{0.81} \approx 1.23 \cdot \text{original resistance}$$

$$\text{New flow} = 0.95^4 \approx 0.81 \cdot \text{original flow}$$

REMEMBER: Afterload represents resistance to arterial flow. Afterload is increased when blood is more viscous and the arterial radius is smaller. Arterial flow is equivalent to cardiac output; therefore, the effect of increased afterload is a reduction in cardiac output.

CONTRACTILITY AND HEART RATE

The last two determinants of cardiac output—contractility and heart rate—are somewhat simpler concepts than preload and afterload. *Contractility* refers to the intrinsic efficiency of the ventricular contraction at any given preload and against any given afterload. With higher contractility, the ejection fraction is higher; with a higher ejection fraction, stroke volume is higher; and with a higher stroke volume, cardiac output is higher.

Factors that increase or decrease contractility are called positive or negative *inotropic* factors. For example, decreased contractility is a key element in heart failure, and a drug called *digitalis* is frequently used to improve contractile function; thus, digitalis is positively inotropic.

We have already looked at *heart rate,* in the formula for cardiac output (stroke volume times heart rate). In general, at any given stroke volume, a higher heart rate means higher cardiac output. However, an extremely high rate may not allow adequate time for ventricular filling; in this circumstance, the heart rate itself limits the stroke volume and thereby limits cardiac output.

Factors that increase or decrease the heart rate (which is mediated by the peripheral nervous system) are called positive or negative *chronotropic* factors. A wide variety of drugs produce chronotropic effects as the main therapeutic action or as a side effect.

REMEMBER: Cardiac output is determined by four factors: *preload* (ventricular myocardial stretching to contain the end-diastolic volume); *afterload* (resistance to the ejection of blood into and through the arteries); *contractility* (the intrinsic pumping efficiency of the ventricles); and *heart rate* (the number of ventricular contractions per minute). In general terms, output is enhanced by preload, contractility, and heart rate, and limited by afterload; however, extremely high values for preload or heart rate can limit or even reduce stroke volume and output.

VOLTAGE AND THE ACTION POTENTIAL

The mechanical events of the cardiac cycle—the atrial and ventricular contractions—are triggered by electrical events. Let's review some general concepts about membrane voltage in the cell at rest (resting potential) and the voltage change associated with an electrical impulse (action potential).

Transmembrane voltage results from a charge gradient across the cell membrane—a difference in electrical charge on either side of the membrane. This gradient is established by the distribution of intracellular and extracellular cations and anions (see Chapter 2).

At rest, the membrane is *polarized;* if the charge gradient is larger than normal, the membrane is *hyperpolarized. Depolarization* refers to a decrease in the magnitude of the gradient, with complete loss of polarization during

Negative Voltage

Transmembrane voltage is reported in millivolts (mV). In cells at rest, the voltage is given as a negative value—for example, −80 mV. The negative value indicates that the gradient across the membrane goes from negative inside to positive outside. A positive value (seen for a tiny fraction of a second during the generation of an electrical impulse) would indicate that the gradient goes from positive inside to negative outside.

In terms of pure numbers, −40 is greater than −80; but in terms of transmembrane voltage, −40 mV denotes a smaller charge gradient than −80 mV. (Note that at 0 mV, there is no gradient at all.) Because voltage changes occur mainly within the negative range, certain casually used phrases may be confusing.

A change from −80 mV to −40 mV (depolarization) produces an *upward* deflection on a tracing of transmembrane voltage over time; but to describe this change as a "rise in voltage" is misleading, because it actually represents a *decrease* in the magnitude of the charge gradient. Conversely, a change from −40 mV to −80 mV (repolarization) produces a *downward* deflection on the tracing; but this "fall in voltage" represents an *increase* in the magnitude of the gradient.

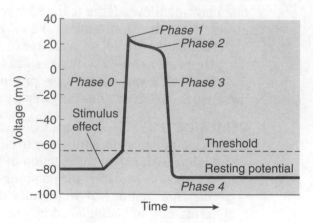

Figure 3.8 Action potential.

the generation of an electrical impulse. *Repolarization* refers to restoration of the normal polarized state of the membrane after depolarization.

Electrical impulses result from ionic shifts—certain extracellular ions move into the cells (influx), while intracellular ions move out of the cells (efflux). The movements of these charged particles across the cell membranes disturb the transmembrane voltage. In this manner, impulses are created and conducted.

The transmembrane voltage in myocardial cells at rest is typically around −80 mV (the value varies in different tissues). A slight depolarization—just enough to nudge the transmembrane voltage to around −65 mV—triggers a dramatic series of events. At that *threshold,* voltage-sensitive sodium channels in the cell membrane suddenly open wide, permitting a tremendous influx of sodium—that is, the main extracellular cation floods inward along its chemical concentration gradient (and also along the electrical gradient; see Chapter 2). Because the charge along the interior of the polarized membrane is negative, the sudden influx of so much positive charge causes rapid depolarization, as shown by the large, almost vertical upward deflection on a voltage tracing (Figure 3.8). This event is *phase 0,* the first part of the action potential. Note that the sodium influx is so great that the depolarization "spike" overshoots the zero point and enters the positive range, peaking at about +20 mV.

Phase 1 represents early repolarization, as shown in Figure 3.8 by the little downward deflection after phase 0.[13] In terms of ionic shifts, phase 1 is caused by an efflux of potassium—the main intracellular cation moves outward along its concentration gradient. As positive charge leaves the cell, the voltage drops back; however, membrane permeability to potassium is still low at this time, so the efflux is relatively slow and the voltage drop is small.

Phase 2 is a plateau, as shown by the flattened segment on the voltage tracing. Phase 2 is caused by an influx of calcium ion; the voltage is almost steady as this slow influx of positive charge counterbalances the slow efflux of potassium that produced phase 1.

---

[13]At this instant, the magnitude of the charge gradient is decreasing as voltage falls from a positive value toward zero. However, the term *repolarization* is used, because this phase is part of the process of restoring the polarized state of the membrane at rest. That is, the membrane at rest is polarized; phase 0 of the action potential is depolarization; and everything that follows is part of repolarization.

Phase 3 represents late repolarization, as shown by the steeper downward deflection, back into the negative range. Phase 3 is caused by an increase in membrane permeability to potassium efflux; as positive charge leaves the cell more quickly, repolarization accelerates.

Phase 4 represents the resting potential, which is again steady at −80 mV. To restore the initial conditions, the sodium that entered the cell in phase 0 and the calcium that entered in phase 2 must be pumped back out, and the potassium that left the cell in phases 1, 2, and 3 must be pumped back in. These counterbalancing ionic shifts take place simultaneously, so there is no significant change in transmembrane voltage—until another stimulus nudges the voltage to the threshold and another action potential begins. (The usual stimulus is depolarization of the adjacent tissue, which is exactly how impulses are conducted and distributed.)

AUTOMATICITY AND CONDUCTION

Electrical activity in the heart normally originates in a specialized patch of right atrial tissue called the *sinoatrial node* (SA node). The unique property of the SA node is *automaticity*—spontaneous generation of periodic electrical impulses. Because of this property, the SA node serves as the heart's primary pacemaker.

In ordinary conductive tissue at rest (that is, during phase 4), sodium is pumped outward at about the same slow rate as it leaks inward along its chemical concentration gradient; likewise, potassium is pumped inward at the same rate as it leaks outward. This balanced state maintains the magnitude of the charge gradient; as a result, the resting potential is stable.

In pacemaker tissue, the cell membrane becomes impermeable to potassium leakage, but the membrane pump mechanism continues to pull the cation inward. This slow but unbalanced influx of positive charge gradually diminishes the magnitude of the charge gradient; as a result, there is no true resting potential. That is, after each action potential, the transmembrane voltage spontaneously approaches the threshold that will trigger the next action potential (Figure 3.9).

The impulse from the SA node is quickly distributed throughout the left and right atria, causing depolarization. Electrical depolarization is the trigger for the mechanical event (atrial contraction). The impulse then arrives at the *atrioventricular node* (AV node), where it is delayed for a split second; this delay allows the atrial contraction to finish before the ventricular contraction

Backup Pacemakers

Like the SA node, the AV node has the property of automaticity—but because its firing rate is slower than that of the SA node, it is depolarized by the SA node impulse before its own action potential arises. However, if the SA node fails, AV node impulses have time to arise and emerge; the heart is then paced by these AV node "escape beats." The AV node thus acts as a slower-rate backup pacemaker. If both nodes fail, nonpacemaker tissue can generate impulses, but at a very slow rate.

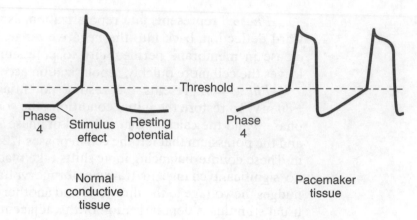

Figure 3.9 In pacemaker tissue, there is no stable resting potential. Even without any stimulus, spontaneous depolarization during phase 4 brings voltage to threshold.

begins (otherwise, the vastly more powerful ventricular contraction would shut the AV valves before filling is complete, thus limiting preload and ventricular efficiency).

From the AV node, the impulse enters the ventricular conductive pathways. As in the atria, the impulse is quickly distributed throughout the left and right ventricles, depolarizing the myocardial tissue and triggering the ventricular contraction on both sides of the heart almost simultaneously (Figure 3.10).

REMEMBER: Transmembrane voltage varies with the charge gradient. Depolarization is a reduction in the magnitude of the gradient; repolarization restores the gradient. When depolarization brings the voltage to a certain threshold, an action potential arises. In pacemaker tissue, the

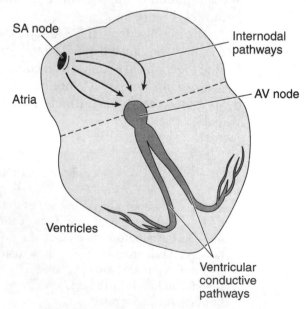

Figure 3.10 General path of depolarization.

voltage spontaneously approaches the threshold after each action potential; the impulses thus generated trigger the atrial and ventricular contractions.

THE ELECTROCARDIOGRAM

Tissue that has been depolarized must be repolarized before the next action potential can arise. Atrial and ventricular depolarization and repolarization are the specific events that are traced on an *electrocardiogram* (ECG or EKG; the device that produces the tracing is the *electrocardiograph*).[14]

The ECG records electrical activity over time on a strip of paper calibrated in millimeters; every fifth line (vertically and horizontally) is dark, creating a grid of squares measuring 5 mm by 5 mm. The paper moves under a stylus that indicates the magnitude of the activity by the vertical height or depth of a deflection; the duration of the event is indicated by the horizontal width of the deflection. With the ECG paper moving at a standard speed of 25 mm · sec^{-1}, 1 mm horizontally represents a time interval of $\frac{1}{25}$ second, or 0.04 second; each 5-mm square represents five times that interval, or 0.2 second (Figure 3.11).

The relationship mm = 0.04 sec is a conversion factor (a ratio equal to 1), which means it can be used as needed in either of its reciprocal forms:

$$\frac{0.04\ \text{sec}}{\text{mm}} = \frac{\text{mm}}{0.04\ \text{sec}} = 1$$

Thus, if a certain feature on the tracing measures 4 mm in width, the duration of that event is calculated as follows:

$$\text{Duration} = 4\ \text{mm} \cdot \frac{0.04\ \text{sec}}{\text{mm}} = 0.16\ \text{sec}$$

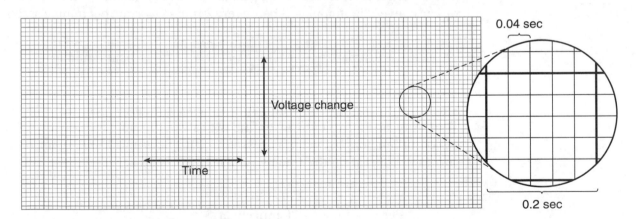

Figure 3.11 ECG paper. Heavy lines create squares 5 mm by 5 mm.

---

[14]Do not confuse an ECG tracing with the voltage tracing of an action potential. The ECG depicts the totality of electrical events everywhere in the heart throughout the cardiac cycle; the action potential tracing depicts a single depolarization in a single cell.

Counting Boxes

If the occurrence interval of a periodic event happens to be a whole number of heavy-line boxes (width = 5 mm), the rate per minute is simply 300 divided by the number of boxes:

ECG Interval	Rate per Minute
1 box	$300 \cdot min^{-1}$
2 boxes	$150 \cdot min^{-1}$
3 boxes	$100 \cdot min^{-1}$
4 boxes	$75 \cdot min^{-1}$
5 boxes	$60 \cdot min^{-1}$
6 boxes	$50 \cdot min^{-1}$

These relationships are based on 5 mm = 0.2 sec. For example, an event that occurs at time intervals equal to 1 box occurs once every 0.2 second, or 5 times per second, or 300 times per minute.

Counting Boxes, Double-Time

Fast events produce narrow ECG deflections at the standard paper speed. Doubling the paper speed spreads the deflections to twice their normal width, which may reveal features and details not otherwise discernible. At a paper speed of 50 mm per second, the time interval represented by a heavy-line box is halved (5 mm = 0.1 sec) and the per-minute rates are doubled (occurrence at every box = 600 per minute, at every second box = 300 per minute, etc.).

The frequency of periodic events can be measured from the horizontal space between successive occurrences. For example, if ventricular depolarization (the electrical correlate to the heartbeat) occurs at intervals of 2.0 cm = 20 mm, we can convert this finding—1 beat every 20 mm—to a rate per minute as follows:

$$\text{Rate} = \frac{\text{beat}}{20 \text{ mm}} \cdot \frac{\text{mm}}{0.04 \text{ sec}} \cdot \frac{60 \text{ sec}}{\text{min}} = 75 \text{ beats} \cdot min^{-1}$$

Since 1 second = 25 mm (the paper speed), 1 minute = $60 \cdot 25 = 1500$ mm. Therefore, an equivalent way to find the rate per minute is to divide the number of millimeters between successive occurrences into 1500. Again, if the occurrence interval on the tracing is 20 mm per beat:

$$\text{Rate} = \frac{1500 \text{ mm} \cdot min^{-1}}{20 \text{ mm} \cdot \text{beat}^{-1}} = 75 \text{ beats} \cdot min^{-1}$$

THE ECG LEADS

Depolarization and repolarization are dynamic processes taking place in the three-dimensional structure of the heart. As we would look at any solid object from different angles or directions, so we must look at the electrical events in the heart from different angles. Each of the 12 leads on a standard ECG is a view of these events as seen along a different directional line.

Don't confuse an ECG *lead* with an *electrode*. An electrode is a sensor device placed on the skin to detect voltage changes caused by depolarization and repolarization. An ECG lead is not a physical device at all; it is a directional line of view created by the machine, utilizing input from the electrodes. Active electrodes are placed on the left arm (LA), right arm (RA) and left leg (LL); the electrode at the right leg is inactive. Regardless of the actual position of the electrode on each limb, the voltage is recorded as if the electrode were placed at the anatomic site of limb attachment to the trunk of the body.

Of the 12 standard ECG leads, six are "limb leads," representing different connections of these electrodes (the machine makes these connections internally). The other six are "precordial leads," obtained by a separate, movable electrode that is sequentially placed at six different positions across the precordium.

To get a sense of the directional lines of view provided by the limb leads, imagine an upside-down equilateral triangle centered directly over the heart, with vertices located approximately at the shoulders and umbilicus (Figure 3.12). This construction is called *Einthoven's triangle.*[15] Its sides show the directional lines for three of the six "limb leads." They are created by the electrocardiograph's internal connections between the active electrodes on the limbs, as follows.

Lead I is created by connecting LA and RA, so the directional line goes horizontally across the body. *Lead II* connects LL and RA, so the directional line goes diagonally down from the right shoulder. *Lead III* connects LL and LA, so the directional line goes diagonally down from the left shoulder.

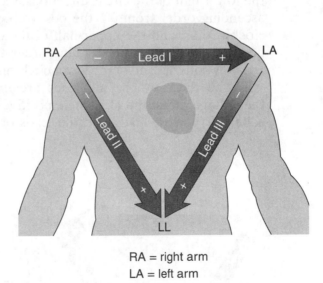

RA = right arm
LA = left arm
LL = left leg

Figure 3.12 Einthoven's triangle. The electrodes can actually be placed anywhere on the limbs.

---

[15]In 1924, Dutch physiologist Willem Einthoven (1860–1927) was awarded the Nobel Prize for Physiology or Medicine.

These leads are called *bipolar leads*, because they show the difference in the voltages recorded at the two connected electrodes. The positive end of each bipolar lead (that is, the positive end of its directional line) is identified by the positive electrode for that lead.

Look again at Figure 3.12. In Lead I, the LA electrode is positive, so the positive end of Lead I is horizontally to the left; the voltage recorded = LA − RA. In Leads II and III, the LL electrode is positive, so the positive end of Lead II is diagonally down and to the left; voltage = LL − RA. The positive end of Lead III is diagonally down and to the right; voltage = LL − LA.

The bipolar limb leads show strong deflections *because* we take these voltage differences. The same electrical activity that is positive in the direction of an ECG lead is negative in the opposite direction, on the other side of the side of the heart. Taking the difference between voltages at the positive and negative electrodes is equivalent to subtracting a negative number—the voltage magnitude on a bipolar lead is significantly greater than what a single electrode would record. (Also, the directional line is different with a single-electrode lead; see below.)

Because parallel lines extend without limit in the same direction, moving a line to a parallel position makes no difference in terms of direction. Therefore, the directional lines for the bipolar leads, corresponding to the sides of Einthoven's triangle, can be repositioned to cross at the center of the triangle—that is, directly over the center of the heart (around the AV node). In this configuration, the positive ends of Leads I, II, and III are designated 0°, 60°, and 120°, respectively (Figure 3.13).

The imaginary circle around the point where the limb leads' directional lines cross is divided into upper and lower halves. Moving clockwise around the lower half of the circle, directions are designated as positive numbers in ascending order from 0° (the positive end of Lead I) to 180°. Continuing clockwise around the upper half of the circle, directions are designated as negative numbers in descending order from −180° back to 0°. Note that ±180° is both the end of the lower circle and the beginning of the upper circle.

The three other limb leads are *unipolar leads,* because they show the voltage measured at a single positive electrode. Unipolar leads are also called V (vector) leads, and the leads obtained by voltage readings at the

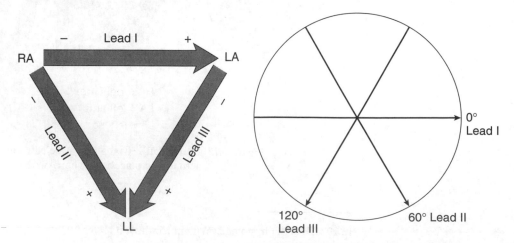

Figure 3.13 Directional lines of the bipolar limb leads (arrows indicate positive ends).

left arm, right arm and left leg (or foot) are designated *VL, VR,* and *VF,* respectively.

In contrast to the directional lines of the bipolar leads, which are created by the connection of two electrodes, the directional lines of the unipolar leads go straight from the center of the heart toward the electrode (that is, toward the anatomic attachment of the limb to the trunk of the body). Thus, the directional line of Lead VL goes diagonally up from the heart toward the left shoulder; Lead VR goes diagonally up from the heart toward the right shoulder; Lead VF goes straight down from the heart toward the left foot.

The voltage readings from a single electrode on a limb are relatively weak. To increase the magnitude of the readings, the ECG machine creates *augmented* leads, designated aVL, aVR, and aVF. Recall that readings on a bipolar lead are obtained as the difference between the voltages recorded at the positive and negative electrodes. With a unipolar limb lead, there is no negative electrode, so voltage in the opposite direction is the average of the voltages recorded at the other two unipolar leads (that is, half their sum). Thus, the differences are obtained as follows:

$$aVL = VL - 0.5(VR + VF)$$

$$aVR = VR - 0.5(VL + VF)$$

$$aVF = VF - 0.5(VL + VR)$$

By taking these differences, the readings are augmented 50%. Here is how it works. With the heart at the center of Einthoven's triangle, the algebraic sum of the voltages recorded at all three vertices is always zero (because voltage read as positive in any given direction is negative in the opposite direction). Therefore:

$$VL + VR + VF = 0$$

We can rearrange this equation to solve for any term. For example:

$$VL = -VR - VF = -(VR + VF)$$

This expression replaces VL in the augmentation formula:

$$aVL = -(VR + VF) - 0.5(VR + VF)$$

$$aVL = -1.5(VR + VF)$$

But if $VL = -(VR + VF)$, then reversing the signs yields $VR + VF = -VL$. Therefore:

$$aVL = -1.5(-VL) = 1.5VL$$

As stated, aVL is 50% greater than VL. It works the same way with aVR and aVF. Like the bipolar limb leads, the augmented unipolar limb leads can be shown crossing at the center of Einthoven's triangle. The electrodes at the left arm, left leg (foot), and right arm identify the positive ends of aVL, aVF, and aVR, which are designated −30°, +90°, and −150°, respectively[16] (Figure 3.14).

The precordial leads are also unipolar or V leads, and they are designated $V_1, V_2, \ldots V_6$. With the movable electrode placed directly over the precordium, the voltage readings are strong; augmentation is unnecessary. The electrode identifies the positive direction of the lead at each of the six positions (Figure 3.15). Note that the limb leads provide directional lines of view in the frontal (coronal) plane of the body, while the precordial leads provide lines of view in the transverse plane (Figure 3.16).

REMEMBER: Each ECG lead views electrical activity within the heart along a different directional line. The directional lines of the six limb leads (I, II, III, aVL, aVR, and aVF) are in the frontal plane; the lines of the six

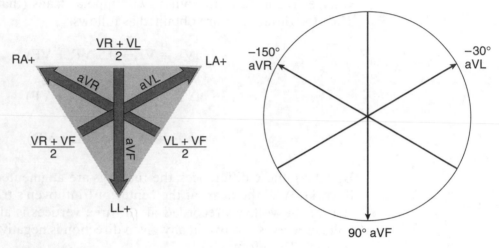

Figure 3.14 Directional lines of the augmented unipolar limb leads (arrows indicate positive ends).

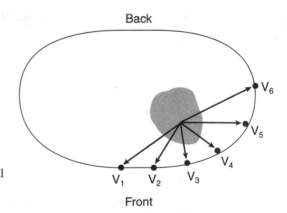

Figure 3.15 Directional lines of precordial leads (seen in transverse section).

[16]Don't get mixed up by the fact that the *positive* ends of aVL and aVR are designated by *negative* degrees.

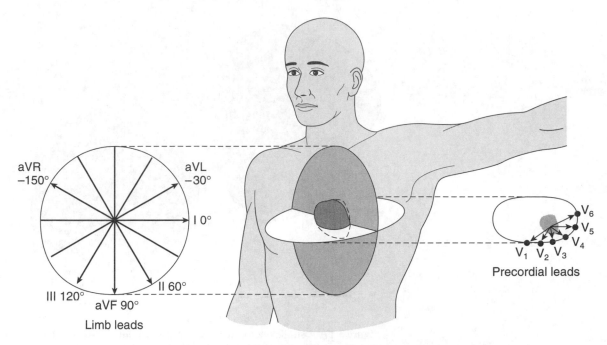

Figure 3.16 Intersecting planes of limb leads (frontal) and precordial leads (transverse).

precordial leads (V_{1-6}) are in the transverse plane. The positive end of each lead is identifed by the positive electrode serving that lead.

CARDIAC VECTOR

During diastole, when no distinct electrical event is taking place, the ECG stylus barely moves as the paper advances beneath it, and the tracing remains at baseline—a steady horizontal line indicating that the electrodes detect no voltage change. When voltage changes caused by depolarization and repolarization are detected, the stylus moves up or down.

Depolarization and repolarization progress as waves of electrical activity through the conductive pathways and the myocardial tissues. *Some portion of the wave of activity lies along each directional line of the ECG, moving toward or away from that lead*—that is, toward or away from the positive end of that lead.

In practical terms, electrical activity moving toward an ECG lead along its directional line causes an upward deflection on the tracing *for that lead;* activity moving away from a lead causes a downward deflection.

The electrical activity in any tiny region of the heart can be represented by a *vector*—an arrow whose length is proportional to the voltage driving the activity, and whose direction denotes the course of the moving wave of activity. Because vectors can be combined, the magnitudes and directions of uncountable vectors from every region can be replaced by a single vector representing overall electrical activity throughout the heart (Figure 3.17).

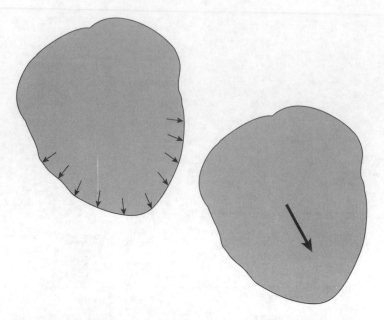

Figure 3.17 A single vector arrow can represent all electrical activity in the heart.

Each ECG lead records the portion of the cardiac vector lying along its own directional line. We can see that portion by constructing a rectangle such that the diagonal is the vector and one side is parallel to the direction of the ECG lead. The length projected along the side parallel to the ECG lead is the portion of the vector recorded by that lead (Figure 3.18).

As shown in Figure 3.18, the vector and its projection along an ECG lead form a right triangle. If the angle between the vector and the lead is α, the vector is the hypotenuse of the triangle and its projection is the side adjacent to α. Therefore:

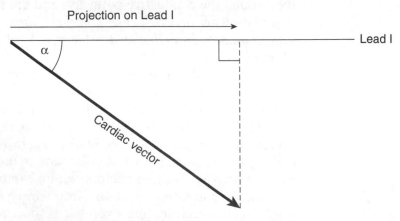

Figure 3.18 The voltage recorded on Lead I is proportional to the length of the side adjacent to angle α in a right triangle whose hypotenuse is the cardiac vector. Since cosine α = adjacent divided by hypotenuse, the projection on Lead I = cardiac vector times cosine α.

$$\text{Cosine } \alpha = \frac{\text{adjacent side}}{\text{hypotenuse}} = \frac{\text{projection on ECG lead}}{\text{vector}}$$

$$\text{Projection on ECG lead} = \text{vector} \cdot \text{cosine } \alpha$$

$$\text{Vector} = \frac{\text{projection on ECG lead}}{\text{cosine } \alpha}$$

For example, if angle $\alpha = 40°$, the projection along that ECG lead is the product of the vector magnitude and the cosine of 40° (approximately 0.766). The ECG lead would record 0.766 of the actual vector voltage, and the vector voltage would be $1/0.766 \approx 1.305$ times the voltage recorded along that lead.

If the cardiac vector is more or less parallel to an ECG lead, the deflection on that lead is large, because the recorded portion is almost the full magnitude of the vector; mathematically, as angle α decreases toward 0°, cos α increases toward 1. If the vector is perpendicular to the lead, the deflection is just a small baseline oscillation, because the recorded portion of the vector is short; as α approaches an absolute maximum of 90°, cos α approaches 0 (Figure 3.19).

The six limb leads may be grouped as three *orthogonal* (perpendicular) pairs, as follows: Leads I and aVF (at 0° and +90°, respectively); Leads II and aVL (at +60° and −30°); and Leads III and aVR (at +120° and −150°). Whichever lead has the largest deflections, its orthogonal partner will have the smallest—because when most of the cardiac vector is recorded along one directional line, very little is recorded along the perpendicular line.

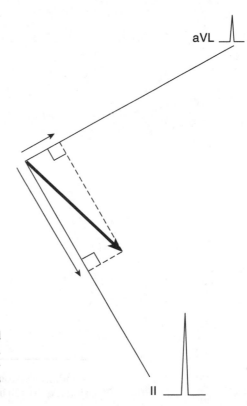

Figure 3.19 The vector is about parallel to Lead II, which shows a large deflection. The same vector is almost perpendicular to Lead aVL, which shows a small deflection.

REMEMBER: On each ECG lead, the magnitude of the deflection is proportional to the voltage driving the activity along that directional line. The deflection is upward if the activity is moving toward the lead, or downward if it is moving away.

VOLTAGE MAGNITUDE

Having looked at voltage in general terms, we now turn to actual measurements.[17] Just as the duration and frequency of electrical events can be measured on the horizontal axis of the ECG tracing (25 mm = 1 second), the magnitude of voltage deflections can be measured on the vertical axis, by the following relationship:

$$10 \text{ mm} = 1 \text{ cm} = 1 \text{ mV}$$

At the beginning of a tracing, a 1-mV spike is generated to test the machine's calibration; the deflection of this test spike should measure exactly 1 cm.

If Lead I shows an upward deflection of 12 mm above baseline, it is recording 1.2 mV of electrical activity moving toward 0°; and if we know that angle α is 20°, the magnitude of the cardiac vector is calculated as follows:

$$\text{Vector voltage} = \frac{\text{voltage on Lead I}}{\text{cosine } 20°} = \frac{1.2 \text{ mV}}{0.94} \approx 1.28 \text{ mV}$$

On Lead III, the same electrical event produces a downward deflection of 2 mm below baseline; it is recording 0.2 mV moving away from 120° (Figure 3.20).

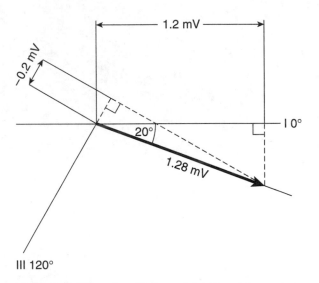

Figure 3.20 When Lead I shows 1.2 mV and the angle to the vector is 20°, the vector magnitude is 1.28 mV. Lead III shows −0.2 mV. (Scale is enlarged from ECG deflections.)

---

[17]Standard electrocardiography does not provide perfectly precise voltage readings. Greater precision requires an invasive procedure in which electrodes are placed directly upon the heart itself.

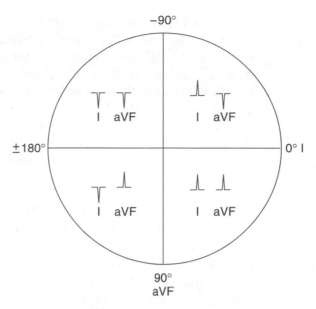

Figure 3.21 Electrical axis by quadrant.

ELECTRICAL AXIS

Each ECG lead is a vector—the direction is that of the lead, and the magnitude (defined by the height or depth of the deflection) is the voltage along that line. *Overall* electrical activity is represented by the cardiac vector arrow. The length of the arrow defines overall voltage, and the direction defines the *electrical axis* of the heart—the overall direction of the wave of electrical activity. Like the ECG limb leads, the electrical axis is measured in degrees on the frontal plane.

A quick way to get the approximate direction of the axis is to look at the deflections in Leads I and aVF; these leads define a quadrant of the complete frontal-plane circle. If the deflections are positive in both of these leads, the axis lies somewhere between 0° and 90°. Likewise, location in each of the other quadrants is identified by the deflections in these two leads (Figure 3.21).

Locating the axis with greater precision is more complex. The limb lead most closely parallel to the electrical axis of the heart should show maximal deflections—yet we do *not* identify the electrical axis by looking for the lead with the largest deflections.[18] Instead, we look for the lead with the most symmetrically positive and negative deflections. A lead showing a relatively small oscillation with deflections of equal magnitude above and below the baseline is called the *isoelectric lead*. The closer a lead is to being perfectly isoelectric, the closer it is to being perfectly perpendicular to the axis, and the closer its orthogonal partner is to being perfectly parallel to the electrical axis. For example, if Lead III is isoelectric, the axis is parallel to Lead aVR. Then if the deflections in aVR are negative, the axis is pointing directly away from aVR; that is, away from −150°, which places the axis at +30°.

---

[18]The problem with looking for the largest deflections is that the bipolar limb leads *inherently* produce slightly larger deflections than the augmented unipolar leads.

Usually, no lead is perfectly isoelectric, so we make a small adjustment. For example, if Lead III is almost isoelectric but slightly more positive than negative, the axis lies slightly less than a right angle from Lead III's 120° heading, perhaps at 35° or 40°. Conversely, if Lead III is slightly more negative than positive, the axis lies slightly more than a right angle from 120°— perhaps at 25° or 20°.[19]

Here is the most accurate way to locate the axis. Trace the net deflection in each of the bipolar leads along its directional line, toward or away from the lead depending on whether the net deflection is positive or negative. (Net deflection is the maximum height above baseline minus the maximum depth below baseline.) Suppose the ECG tracing shows the following net deflections: +12 mm (corresponding to +1.2 mV) in Lead I, +8 mm (+0.8 mV) in Lead II, and −4 mm (−0.4 mV) in Lead III. From these measured positions, construct perpendiculars to intersect at a single point (Figure 3.22). A line drawn from the origin to this triple intersection point is the cardiac vector arrow. Its direction (measured by protractor) is the electrical axis of the heart. In this case, the axis is about +10°. Its magnitude can be determined by its length (measured by ruler) or by the cosine method shown earlier:

$$\text{Vector voltage} = \frac{\text{voltage on Lead I}}{\text{cosine } 10°} \approx \frac{1.2 \text{ mV}}{0.985} \approx 1.22 \text{ mV}$$

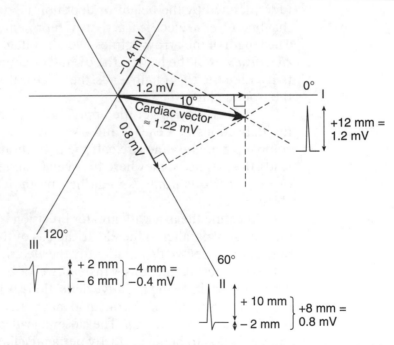

Figure 3.22 Axis determination. From the net deflections along the bipolar leads, perpendiculars intersect at a point that defines the axis and the vector voltage. (Tracings are shown actual size; scale is enlarged to show projected lengths on the leads and the constructed perpendiculars.)

---

[19]The ECG machine computes the axis automatically, but clinicians often do their own estimations anyway—usually by this method of finding the isoelectric lead.

The three constructed perpendiculars will always intersect at a single point. The underlying principle is called *Einthoven's law,* which states that the net deflection (and corresponding voltage) in Lead II is equal to the sum of the net deflections (and corresponding voltages) in Leads I and III.[20] Likewise, the perpendiculars from Leads I and III will always meet at an intersection point with the perpendicular from Lead II.

Let's confirm it, using the given ECG findings. In terms of net deflections, $12 \, \text{mm} + (-4) \, \text{mm} = 8 \, \text{mm}$; in terms of voltage, $1.2 \, \text{mV} + (-0.4) \, \text{mV} = 0.8 \, \text{mV}$.

COMPONENTS OF THE ECG TRACING

Now we are ready to look at the components of an ECG tracing. Considering the anatomic orientation of the heart and the general direction of depolarization from the SA node through the atria and then the ventricles, Lead II often (but not always) provides the best overall view of electrical events; that is, the electrical axis is normally somewhere near Lead II, which therefore shows relatively large positive deflections. (However, look again at Figure 3.22; in that case, the axis is at 10° and Lead I would provide the best overall view.)

As seen in Lead II, the depolarization of the atria produces an upward deflection called the *P wave.* It is a relatively small deflection, because the atria are structurally thin and the amount of tissue being depolarized is small. Recall that when the impulse reaches the AV node, it is delayed (to let the atrial contraction finish before the ventricles contract). During that split-second delay, there is no electrical activity, so the tracing is flat.

---

[20]Einthoven's law is based on a unique property of the 60° angle (the angle between the bipolar limb leads). If n is some angular distance from 60°, the sum of the cosines of $(60 + n)°$ and $(60 - n)°$ is equal to the cosine of $n°$. If the axis is 70°, it is $60 + 10 = 70°$ clockwise from Lead I, $60 - 10 = 50°$ counterclockwise from Lead III, and 10° clockwise from Lead II. With voltage obtained as the product of the vector voltage and the cosine of the angle between the vector and the lead, apply Einthoven's law. *Equate the combined voltages on Leads I and III with the voltage on Lead II:* (vector voltage · cos 70°) + (vector voltage · cos 50°) = vector voltage · cos 10°. *Factor out vector voltage:* cos 70° + cos 50° = cos 10°. *Check by plugging in the numeric values of these cosines:* $0.342 + 0.643 = 0.985$. The rule is, "Wherever the axis lies, Einthoven's law applies."

When the impulse enters the ventricular conductive system, a large mass of tissue is depolarized at great speed, producing a tall but narrow deflection—the *QRS complex*. A *Q wave* indicates a *downward* deflection as the first part of the complex; normally, Lead II shows little or no Q wave. The main part of the complex is the tall upward deflection—the *R wave*. A downward deflection following the R wave is an *S wave* (not the descent from the peak of the R wave, but a deflection below the baseline after the R wave). The interval from the beginning of the P wave to the beginning of the QRS complex (whether or not there is a Q wave) is called the *P-R interval*.

A brief period of quiescense follows the QRS complex before the ventricular repolarization wave appears; the deflection representing this event is called the *T wave*. Note that it is shorter but wider than the R wave, indicating that repolarization is a slower, less orderly process than depolarization.[21] Atrial repolarization is not discernible at all; it occurs at about the same time as ventricular depolarization, and the tiny deflection it would cause is completely lost in the vastly larger QRS complex.

Each of these features of the ECG tracing is evaluated in terms of magnitude and duration (Figure 3.23). On a 12-lead tracing, the deflections in Lead II look similar to those in Lead aVR—but upside down. This finding is normal, because electrical activity that is moving generally toward Lead II is moving generally away from Lead aVR; as shown in Figure 3.16, these leads are 150° apart (not quite directly opposite from each other).

Note that while the direction (positive versus negative) and magnitude of the voltage readings vary from lead to lead, the timing of events (duration and frequency) is the same on every lead.

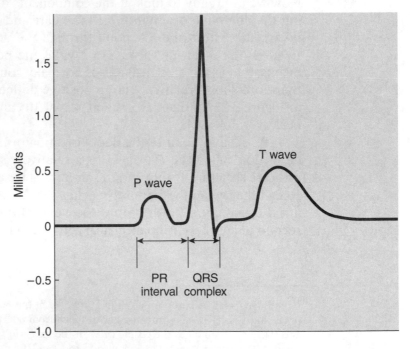

Figure 3.23 Components of normal ECG tracing (enlarged scale).

---

[21]Because all the ventricular tissue that was depolarized must be repolarized, the areas under the R wave and T wave should be mathematically equal.

1. A patient's arterial blood pressure is measured at 180/110 mm Hg (systolic/diastolic). What is the mean arterial pressure (MAP)? How is MAP affected if the pressure is reduced to 165/110 mm Hg? To 180/95 mm Hg?

2. What is the pulse pressure in a patient with a systolic pressure of 135 mm Hg and MAP of 95 mm Hg?

3. As blood goes from an artery of a certain inner diameter to an artery of half that diameter, does resistance increase or decrease? By what proportion? Does flow increase or decrease? By what proportion?

4. Convert a pressure of 25 cm H_2O to units of mm Hg and also to pascals (Pa). Convert 25 mm Hg to cm H_2O and Pa. Convert 3000 Pa to mm Hg and cm H_2O.

5. What is the stroke volume in a patient with a heart rate of 68 beats/minute, an arterial oxygen content of 18 mL/dL, a venous oxygen content of 14 mL/dL, and an oxygen consumption rate of 240 mL/minute?

6. Compute the cardiac output (in liters per minute) in a patient with a blood pressure of 130/85 mm Hg and peripheral resistance of 1600 dyne · sec · cm^{-5}.

7. What is the body surface area if stroke volume is 55 mL, heart rate is 80 beats/minute, and cardiac index is 2.8 liters/minute per square meter?

8. In Patient A, cardiac output is 5.0 liters/minute and heart rate is 76 beats/minute. In Patient B, output is 4.5 L/min, heart rate is 78/minute. Both patients have the same end-diastolic volume. All other things being equal, is it possible to predict which patient has the higher filling pressure?

9. At standard ECG paper speed (25 mm/second), the width of the QRS complex is 0.6 cm. What is the duration of this event (in seconds)? If these QRS complexes occur at intervals of exactly 1.5 heavy-line boxes, what is the heart rate?

10. If the electrical axis of the heart is 15° and the voltage on Lead II is +1.4 mV, predict the voltage readings on Leads I and III. The following cosine values are provided:

cos 0° = 1.0	cos 105° ≈ −0.259
cos 15° ≈ 0.966	cos 120° = −0.5
cos 30° ≈ 0.866	cos 135° ≈ −0.707
cos 45° ≈ 0.707	cos 150° ≈ −0.866
cos 60° = 0.5	cos 165° ≈ −0.966
cos 75° ≈ 0.259	cos 180° ≈ −1.0
cos 90° = 0	

Answers and explanations at end of book.

Air, Blood, and Respiratory Function

Chapter Outline

Mathematical Tools Used in This Chapter

The respiratory system aerates blood, which means bringing in oxygen and removing carbon dioxide. Unaerated blood is pumped through the lungs by the right side of the heart; the aerated blood then returns to the left side of the heart.[1]

---

[1]The term *aerated* refers to systemic arterial blood, which is fully primed with oxygen from the lungs; that is, oxygenated blood. *Unaerated* refers to systemic venous blood, which has delivered much of its oxygen to the body cells. Venous blood is *not* totally deoxygenated; it still carries a significant amount of oxygen, as well as the carbon dioxide it has picked up from the cells.

OVERVIEW

Before we approach the mathematical expression of respiratory function, a brief review of pertinent anatomy and physiology is useful. The movement of air into and out from the lungs follows a pressure gradient. When the pressure inside the lungs drops below the pressure outside the body, air is pulled inward; when the inside pressure rises above the outside pressure, air is pushed outward.

Inspiration is due mainly to movement of the diaphragm. From its arched, at-rest position, downward contraction of the diaphragm expands the intrathoracic space and lowers the intrapulmonary pressure, pulling air inward and expanding the lungs. Expiration is mostly passive, due mainly to the elasticity of the lungs. From their inflated state, recoil contraction raises the intrapulmonary pressure, pushing air outward and allowing the diaphragm to return to its at-rest position.

The lungs are encased in a doubled-over sheet of tissue—the *pleurae*. The outer layer (the parietal pleura) is attached to the interior surfaces of the chest wall. The inner layer (the visceral pleura) is attached to the exterior surfaces of the lungs. The potential space between the two layers is called the *pleural space*.

Negative Pressure

Normally, the pleural space is closed. The parietal and visceral pleurae lie against each other, gliding smoothly over each other by virtue of a thin film of lubricant fluid between them. The pleurae are held close together by "negative pressure." Since the pressure in a perfect vacuum would be 0 mm Hg, negative pressure is a relative measure, not an absolute value.

Pressures within the respiratory system are given as the *difference* from atmospheric pressure. "Negative pressure in the pleural space" means that the intrapleural pressure is lower than atmospheric pressure; $-N$ mm Hg means N mm Hg is lower than atmospheric pressure.

Negative pressure is crucial to respiratory function. As long as the intrapleural pressure (pressing inward upon the lungs) is lower than the intrapulmonary pressure (air in the lungs, expanding the lungs outward), the lungs occupy the full volume of the thoracic cavity throughout the respiratory cycle. As the diaphragm descends and the rib cage expands, the volume of the thoracic cavity increases, transiently pulling the parietal pleura away from the visceral pleura and further lowering the intrapleural pressure. The result: the lungs expand, the pleural space remains closed, intrapulmonary pressure drops, and air is pulled inward. This mechanism prevents the lungs from collapsing under their own contractile elasticity.

If the chest wall is punctured, outside air enters the pleural space; likewise, if an area of lung ruptures outward, intrapulmonary air is released into the pleural space. In either case, the inward pressure upon the lungs may approach or even surpass the outward pressure from within the lungs. Without negative pressure in the pleural space, the lung on the involved side will partially or totally collapse—a condition called *pneumothorax*. Then the pleural space is open, which means that there is a measurable gap between the visceral pleura covering the collapsed lung and the parietal pleura lining the inside of the chest wall.

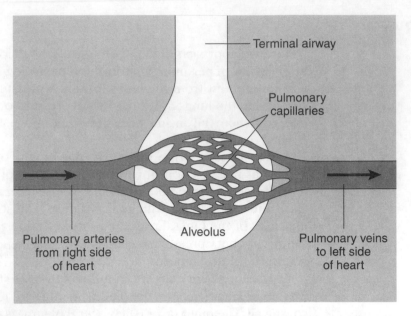

Figure 4.1 Schematic of alveolus surrounded by pulmonary capillaries.

During inhalation, air moves through the upper airways (nose, mouth, trachea) and enters a branching "tree" of narrowing passageways within each lung. These passageways terminate in the *alveoli*—a few hundred million microscopic sacs surrounded by capillaries of the pulmonary circulation (Figure 4.1).

Gas exchange takes place across the ultrathin membranous structures between air in the alveoli and blood in the surrounding capillaries. This boundary is the *interface* between air and blood in the body. The aeration of blood occurs as inhaled oxygen in the alveoli diffuses across the interface and enters the blood in the capillaries; at the same time, carbon dioxide in the blood diffuses across in the opposite direction, leaving the capillaries and entering the alveoli to be exhaled.

Because of the large number of branchings between the mainstem bronchus and the alveoli, the combined alveoli in the two lungs represent a large surface area, which facilitates efficient diffusion and exchange of gases.

AIR VOLUMES IN THE RESPIRATORY SYSTEM

If a healthy man inhales as deeply as possible, the volume of air in his lungs would be around 6 liters; for a woman, it would be around 4.3 liters. This volume—the amount of air in the lungs after maximum forced inhalation—is called the *total lung capacity*. All other measures of respiratory volume are fractions of the total lung capacity (Figure 4.2).

In a healthy adult at rest, each normal breath moves approximately 500 milliliters of air in and out of the lungs. The total amount of air in the lungs rises and falls by this *tidal volume* during normal breathing.

The difference between total lung capacity and the peak of the tidal volume—that is, the *additional* volume that can be forcibly inhaled after normal inspiration—is called the *inspiratory reserve volume*. For men, the inspiratory reserve volume is around 3.3 liters; for women, around 2.0 liters.

Chapter 4 / Air, Blood, and Respiratory Function

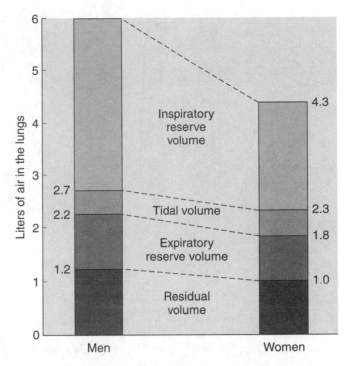

Figure 4.2 Respiratory volumes.

Even after a maximum-effort forced expiration, a certain volume of air remains in the lungs. This minimum volume is called the *residual volume*. For men, the residual volume is around 1.2 liters; for women, around 1.0 liter.

The difference between residual volume and the trough of the tidal volume—that is, the *additional* volume that can be forcibly exhaled after normal expiration—is called the *expiratory reserve volume*. For men, the expiratory reserve volume is around 1.0 liter; for women, around 0.8 liter. A continuous tracing of the volume of air in the lungs over time shows these measures dynamically (Figure 4.3).

The difference between the total lung capacity (the volume at the top of the inspiratory reserve) and the residual volume (at the bottom of the expiratory reserve) is called the *vital capacity*. For men, vital capacity is around 4.8 liters; for women, around 3.3 liters.

The total volume that can be forcibly inhaled after normal expiration is the *inspiratory capacity,* which is around 3.8 liters in men, and 2.5 liters in women. Note that the inspiratory capacity is equal to the tidal volume plus the inspiratory reserve volume.

The volume of air remaining in the lungs after normal expiration is the *functional residual capacity,* which is around 2.2 liters in men, and 1.8 liters in women. The functional residual capacity is equal to the residual volume plus the expiratory reserve volume; it is the volume of air in the lungs when the diaphragm is in its at-rest position.

One last compartment to consider is the *anatomic dead space*—inhaled air that never reaches the alveoli but is left in the passageways leading to the alveoli at the end of inspiration. (In contrast, the residual volume is air left in the alveoli and the passageways after forced expiration.) Dead-space air is the last air inhaled and the first air exhaled; and because it does not reach the alveolar-capillary interface, it is exhaled with no real change in its gaseous

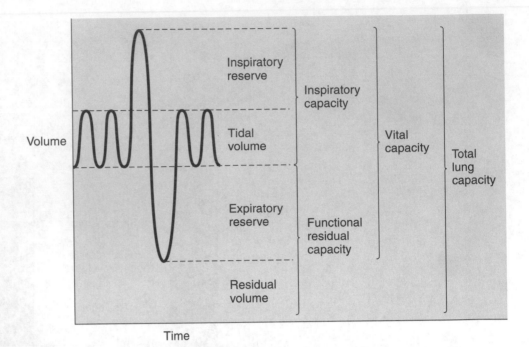

Figure 4.3 Spirometric terminology.

composition. Typically, the dead-space volume in milliliters is very roughly equal to the lean body weight in pounds.

PULMONARY FUNCTION TESTING

Multiplying the tidal volume by the respiratory rate at rest gives the *pulmonary ventilation* (respiratory volume per minute)—the total volume of air moved in 1 minute of normal breathing. At typical values in a healthy individual:

$$\text{Pulmonary ventilation} = \text{tidal volume} \cdot \text{respiratory rate}$$

$$= 0.5 \frac{\text{L}}{\text{breath}} \cdot 12 \frac{\text{breaths}}{\text{min}} = 6 \text{ L} \cdot \text{min}^{-1}$$

Multiplying the difference between the tidal volume and the dead space by the respiratory rate gives the *alveolar ventilation*—the total volume of air undergoing gas exchange in 1 minute. If the dead-space volume is 140 mL = 0.14 L:

$$\text{Alveolar ventilation} = (\text{tidal volume} - \text{dead space}) \cdot \text{respiratory rate}$$

$$\approx 0.36 \frac{\text{L}}{\text{breath}} \cdot 12 \frac{\text{breaths}}{\text{min}} \approx 4.3 \text{ L} \cdot \text{min}^{-1}$$

In addition to measuring ventilation during normal breathing, pulmonary function testing measures the *maximum* volume and rate of air movement. The procedure involves forced inspiration followed by forced expiration—from the top of the inspiratory reserve to the bottom of the expiratory reserve, in one continuous effort.

The total volume of air that can be exhaled during this maneuver is called the *forced vital capacity*. Recall that vital capacity is the difference between total lung capacity and residual volume; forced vital capacity (FVC) is the measurement of that difference during one continuous, maximum expiratory effort. FVC is the total volume exhaled, regardless of how long the maneuver takes.

With respect to the rate of expiration, the *forced expiratory volume in one second* (FEV_1) gives the volume exhaled in the first second of the FVC maneuver. FEV_1/FVC is the ratio of the volume exhaled in the first second to the total volume exhaled in the completed maneuver. In a healthy adult, the ratio is 0.8 or higher; that is, at least 80% of FVC is exhaled in the first second.

The *peak expiratory flow rate* is the maximum rate achieved at any time during the FVC maneuver. If we divide FVC into quartiles, the *mid-expiratory flow* from the 25th to the 75th percentile of volume (MEF_{25-75}) is the airflow rate for that middle portion of FVC (Figure 4.4).

Lung *compliance* reflects the stretchability versus stiffness of the lung tissue, as defined by the relationship between intrapulmonary air volume and intrapleural negative pressure during respiration (Figure 4.5). Compliance is measured as the slope of the line connecting the points representing volume and pressure after inspiration and expiration. An abnormally steep slope indicates increased compliance (airway and alveolar distention and flaccidity); this condition is characteristic of *emphysema*. Conversely, an abnormally shallow slope indicates decreased compliance (excessive stiffness); this condition is characteristic of a variety of diseases, such as *interstitial fibrosis*.

REMEMBER: The maximum *volume* of air exhaled in one continuous effort is the forced vital capacity; the *rate* of exhalation during this maneuver

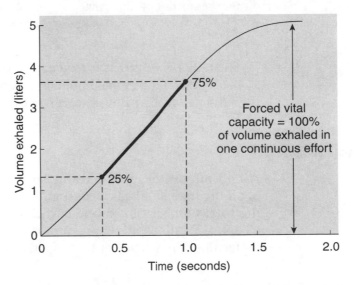

Figure 4.4 Mid-expiratory flow rate is the slope of the marked line segment. If the volume change is $3.6 - 1.2 = 2.4$ L, and the time interval is $1.0 - 0.4 = 0.6$ sec, the flow rate is 2.4 L/0.6 sec = 4 L/sec.

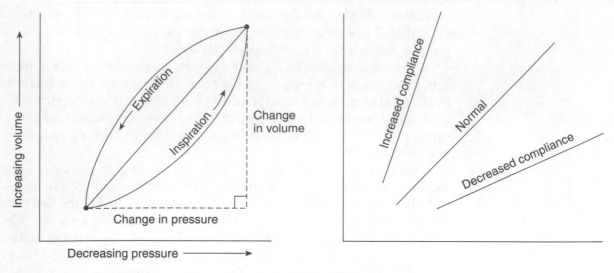

Figure 4.5 Volume-pressure loop during respiration. Compliance is the slope of the straight line connecting the tips of the loop. Increased compliance means the lung is flaccid; it takes very little pressure change to achieve a large volume change. Decreased compliance means the lung is stiff; it takes a large pressure change to achieve significant volume change.

Two General Categories of Ventilatory Impairment

In *obstructive* pulmonary conditions such as asthma or chronic bronchitis, the airways are narrowed and the airflow rate is reduced; so the characteristic finding on pulmonary function testing is a low FEV_1. (Air volume, as reflected in FVC, may be normal or abnormal.)

In *restrictive* pulmonary conditions such as chest wall immobility or movement disorders involving the diaphragm, the total air volume is reduced; so the characteristic finding is a low FVC. (Airflow rate, as reflected in FEV_1, may be normal or abnormal.)

can be measured as forced expiratory volume in 1 second, peak expiratory flow rate, or maximum mid-expiratory flow rate. Obstructive conditions reduce airflow rate; restrictive conditions reduce total air volume.

PARTIAL PRESSURE

Air is a mixture of gases, and atmospheric pressure is the sum of the pressures of all the individual gases present. Just as osmotic pressure is determined by the total number of osmotic particles *of any type* per liter of solution (see Chapter 2), atmospheric pressure is determined by the total number of gas molecules *of any type* per liter of air, as shown in the following relationship:[2]

---

[2]The gas constant (8.314 joules · kelvin^{-1} · mole^{-1}) relates pressure to the number of moles of gas held at a certain Kelvin-scale temperature within a certain volume. Pressure is expressed as joules per unit-volume (see Chapter 1). Note that the *type* of gas is not part of the equation.

Chapter 4 / Air, Blood, and Respiratory Function

$$Pressure = \frac{moles \cdot absolute\ temperature \cdot gas\ constant}{volume}$$

If total atmospheric pressure is the sum of the individual gas pressures and each gas represents a certain percentage of the atmosphere, the *partial pressure* (P) of any individual gas is simply that percentage of the total pressure. Since atmospheric pressure[3] at sea level is 760 mm Hg and nitrogen accounts for 78% of the atmosphere, the partial pressure of nitrogen (P_{N_2}) is obtained as follows:

$$P_{N_2} = 760 \cdot 0.78 \approx 593\ mm\ Hg$$

Oxygen accounts for 21% of the atmosphere; this value is called the *fraction of inspired oxygen* (Fio_2). Therefore, its partial pressure is obtained as follows:

$$P_{O_2} = 760 \cdot 0.21 \approx 160\ mm\ Hg$$

Small amounts of carbon dioxide, water vapor, and other gases account for the remaining 1% of the atmosphere and the last 7 mm Hg of atmospheric pressure.

Now that we know the meaning of partial pressure *in a mixture of gases* (the alveolar side of the interface), we must look at the meaning of partial pressure *in a liquid such as blood* (the capillary side of the interface).

The Rate of Gas Exchange

The rate at which gas molecules cross the alveolar-capillary interface varies directly with two general factors and two gas-specific factors:

- The total surface area of the interface.
- The thinness of the interface.
- The intrinsic permeability of the interface to *that gas.*
- The transinterface concentration gradient for *that gas.*

Diffusion is fastest when the interface is large in surface area but thin; and when the interface is highly permeable to the gas, and there is a large difference in concentration across the interface.

Interface permeability is a constant for each gas. (Note that the chemical nature of a gas is irrelevant in terms of air pressure, but crucial in terms of membrane permeability.)

Thus, the concentration gradient across the interface is the main variable. The larger the gradient, the faster the rate of crossing (going from the side at higher concentration to the side at lower concentration). *Concentrations in air and in blood are both measured as partial pressure.*

---

[3]One atmosphere (1 atm) $\approx$ 760 mm Hg = 101,325 Pa (the SI unit of pressure; see Chapter 1). The *torr* is distinct from mm Hg in definition but virtually interchangeable in value—that is, 1 torr = 1 mm Hg. All of these measures of atmospheric pressure at sea level are equivalent to the value traditionally taught to schoolchildren: 14.7 lb $\cdot$ in^{-2}.

If a mixture of gases is held above a volume of liquid in a closed system, some portion of each gas will dissolve into the liquid to a point of equilibrium. For each gas, the amount that dissolves is directly proportional to its partial pressure in the gas mixture (a variable) and its solubility in that liquid (a constant; an intrinsic characteristic of the gas). At a higher partial pressure and with a higher solubility coefficient, more of the gas will dissolve into the liquid; and vice versa.

For a given gas with its characteristic solubility coefficient, the partial pressure in the gas mixture determines the amount that dissolves into the liquid; and the amount that dissolves into the liquid implies its partial pressure in the gas mixture. For example, if the partial pressure of carbon dioxide in air is 40 mm Hg, the equilibrium concentration in the blood is around 1.2 mmol/L. Then if the blood concentration of dissolved carbon dioxide is 1.2 mmol/L, we say that the partial pressure of carbon dioxide in the blood is 40 mm Hg.

REMEMBER: The partial pressure of a dissolved gas in the blood is the partial pressure in a gas mixture that would yield the measured concentration of that gas in the blood.

GAS EXCHANGE

To see how gas exchange occurs, let's start with inhaled air *approaching* the alveoli. Air becomes saturated with water vapor as it passes through the upper air passages and the multiple bronchial branchings to reach the alveoli. So the partial pressure of water vapor rises significantly, to around 47 mm Hg, while the partial pressures of the other gases drop slightly from their characteristic values in the ambient atmosphere. For oxygen, at $Fio_2 = 21\%$:

$$Po_2 = (760 - 47) \cdot 0.21 = 713 \cdot 0.21 \approx 150 \text{ mm Hg}$$

The air *in* the alveoli is lower in oxygen (which is diffusing into the blood) and higher in carbon dioxide (which is diffusing out of the blood).

These gas exchanges take place *constantly*. Do *not* think in terms of a load of air inhaled, undergoing gas exchange, and then exhaled, with the alveoli empty and nothing happening between breaths. Because of the negative pressure in the pleural space, the alveoli are *always* filled with air. Oxygen continually crosses the interface from the alveolar air to the capillary blood, and carbon dioxide continually crosses in the opposite direction. *The purpose of the cyclic movement of air in and out of the lungs is to replenish the alveolar air.* With each breath, inhaled air mixes with alveolar air, replacing the oxygen that has entered the blood and removing the carbon dioxide that has come out from the blood.

Thus, respiration maintains the partial pressure of each gas in the alveoli within an optimal steady-state range (Figure 4.6). With alveolar oxygen held at around 100 mm Hg, oxygen crosses the interface at a rate that raises oxygen tension from 40 mm Hg in unaerated blood to 100 mm Hg after aeration.[4] With carbon dioxide crossing the interface at a rate that drops carbon dioxide tension from 45 mm Hg in unaerated blood to 40 mm Hg after aeration, alveolar carbon dioxide is held at around 40 mm Hg.

The rate-limiting factor in gas exchange is the diffusing capacity of oxygen. Not only is the alveolar-capillary interface more permeable to carbon dioxide than to oxygen, but the blood itself soon becomes saturated with oxygen (see the following section). At a certain respiratory rate, oxygen is diffusing *into* the blood as fast as it can go. However, carbon dioxide continues diffusing *out* of the blood faster and faster as the respiratory rate rises still higher.

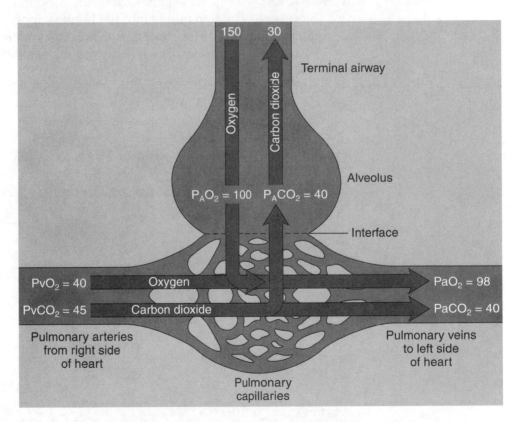

Figure 4.6 Gas exchange at the alveolar-capillary interface.

---

[4]The lung tissue itself needs a blood supply. Blood from the left side of the heart arrives via the bronchial arteries, and most of it returns to the right side via the bronchial veins. However, some of this unaerated blood returns to the *left* side via the *pulmonary* veins, mixing with aerated blood from the pulmonary capillaries. Because of this *physiologic shunt*, P_aO_2 (oxygen tension in the left side of the heart) is slightly lower than P_AO_2 (partial pressure of alveolar oxygen, reflecting oxygen tension in the pulmonary capillaries, before the mixing occurs).

REMEMBER: Carbon dioxide diffuses across the alveolar-capillary interface more easily than oxygen does. Even when oxygen tension reaches its peak, carbon dioxide tension continues to drop with increased respiration.

OXYGEN IN THE BLOOD

Oxygen is carried in the blood in two forms. Of the total amount of oxygen in arterial blood, about 99% is bound to hemoglobin in red blood cells, while 1% is dissolved in the plasma (and measured as P_aO_2). Gas exchange directly involves dissolved oxygen; hemoglobin-bound oxygen is affected secondarily.

The diffusion of alveolar oxygen across the interface transiently raises the oxygen tension in the plasma; as a result, some of the dissolved oxygen enters the red blood cells and becomes bound to hemoglobin. At the systemic capillaries, the process runs in the opposite direction. The delivery of dissolved oxygen to the tissues via the interstitial fluid transiently lowers the oxygen tension in the plasma; as a result, hemoglobin-bound oxygen is released into the plasma (Figure 4.7).

REMEMBER: The main supply of oxygen is bound to hemoglobin in red blood cells. Oxygen moves into the bound supply at the pulmonary capillaries, and out of the bound supply at the systemic capillaries; in both cases, the transfer occurs via dissolved oxygen in the plasma.

But how does hemoglobin "know" when to bind and release oxygen?

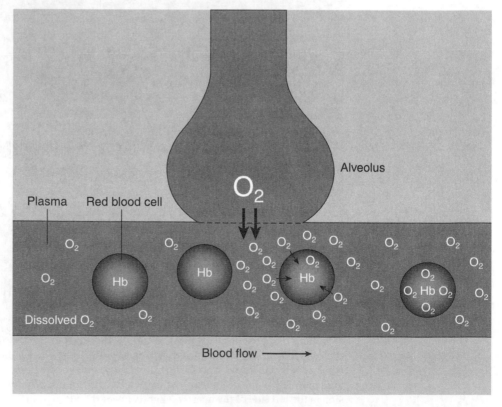

Figure 4.7 Uptake of inhaled oxygen by hemoglobin.

THE OXYGEN-HEMOGLOBIN DISSOCIATION CURVE

Each hemoglobin molecule can bind up to four oxygen molecules. Starting with all four binding sites unoccupied and available, the successive binding of each oxygen molecule increases hemoglobin's affinity for the next; that is, the first binding accelerates the second, which accelerates the third, which accelerates the fourth. Fully oxygenated hemoglobin is called *oxyhemoglobin*.[5]

Conversely, starting with all four sites occupied, the successive unbinding of each oxygen molecule decreases hemoglobin's affinity for the remainder; that is, the first release accelerates the second, which accelerates the third, which accelerates the fourth (Figure 4.8).

If whole blood is allowed to equilibrate with pure oxygen at high pressure, the largest possible amount of the oxygen will dissolve into the blood,

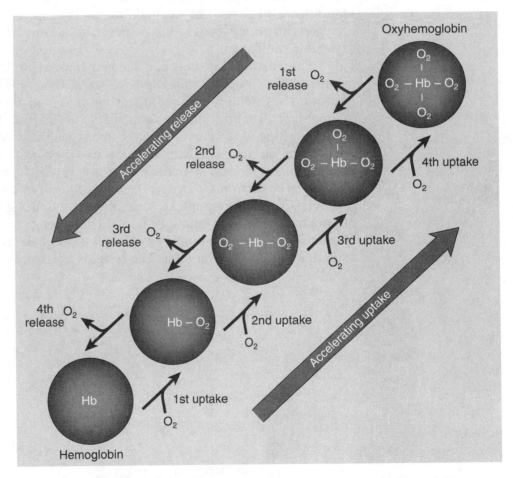

Figure 4.8 Successive uptake and successive release of oxygen by hemoglobin.

---

[5]Oxyhemoglobin gives arterial blood its characteristic bright red color, while hemoglobin carrying much less oxygen gives venous blood its duskier red color. Beware of the widely held myth that venous blood is blue but somehow instantly turns dark red when exposed to air; this notion probably arose because veins seen through the skin look bluish against the surrounding tissues.

virtually all hemoglobin binding sites will be occupied with oxygen, and saturation will be 100%. Blood aeration in a healthy individual breathing normal air at normal pressure should produce an arterial oxygen saturation of around 97%.

At the pulmonary capillaries, oxygen follows its concentration gradient and crosses the alveolar-capillary interface to enter the plasma as a dissolved gas. The resulting rise in oxygen tension in the blood initiates the binding of oxygen to hemoglobin—and the beginning of that process accelerates its continuation. The accelerating uptake of oxygen by hemoglobin explains how such a large amount of oxygen can be stored so quickly as blood passes through the pulmonary capillaries.

Conversely, at the systemic capillaries, oxygen leaves the plasma and enters the interstitial fluid to reach the body cells. The resulting drop in oxygen tension in the plasma initiates the unbinding of oxygen from hemoglobin—and the beginning of that process accelerates its continuation. The accelerating release of oxygen by hemoglobin explains how such a large amount of oxygen can be delivered to the tissues so quickly as blood passes through the systemic capillaries.

The relationship between oxygen tension and oxygen saturation is shown in the *oxygen-hemoglobin dissociation curve* (Figure 4.9). As oxygen tension rises steadily from zero (a state that is incompatible with life), the percentage of saturation increases slowly at first but then accelerates. Note that the slope of the curve—change in oxygen saturation divided by change in oxygen tension—becomes steepest over the range of partial pressures between 15 mm Hg and 40 mm Hg. *The steeper the slope, the larger the change in oxygen saturation for any given change in oxygen tension.*

The accelerating uptake soon reaches a limiting factor—as each additional molecule of oxygen is bound, the number of available binding sites is diminished. When the binding process is at its peak rate, the number of sites still open is rapidly dwindling. The result is a slowdown in the rate of further

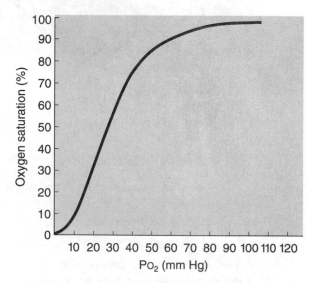

Figure 4.9 Oxygen-hemoglobin dissociation curve.

Chapter 4 / Air, Blood, and Respiratory Function

binding, as shown by the way the curve flattens out going toward the right side. The percentage of saturation is still rising as oxygen tension rises, but at a much slower rate.

Look again at Figure 4.9. At the pulmonary capillaries, as oxygen tension rises from a pre-aeration value of 40 mm Hg up to about 70 mm Hg, the slope of the curve is steep and saturation rises sharply, from about 75% up to 90%. However, as oxygen tension continues rising, up to about 100 mm Hg, the curve is much flatter and saturation rises only another 7% or so.

At the systemic capillaries, the process runs in the opposite direction. As oxygen delivery to the tissues begins and oxygen tension falls from a fully aerated value of 100 mm Hg down to approximately 70 mm Hg, the curve is relatively flat and saturation decreases slowly. However, as oxygen tension continues falling, down to about 40 mm Hg, the slope of the curve is much steeper and saturation drops off more rapidly.

REMEMBER: At the pulmonary capillaries, a large amount of oxygen is quickly taken up and bound by hemoglobin; at the systemic capillaries, a large amount of oxygen is quickly released for delivery to the body tissues.

The oxygen-hemoglobin dissociation curve shown in Figure 4.9 reflects normal physiologic conditions of acid-base balance and temperature. The entire curve is shifted leftward under conditions of increasing alkalinity and/or falling temperature; and rightward under conditions of increasing acidity and/or rising temperature (Figure 4.10).

Now observe how a leftward or rightward shift affects oxygen saturation. At any given oxygen tension, the percent saturation is higher when the curve is shifted leftward, and lower when the curve is shifted rightward. In any position, the curve functions in the same way because the shape of the curve is essentially the same; that is, an increase in oxygen tension increases

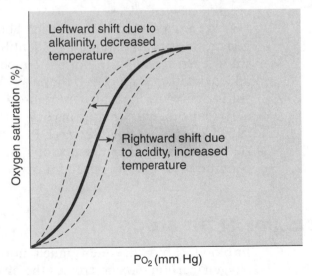

Figure 4.10 The position of the oxygen-hemoglobin dissociation curve may shift left or right, but the shape of the curve is the same.

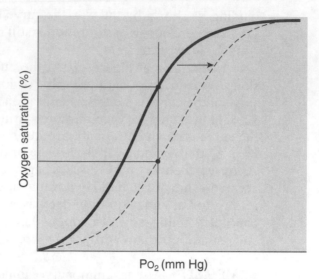

Figure 4.11 Rightward shift of oxygen-hemoglobin dissociation curve. At any given oxygen tension, the percentage of saturation is lower.

binding affinity and saturation, and a decrease in oxygen tension decreases binding affinity and saturation (Figure 4.11).

This mechanism is physiologically efficient. Increased cellular activity may create a buildup of carbon dioxide, and an increase in carbon dioxide tension means an increase in acidity (see Chapter 6). Likewise, increased muscular activity creates a buildup of heat. These conditions shift the dissociation curve rightward. So at the same oxygen tension, oxygen saturation is lower; that is, oxygen has been released. And at any position of the curve, the beginning of oxygen release accelerates further release. In short, when more oxygen is needed to meet increased metabolic needs, the conditions favor increased oxygen release.

At any position of the oxygen-hemoglobin dissociation curve, the oxygen tension at 50% saturation is identified as P_{50}. When the curve is at the position corresponding to normal temperature and acid-base balance, P_{50} is approximately 30 mm Hg. If the curve shifts leftward, P_{50} is lower; if the curve shifts rightward, P_{50} is higher (Figure 4.12).

One other factor affecting oxygen binding to hemoglobin is a compound called *2,3-diphosphoglycerate* (DPG). In red blood cells, DPG promotes the release of oxygen from hemoglobin. Logically, DPG levels rise in just those conditions in which the need for oxygen delivery to the tissues is increased.

CARBON DIOXIDE IN THE BLOOD

In blood, carbon dioxide is much more soluble than oxygen. Compared to oxygen, carbon dioxide crosses the alveolar-capillary interface more easily and dissolves into plasma more extensively.

However, the majority of carbon dioxide is carried in other forms. Both in the plasma and in the red blood cells, some carbon dioxide forms compounds with the amino groups of protein molecules. But something else happens in the red blood cells—an enzyme called *carbonic anhydrase* causes

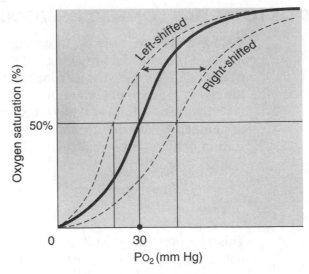

Figure 4.12 P_{50} is the oxygen tension at which saturation = 50%. P_{50} is lower when the oxygen-hemoglobin dissociation curve shifts leftward, and higher when the curve shifts rightward. Note that at any one of these three partial pressures, saturation is higher for the left-shifted curve, and lower for the right-shifted curve.

rapid hydration of carbon dioxide to form carbonic acid, which undergoes partial dissociation into hydrogen ions and bicarbonate ions:

$$H_2O + CO_2 \rightleftarrows H_2CO_3 \rightleftarrows H^+ + HCO_3^-$$

The bicarbonate ions reenter the plasma, while the hydrogen ions are bound up by hemoglobin.[6] This process is a buffering action; as explained in Chapter 6, a buffer is a system that prevents large changes in acid-base balance. If the hydrogen ions formed from the hydration of carbon dioxide were all released into the plasma, the plasma would become more acidic; instead, the hydrogen ions thus formed are bound up by hemoglobin—and the less oxygen is bound to hemoglobin, the more efficiently hemoglobin can bind hydrogen ions.

Again, note the efficiency of the system. Venous blood contains less oxygen and more carbon dioxide than arterial blood, and a higher concentration of carbon dioxide means a higher concentration of hydrogen ions. So venous blood has more hydrogen ions; but because it also has less oxygen, the hemoglobin is more efficient at binding hydrogen ions and thus preventing significant acidification.

REMEMBER: In red blood cells, carbon dioxide undergoes hydration to form carbonic acid, which dissociates into hydrogen ions (bound by hemoglobin) and bicarbonate ions (released back into the plasma).

---

[6]Bicarbonate ions leaving the red blood cells are replaced by chloride ions that enter the cells; because of this "chloride shift," there is no net change in electric charge.

MEASURES OF TOTAL GAS CONTENT IN BLOOD

We have seen that dissolved oxygen and carbon dioxide are measured in terms of their partial pressures, but that a greater amount of each gas is carried in other forms. Now let's look at the actual total amounts of each gas in the blood (at normal temperature and normal acid-base conditions).

In arterial blood at $P_aO_2 \approx 100$ mm Hg (97% oxygen saturation), 1 gram of hemoglobin carries around 1.3 mL of oxygen. If the hemoglobin concentration in blood is normal at 15 g/dL:

$$\text{Total arterial oxygen} = 1.3 \, \frac{mL}{g} \cdot 15 \, \frac{g}{dL} \approx 20 \, mL \cdot dL^{-1}$$

This measurement—a volume of gas per deciliter of blood—is sometimes called *volume percent* (in this usage, percent means per deciliter; see Chapter 2).

In venous blood at $P_vO_2 \approx 40$ mm Hg (75% saturation), 1 gram of hemoglobin carries 1 mL of oxygen. Then at the same hemoglobin concentration:

$$\text{Total venous oxygen} = 1 \, \frac{mL}{g} \cdot 15 \, \frac{g}{dL} \approx 15 \, mL \cdot dL^{-1}$$

The reduction in oxygen content going from arterial to venous blood is called the *A-V difference*. At the preceding values, 20 mL/dL − 15 mL/dL = 5 mL of oxygen delivered to the tissues for each deciliter of blood that passes through the systemic capillaries; most of this amount is oxygen that had been bound by hemoglobin.

Now let's look at total carbon dioxide content: 1 dL of arterial blood at $P_aCO_2 \approx 40$ mm Hg carries almost 50 mL of carbon dioxide, and 1 dL of venous blood at $P_vCO_2 \approx 46$ mm Hg carries almost 53 mL.

Note that while P_aO_2 is normally 2.5 times greater than P_aCO_2 (100 mm Hg versus 40 mm Hg), the total volume of carbon dioxide in arterial blood is 2.5 times greater than the total volume of oxygen (50 mL/dL versus 20 mL/dL). Recall that carbon dioxide is far more soluble than oxygen and thus dissolves into plasma more extensively at a lower partial pressure. In addition, dissolved carbon dioxide represents just a small part of "total CO_2"—most of the carbon dioxide is carried as bicarbonate and compounds formed with protein amino groups (Figure 4.13). At $P_aCO_2 = 40$ mm Hg, the *molar* concentration of dissolved carbon dioxide itself is only 1.2 mmol/L, but the molar concentration of total CO_2 is around 23 mmol/L; this latter value corresponds to the *volume* concentration 50 mL/dL.

MEASURES OF RED BLOOD CELLS AND HEMOGLOBIN

Hemoglobin in red blood cells plays a key role in gas transport, binding 99% of the oxygen in arterial blood and most of the hydrogen ions formed by the hydration of carbon dioxide in venous blood. The *complete blood count* (CBC) offers a standard set of measurements of red blood cells and their hemoglobin content.

The counts of *erythrocytes, leukocytes,* and *thrombocytes* (red cells, white cells, and platelets) are each given as the number per microliter of

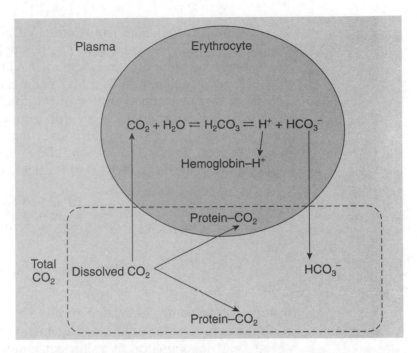

Figure 4.13 Dotted line encompasses various forms in which carbon dioxide is carried in blood. Dissolved gas in plasma (measured as P_{CO_2}) is a small part of total CO_2.

whole blood. The *hematocrit* is the decimal or percent fraction of the volume of whole blood represented by cellular elements (red cells, outnumbering white cells 1000 to 1, account for almost all of the cellular volume); the remainder of the whole-blood volume is plasma. The hemoglobin concentration in whole blood is reported in units of grams per deciliter.

The CBC also includes three *erythrocyte indices*—direct measurements of red blood cells. The *mean corpuscular volume* (MCV) is the average volume of an individual red cell. The hematocrit fraction of 1 microliter of whole blood is the combined volume of *all* the cellular elements (mostly erythrocytes) in 1 microliter; then to find MCV (the volume of *one* erythrocyte), divide the hematocrit by the number of erythrocytes in 1 microliter. If the hematocrit is $45\% = 0.45$ and the erythrocyte count is 5 million per microliter, MCV (in femtoliters; 1 fL $= 10^{-15}$ L) is calculated as follows:

$$MCV = \frac{\text{hematocrit}}{\text{erythrocyte count}} = \frac{0.45}{5 \cdot 10^6 \, \mu L^{-1}} = \frac{0.45 \, \mu L}{5 \cdot 10^6}$$

$$= \frac{0.45 \cdot 10^{-6} \, L}{5 \cdot 10^6} = 0.09 \cdot 10^{-12} \, L = 90 \cdot 10^{-15} \, L = 90 \, \text{fL}$$

As a shortcut, multiply the decimal hematocrit by 1000 (or the percent hematocrit by 10), and then divide by the number of millions of red cells per microliter:

$$MCV = \frac{0.45 \cdot 1000}{5} \quad or \quad \frac{45 \cdot 10}{5} = \frac{450}{5} = 90 \, \text{fL}$$

The *mean corpuscular hemoglobin* (MCH) is the average amount of hemoglobin per individual red cell. The hemoglobin concentration in whole blood is the total amount of hemoglobin in *all* the erythrocytes in the volume of blood measured; then to find MCH (the amount in *one* erythrocyte), divide the hemoglobin concentration by the number of erythrocytes in that volume; the volume units in the hemoglobin concentration and the erythrocyte count are different, so we will be converting one to the other. If the hemoglobin concentration is 15 g/dL and the erythrocyte count is 5 million/μL, MCH (in picograms; 1 pg $= 10^{-12}$ g) is calculated as follows:

$$\text{MCH} = \frac{\text{hemoglobin concentration}}{\text{erythrocyte count}} = \frac{15 \text{ g} \cdot \text{dL}^{-1}}{5 \cdot 10^6 \, \mu\text{L}^{-1}} = \frac{15 \cdot 10^{-5} \text{ g} \cdot \mu\text{L}^{-1}}{5 \cdot 10^6 \, \mu\text{L}^{-1}}$$

$$= \frac{15 \cdot 10^{-5} \text{ g}}{5 \cdot 10^6} = 3 \cdot 10^{-11} \text{ g} = 30 \cdot 10^{-12} \text{ g} = 30 \text{ pg}$$

Multiplying the hemoglobin concentration by 10^{-5} converts its volume unit to μL^{-1} (the same as in the erythrocyte count), so that μL^{-1} cancels out of the numerator and denominator. As a shortcut, multiply the hemoglobin concentration by 10 and then divide by the number of millions of red cells per microliter:

$$\text{MCH} = \frac{15 \cdot 10}{5} = 30 \text{ pg}$$

The *mean corpuscular hemoglobin concentration* (MCHC) is the weight concentration of hemoglobin in the red cells (which is, of course, where hemoglobin is found), rather than in whole blood. To find MCHC, divide the hemoglobin concentration in whole blood by the hematocrit. If the hemoglobin concentration is 15 g/dL and the decimal hematocrit is 0.45, MCHC (in g/dL) is calculated as follows:

$$\text{MCHC} = \frac{\text{hemoglobin concentration}}{\text{hematocrit}} = \frac{15 \text{ g} \cdot \text{dL}^{-1}}{0.45} \approx 33 \text{ g} \cdot \text{dL}^{-1}$$

If the denominator is the percent hematocrit, the numerator is multiplied by 100:

$$\text{MCHC} = \frac{(15 \text{ g} \cdot \text{dL}^{-1}) \cdot 100}{45} \approx 33 \text{ g} \cdot \text{dL}^{-1}$$

MCHC is occasionally reported as a percentage rather than as a weight-per-volume concentration—that is, 33% rather than 33 g/dL. The computation is performed in the same way and yields the same numeric value, but it is interpreted as the percentage of the total red cell mass represented by hemoglobin.

VENTILATION-PERFUSION RELATIONSHIPS

So far, we have looked at gas exchange across the alveolar-capillary interface in theoretical terms. To estimate how efficiently aeration is really taking place, we can compare the partial pressure of *alveolar* oxygen ($P_{A}O_2$, calculated from atmospheric data) and the *arterial* oxygen tension (P_aO_2, measured directly). Nor-

A Deadly Bond

The binding of oxygen and hemoglobin provides an efficient system for carrying oxygen in the blood and delivering it to the body tissues. *Carbon monoxide (CO)* is a poison gas that exerts its deadly effects by interfering with the normal interaction of oxygen and hemoglobin.

Hemoglobin binds CO with enormous strength—more than 200 times its affinity for oxygen—forming *carboxyhemoglobin*. Because of this preferential binding to CO, the affected hemoglobin will not carry oxygen. If the exposure to CO is limited, there may be enough unaffected hemoglobin to carry sufficient oxygen to sustain life.

Treatment with pure oxygen at high pressure will gradually break down carboxyhemoglobin so that the CO can be eliminated from the body.

mally, aeration produces near-equilibrium between these oxygen levels, so the alveolar-arterial gradient (A-a oxygen gradient $= P_AO_2 - P_aO_2$) should be small.

To calculate P_AO_2, we must make adjustments to the air mix of the ambient atmosphere. Nitrogen accounts for most of the air, but it does not undergo exchange; so the only adjustments to be made involve water vapor and carbon dioxide. Both of these gases are minor components in the ambient atmosphere, but their partial pressures are significant in the alveolar air.

Recall how the partial pressure of oxygen is reduced by the rising partial pressure of water vapor in the air approaching the alveoli. We subtract that water vapor pressure (around 47 mm Hg) from the overall pressure (760 mm Hg), and then, as usual, multiply by the Fio_2 (around 21%):

$$(760 - 47) \cdot 0.21 = 713 \cdot 0.21 \approx 150 \text{ mm Hg}$$

Since carbon dioxide is constantly diffusing out of the blood and into the alveolar air, we subtract its partial pressure from the overall alveolar pressure. The estimated partial pressure of carbon dioxide in the alveolar air is P_aco_2 multiplied by the factor 1.2. Therefore:

$$P_AO_2 = [(\text{atmospheric pressure} - H_2O \text{ pressure}) \cdot Fio_2] - (P_aco_2 \cdot 1.2)$$

With the same conditions as before, and $P_aco_2 = 40$ mm Hg:

$$P_AO_2 = ([760 - 47] \cdot 0.21) - (40 \cdot 1.2)$$

$$= (713 \cdot 0.21) - 48 \approx 150 - 48 = 102 \text{ mm Hg}$$

Note that this *predicted* partial pressure of alveolar oxygen (102 mm Hg) is close to the *measured* arterial oxygen tension in a healthy individual (100 mm Hg), just as it should be—and so the gradient is small:

$$\text{A-a oxygen gradient} = P_AO_2 - P_aO_2$$

$$= 102 - 100 = 2 \text{ mm Hg}$$

At high altitude, atmospheric pressure is lower; therefore, the partial pressure of oxygen in the alveolar air and oxygen tension in the arterial plasma are both lower than normal. However, the A-a gradient (the difference between these two values) should still be small.

In contrast, a large gradient means that something is interfering with the aeration of blood. The problem may involve *ventilation* (the ability of air to reach the alveolar-capillary interface via the air passages in the lungs) or *perfusion* (the ability of blood to reach the interface via the pulmonary vasculature). The problem may also lie at the interface itself, in terms of its total surface area and/or permeability (Figure 4.14).

An obstruction in the airways (a ventilation defect) creates a functional "right-to-left shunt." Blood is perfusing the pulmonary capillaries normally, but air is not reaching all the alveoli. As a result, some of the unaerated blood coming from the right side of the heart returns to the left side of the heart without having undergone the normal gas exchanges. Causes of airway obstruction include inhaled foreign bodies and bronchogenic carcinoma.

Conversely, an obstruction within the pulmonary vasculature (a perfusion defect) represents "wasted ventilation." Air is reaching the alveoli normally, but blood is not perfusing the pulmonary capillaries. With blocked pulmonary circulation, the nonperfused alveoli represent a *physiologic dead space*. Like the anatomic dead space, the physiologic dead space refers to air in a region of the lung where gas exchange is not taking place. The classic cause of circulatory obstruction in the lungs is pulmonary embolism.

Such conditions are called *ventilation-perfusion defects*. Various clinical tests can be used to assess the quality of ventilation (V) and perfusion (Q); the status of these functions is then defined by the "V/Q ratio." Obviously,

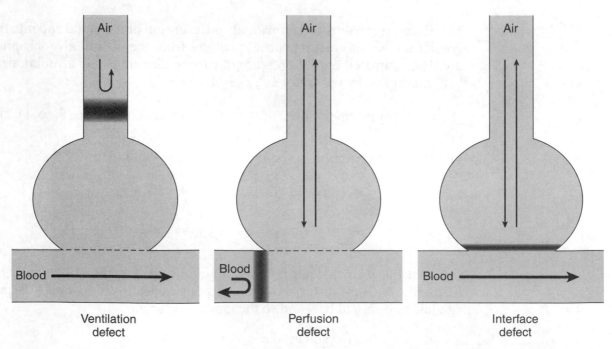

Figure 4.14 In a ventilation defect, blood is shunted back to the heart without adequate aeration. In a perfusion defect, ventilation is "wasted." In an interface defect, air and blood may both reach the interface, but gas exchange is impaired.

both ventilation and perfusion must be intact for the aeration of blood to proceed normally.

QUIZ

1. An adult has a total lung capacity of 5.4 L, a residual volume of 0.9 L, a functional residual capacity of 2.0 L, and a tidal volume of 500 mL. What is the inspiratory reserve volume? The expiratory reserve volume? The vital capacity?

2. On pulmonary function testing, a man's forced vital capacity (FVC) is 4.4 L and his forced expiratory volume in one second (FEV_1) is 2.8 L. What is the FEV_1/FVC ratio? What do these findings suggest?

3. If the mid-expiratory flow rate from the 25th to the 75th percentile (MEF_{25-75}) is 4 L/sec and the actual time for this interval is 0.6 sec, what is the FVC? (Hint: first calculate the volume between the 25th and 75th percentiles of FVC, and then the entire FVC.)

4. A patient has a P_aO_2 of 96 mm Hg and a P_aCO_2 of 42 mm Hg. At normal sea-level atmospheric conditions, what is the alveolar-arterial (A-a) oxygen gradient?

5. A patient is breathing a special mix of air containing 50% oxygen at normal atmospheric pressure. What is the partial pressure of oxygen in this air? If P_aO_2 is measured at 78 mm Hg and P_aCO_2 is 54 mm Hg, what is the A-a oxygen gradient?

6. Using the oxygen-hemoglobin dissociation curve shown in Figure 4.9, estimate the oxygen saturation in arterial blood at $P_aO_2 = 60$ mm Hg, $P_aO_2 = 70$ mm Hg, and $P_aO_2 = 80$ mm Hg. (Assume normal temperature and acid-base balance.)

7. If an individual has a disorder that causes an abnormally large amount of carbon dioxide to be retained in the blood, would the oxygen saturation at any given oxygen tension be higher than normal, lower than normal, or unaffected?

8. A patient has a hemoglobin concentration of 12 g/dL and an arterial oxygen tension of 98 mm Hg. What is the total *volume* of oxygen in the arterial blood? Assuming that oxygen saturation in venous blood is 75%, estimate the rate of oxygen delivery to the tissues.

9. A patient has one lung surgically removed. What effects might be expected in terms of FVC, FEV_1, lung compliance, P_aO_2, and P_aCO_2?

10. Calculate the MCV, MCH, and MCHC in a patient whose hematocrit is 38%, erythrocyte count is 4.6 million/μL, and hemoglobin concentration is 12 g/dL.

Answers and explanations at end of book.

chapter 5

Renal Function

Chapter Outline

Overview
The Formation of Urine
The Medullary Osmotic Gradient
The Glomerular Filtration Rate
Clearance
The Clearance Formula
Assessing GFR
Volume Considerations Related to GFR
Transport Maximum
Renal Control of pH

Mathematical Tools Used in This Chapter

Negative exponents (Appendix C)
Dimensional units with negative exponents (Appendix D)
Unit conversions (Appendix E)
Logarithms and logarithmic scales (Appendix F)

The kidney is one of the body's main systems for clearing waste products from the blood and eliminating foreign substances.[1]

OVERVIEW

Before approaching the mathematical aspects of renal function, we must review some basic anatomy and physiology. Seen in cross section, the dark outer area of the kidney is the renal *cortex*. The lighter-colored "pyramids" beneath the cortex form the renal *medulla* (Figure 5.1).

---

[1]The lungs and liver also eliminate wastes and foreign substances. The kidney's role in controlling blood pressure is addressed briefly in Chapter 3; its other functions are deferred to a comprehensive course in physiology.

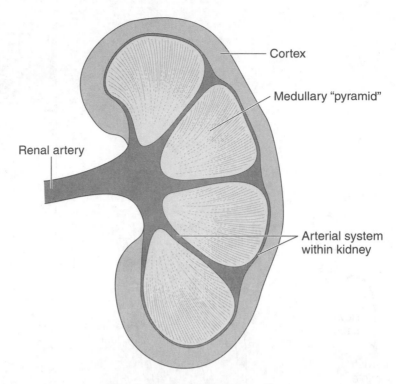

Figure 5.1 Kidney in cross section.

Blood reaches the kidneys via the renal arteries, which branch off on either side from the abdominal aorta. Within the kidney, blood vessels run between and around the medullary pyramids. Tiny branches going into the cortex give rise to vessels called *afferent arterioles* (afferent means "going in"). The afferent arteriole enters the *nephron,* the functional unit of the kidney (Figure 5.2). Each kidney has about a million nephrons, and renal function represents the combined action of all the nephrons in both kidneys.

The beginning of the nephron is a rounded receptacle called *Bowman's capsule.* As the afferent arteriole enters Bowman's capsule, it forms a bunched-up capillary called the *glomerulus.* Like any other capillary, the glomerulus acts like a sieve, allowing water and solutes to filter through, but not proteins or cells. This *glomerular filtrate* is chemically similar to an ultrafiltrate of serum. As this fluid moves down the nephron tubule, it is called *tubular fluid.*

From this point on, we must follow two parallel paths—the unfiltered blood that is still in the glomerulus, and the tubular fluid that has filtered out from the glomerulus and is moving down the nephron tubule. The unfiltered blood leaves Bowman's capsule by way of the *efferent arteriole* ("going out"), which is the continuation of the glomerulus. The efferent arteriole forms a network of *peritubular capillaries* that surround and accompany the convoluted portions of the tubule[2] and the *Loop of Henle.* These peritubular vessels join to form the renal veins, which leave the kidneys and return the blood to the general circulation.

---

[2]The proximal and distal convolutions are usually called the "proximal convoluted tubule" and "distal convoluted tubule." However, such terminology could be misinterpreted as meaning that they are separate structures, when in fact they are segments of the same structure.

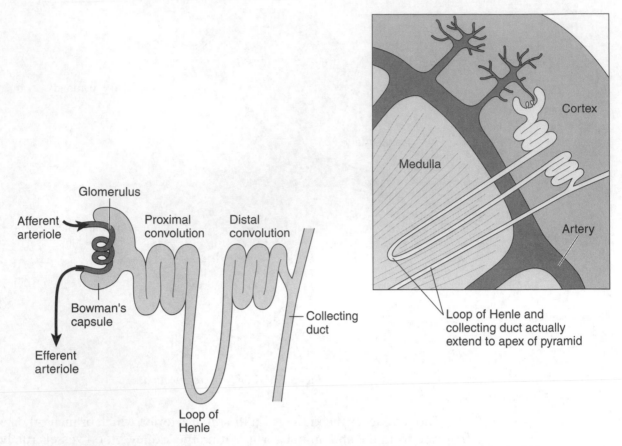

Figure 5.2 The nephron, shown schematically; inset shows position of a nephron in the kidney.

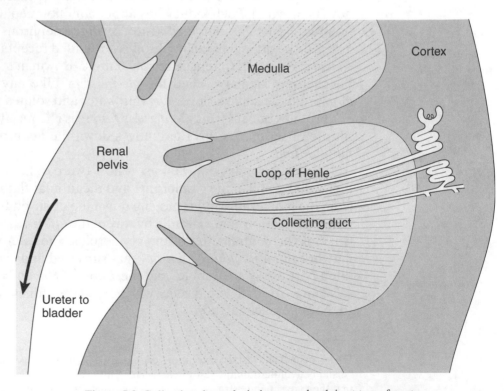

Figure 5.3 Collecting ducts drain into renal pelvis at top of ureter.

Meanwhile, the tubular fluid undergoes chemical changes and water reabsorption as it passes through the proximal convolution, the Loop of Henle, and the distal convolution. These processes involve the movement of molecules between the tubular fluid and the blood in the peritubular capillaries. The fluid coming out of the distal convolution is urine. Collecting ducts convey urine to the renal pelvis, which is the expanded top of the ureter (Figure 5.3).

REMEMBER: Blood reaching the glomerulus is split into two pathways: a serum-like filtrate enters the nephron tubule, while the unfiltered blood enters the peritubular capillaries.

THE FORMATION OF URINE

The formation of urine for excretion involves *chemical changes* and *volume adjustments*. The chemical changes occur in the proximal and distal convolutions of the tubule, as solutes and water move from the nephron to the accompanying peritubular vessels, or vice versa. (These structures, along with Bowman's capsule, give the renal cortex its dark, granular appearance.)

The glomerular filtrate in the nephron tubule and the unfiltered blood in the peritubular capillaries both contain a combination of useful substances and waste products. As the tubular fluid passes through the nephron, the useful substances (such as glucose and sodium) are *reabsorbed* from the tubule into the peritubular capillaries for return to circulation. At the same time, waste products in the unfiltered blood are *secreted* from the peritubular capillaries into the tubule for excretion (Figure 5.4). As a result of reabsorption

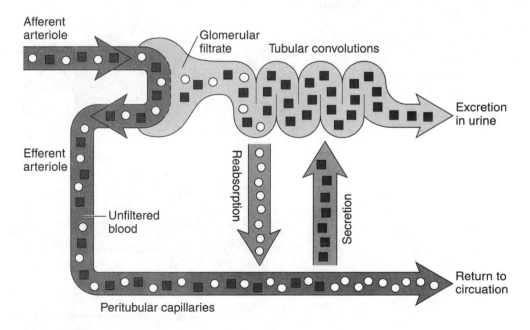

O Useful substance

■ Waste product

Figure 5.4 Reabsorption and secretion, shown schematically.

and secretion, the useful substances and waste products that were together in the blood reaching the kidney are largely separated from each other—useful substances are in the peritubular blood for return to the general circulation, while wastes are in the tubular fluid for excretion via the urine.

Volume adjustment refers to the recovery of water from the tubular fluid. Water recovery is an osmotic process (see Chapter 2). As osmotically active solutes are reabsorbed from the fluid passing through the proximal and distal convolutions, water follows passively. The final volume adjustment occurs in the collecting ducts. (These ducts, along with the Loop of Henle, give the renal medulla its linear appearance.)

The collecting ducts pass through a rising osmotic gradient in the renal medulla (see the following section). As fluid passes down the length of the collecting duct, a variable amount of water is drawn outward through pores in the duct by this osmotic gradient. The extracted water is carried away by the *vasa recta* (the peritubular capillary accompanying the Loop). Depending on the volume of water extracted, the urine may be relatively dilute or highly concentrated.

THE MEDULLARY OSMOTIC GRADIENT

The osmotic gradient in the renal medulla is created and maintained by the Loop of Henle and the vasa recta. The descending limb of the Loop receives fluid from the proximal convolution of the tubule and carries it down toward the apex of the pyramid. There, the Loop makes a hairpin turn, and the ascending limb carries the fluid back up to the distal convolution. So the descending and ascending limbs of the Loop of Henle run in opposite directions right next to each other—and the outflow in the ascending limb directly affects the inflow in the descending limb. This mechanism is called a *countercurrent multiplier* (Figure 5.5).

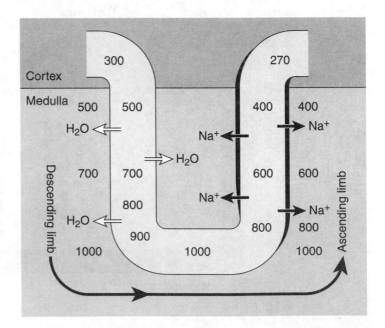

Figure 5.5 Countercurrent multiplier in Loop of Henle. Numbers show osmolality. Sodium is pumped out of ascending limb. Osmosis draws water out of descending limb.

Chapter 5 / Renal Function

The descending limb of the Loop is thin and highly permeable to water. Therefore, the osmolality must be equal inside and outside the descending limb at any point along its length. In contrast, the ascending limb is thick and impermeable, and it continually pumps sodium ions outward (with chloride ions following passively). This outpumping of sodium raises the osmolality outside the Loop.

Water is drawn out of the permeable descending limb by the rising osmolality outside the Loop, so the concentration keeps increasing in the fluid going down the descending limb—and, thanks to osmosis, *the rising osmolality going toward the apex of medullary pyramids is parallel inside and outside the Loop.* However, water cannot follow the sodium being pumped outward from the impermeable ascending limb, so the concentration keeps decreasing in the fluid going up the ascending limb.

REMEMBER: The osmotic concentration in the medullary pyramids increases enormously toward the apices.

This osmotic gradient is utilized in the final adjustment of urinary volume. The collecting ducts have pores that can be opened or closed. If the pores are closed, water is kept inside; if the pores are open, water is drawn outward by the osmotic gradient, and returned to circulation via the vasa recta.

Antidiuretic hormone (ADH; also called *vasopressin*) causes the collecting duct pores to open. Water is then reabsorbed, resulting in a smaller volume of concentrated urine. Without ADH, the collecting duct pores stay closed and water reabsorption is restricted, resulting in a larger volume of dilute urine.

ADH is released from the brain in response to a rise in the osmotic concentration in body fluids or a severe decline in total body water volume. In these circumstances, the body will try to boost its water content through increased intake (by creating the sensation of thirst) and decreased output (by releasing ADH). As a result, the body's water content rises and osmolality falls back toward normal levels. If the osmotic concentration falls below normal, ADH is *not* released and the collecting duct pores stay closed, causing diuresis; the body's water content falls and osmolality rises back toward normal levels.

THE GLOMERULAR FILTRATION RATE

Renal function begins with and depends on filtration at the glomerulus, and the glomerular filtration rate (GFR) is the most important indicator of overall renal function. So we must take a closer look at glomerular filtration and the mathematical expression of its rate.

In the nephron, the glomerulus is surrounded by a porous membrane through which fluid is filtered. According to the Starling Hypothesis of fluid movement at the capillary level, hydrostatic or hydraulic pressure pushes fluid outward while colloid osmotic pressure from plasma proteins pulls fluid back inward (see Chapter 2). Since the glomerulus is a capillary, this principle applies. The pressure inside the glomerulus is much higher than the pressure outside (in the space inside Bowman's capsule). As a result, the outward push is very effective. The inward pull depends on the protein concentration in the unfiltered blood inside the glomerulus. The net pressure—outward—is obtained as the outward push minus the inward pull. Two other factors are the combined surface area of the glomerular membranes and the permeability of the membrane; the larger and more permeable the membrane, the more efficient filtration will be.

These factors are represented in the following proportionality:

$$\text{GFR} \propto \text{area} \cdot \text{permeability} \cdot \text{net pressure}$$

In a healthy young adult, with two million functioning nephrons, a typical value for GFR might be 120 mL/min.[3] Aging is associated with a gradual loss of functioning nephrons, which represents a decrease in the combined surface area of the glomerular membranes. For awhile, the GFR is maintained as intraglomerular pressure increases and pushes out more fluid; that is, there are fewer nephrons, but those that remain are filtering larger volumes. Eventually, however, as age and disease continue to take their toll, this mechanism will no longer compensate for the ongoing loss of nephrons, and GFR will start to fall. Even so, GFR must decrease significantly before any clinical consequences become evident.

CLEARANCE

Measuring GFR is a challenge. There is a totally impractical method that happens to be absolutely accurate, and some easier methods that are variably inaccurate. All of these methods are based on the concept of *clearance*—the rate at which blood is cleared of a specific substance.[4]

> **REMEMBER:** Clearance is measured as the *volume of blood* cleared of a certain substance per minute; *not* as the amount of substance removed from the blood per minute.

Clearance of a specific substance depends not only on the rate at which it is filtered out of the plasma, but also on any reabsorption or secretion that takes place. If there is no reabsorption or secretion, then the rate at which the blood is cleared is the same as the rate of filtration.

However, different substances undergo varying degrees of reabsorption or secretion. For a substance that is *totally reabsorbed,* whatever is filtered goes back into the circulation; therefore, clearance = 0 (Figure 5.6). For a substance that is *partially reabsorbed,* some of what is filtered goes back into the circulation but some does get excreted; therefore, GFR > clearance > 0 (Figure 5.7). For a substance that is *totally secreted,* whatever escapes filtration gets excreted anyway; therefore, clearance = total renal plasma flow (Figure 5.8). For a substance that is *partially secreted,* some of what escapes filtration gets excreted anyway but some goes back into the

---

[3]GFR is typically about 15% to 20% higher in men than in women.

[4]Because clearance is a rate, the phrases *clearance rate* and *rate of clearance* are redundant. It is like saying, "rate of GFR," which would mean "the rate of the glomerular filtration rate."

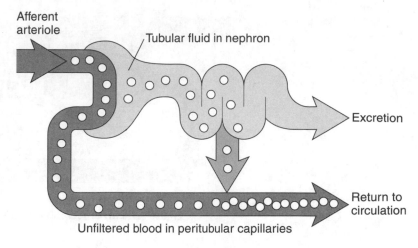

Figure 5.6 Total reabsorption; clearance = 0.

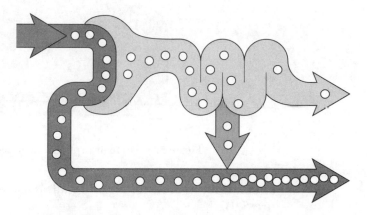

Figure 5.7 Partial reabsorption; GFR > clearance.

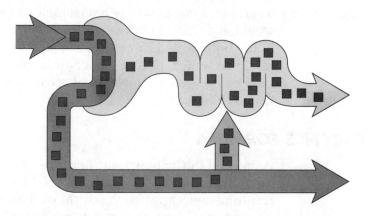

Figure 5.8 Total secretion; clearance = total renal plasma flow.

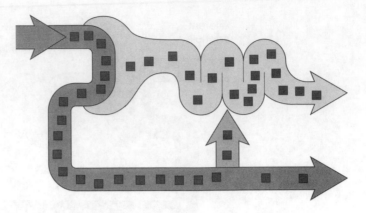

Figure 5.9 Partial secretion; clearance > GFR.

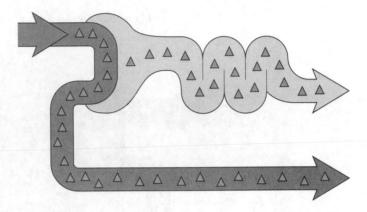

Figure 5.10 No reabsorption and no secretion; clearance = GFR.

circulation; therefore, total renal plasma flow > clearance > GFR (Figure 5.9).

It follows, then, that for a substance that is *neither reabsorbed nor secreted at all,* whatever is filtered gets excreted and whatever escapes filtration goes back into the circulation; therefore, clearance = GFR (Figure 5.10).

The general strategy for determining GFR is to measure the clearance of a substance that is neither reabsorbed nor secreted. Ideally, such a substance would undergo no reabsorption and no secretion at all, and its measurement in the body would be easy and inexpensive. Unfortunately, there is no such ideal substance. Before we consider various less-than-ideal options, let's look at the general method of measuring the clearance of any substance from the blood.

THE CLEARANCE FORMULA

As we see in Chapter 2, introducing a known quantity of substance into an unknown amount of water and then dividing the quantity introduced by the resulting concentration yields the volume of water present. For example, if introducing 20 grams of substance yields a solution whose concentration is 5 g/L, the amount of water that must be present is calculated as follows:

$$\frac{20\,\text{g}}{5\,\text{g}\cdot\text{L}^{-1}} = 4\,\text{L}$$

If each liter contains 5 grams of the substance, then it takes 4 liters to account for the 20 grams of substance introduced.

We employ the same concept in the clearance formula. In the numerator, the product of the urinary concentration of the substance and the urine formation rate is the total quantity of substance excreted. For example, if the urinary concentration is 1.1 mg/mL and the 24-hour urinary volume is 1200 mL, the rate at which the substance appears in the urine is:

$$1.1\,\frac{mg}{mL} \cdot 1200\,\frac{mL}{day} = 1320\,mg \cdot day^{-1}$$

The denominator is the serum or plasma concentration of that substance. If the concentration is 0.8 mg/dL:

$$\frac{1320\,mg \cdot day^{-1}}{0.8\,mg \cdot dL^{-1}} = 1650\,dL \cdot day^{-1}$$

If each deciliter of serum contains 0.8 mg of substance, then it takes the clearance of 1650 deciliters of serum to account for the 1320 milligrams of substance collected. Again, a known quantity of substance is divided by its concentration in fluid to determine the volume of fluid that would account for that quantity.

We use conversion factors so that the clearance is in units of mL/min rather than dL/day. At 100 milliliters per deciliter and 1440 minutes per day:

$$1650\,\frac{dL}{day} \cdot \frac{100\,mL}{dL} \cdot \frac{day}{1440\,min} \approx 115\,mL \cdot min^{-1}$$

In the unit-conversion, multiplying by 100 and dividing by 1440 is the same as dividing by 14.4 or multiplying by the reciprocal of 14.4, which is just over 0.069. Rounding off that reciprocal to 0.07, the general formula for clearance is:

$$\frac{urine\ concentration\ \left(\dfrac{mg}{mL}\right) \cdot urine\ volume\ \left(\dfrac{mL}{day}\right) \cdot 0.07}{serum\ concentration\ \left(\dfrac{mg}{dL}\right)}$$

In the preceding example:

$$\frac{1.1\,\dfrac{mg}{mL} \cdot 1200\,\dfrac{mL}{day} \cdot 0.07}{0.8\,\dfrac{mg}{dL}} \approx 116\,mL \cdot min^{-1}$$

The conversion constant 0.07 yields the correct clearance of the substance in units of mL/min *if* the urine volume (in mL) represents a 24-hour collection and *if* the urine and serum concentrations of the substance are given in units of mg/mL and mg/dL, respectively; otherwise, a different conversion factor would be needed. The substance whose clearance provides the most useful estimate of GFR is called *creatinine,* and creatinine is measured in just those units.

The more generalized expression of clearance (without any specific unit-conversion constant) is usually abbreviated as follows:

$$Cl = \frac{U \cdot V}{P}$$

To obtain the clearance (Cl) of a given substance, multiply the concentration of that substance in the urine (U) by the urinary formation rate based on a timed urinary volume (V); this product is the total quantity of substance collected in a certain time interval. Dividing that quantity by the concentration of the substance in the serum or plasma (P) gives the volume of serum or plasma cleared that would account for that quantity.

ASSESSING GFR

Looking at the clearance formula, we can see why the clearance of a totally reabsorbed substance like glucose is zero—since the normal urinary concentration of glucose is zero, the entire numerator is zero, and therefore the quotient is zero.

REMEMBER: The main clinical use of clearance is in estimating GFR; for that purpose, the substance whose clearance is measured must be one that is neither reabsorbed from nor secreted into the tubular fluid.

 The renal clearance of a substance called *inulin* is the "gold standard" for assessing GFR. (Don't confuse inulin with insulin; these substances are totally unrelated.) Inulin undergoes no reabsorption and no secretion, so its clearance is exactly equal to GFR. However, inulin requires meticulous preparation and intravenous administration, and measuring it in body fluids is quite difficult. Inulin clearance is sometimes used in nephrology research, but it is impractical as a routine method of assessing GFR in the clinical setting.

A more practical alternative is creatinine, an endogenous substance (a metabolite of creatine, from muscle tissue) whose measurement in body fluids is routine. Creatinine is not reabsorbed at all, but there is some degree of secretion from the peritubular capillaries into the nephron tubule; consequently, the clearance of creatinine is greater than the true GFR (look again at Figure 5.9). The numbers used in the calculation are realistic values for the urinary and serum concentrations of creatinine in a healthy adult.

The extent to which creatinine is secreted varies *inversely* with the true GFR. With a high true GFR, little is secreted, and clearance is only slightly above true GFR. With a low true GFR, more is secreted, and clearance may be significantly higher than true GFR. It is ironic—the method works best when there is no cause for concern, and more poorly as renal function declines.

Note that the inaccuracy is *not* in the determination of creatinine clearance itself; if the correct numbers are plugged in, the formula will yield a correct result. The question is how well creatinine clearance reflects the true GFR. When serum creatinine is high and the true GFR is low, a perfectly accurate value for creatinine clearance will be a gross overestimation of GFR. (Another problem is that serum creatinine levels also vary with alterations in muscle-tissue metabolism and age-related decreases in muscle mass.)

Perhaps the most common method of assessing GFR in the clinical setting is the *Cockroft-Gault formula,* whose main advantage is simplicity (the only laboratory value required is the serum creatinine level).[5] By this formula, creatinine clearance—which is, itself, just an estimate of true GFR—is estimated from the patient's serum creatinine level in mg/dL, age in years, and ideal (lean) weight in kilograms:

$$\text{Clearance}_{\text{creatinine}} \approx \frac{(140 - \text{age}) \cdot \text{ideal weight}}{72 \cdot [\text{creatinine}]_{\text{serum}}}$$

For women, an additional factor appears in the numerator—multiplying by the constant 0.85 adjusts for the lower creatinine clearance and GFR seen in women. Note that the estimated clearance varies directly with ideal weight, and inversely with age and serum creatinine. That is, the value is lower in older people with a smaller lean mass and a higher serum creatinine level; and higher in younger people with a larger lean mass and a lower serum creatinine.

The Cockcroft-Gault formula yields a reasonably accurate estimate of creatinine clearance in most situations, although it may be less accurate in elderly patients with severe renal impairment.

REMEMBER: Creatinine secretion increases as serum creatinine rises. Thus, as creatinine clearance falls in a patient with progressive renal disease, the true GFR is falling faster.

GFR: The Next Generation

Two newer methods of determining GFR utilize iodine-based contrast agents that are neither reabsorbed nor secreted; therefore, GFR is accurately represented by their clearance.

Iohexol is administered intravenously. Plasma samples obtained at set intervals are irradiated and then assayed. The level of induced radioactivity in each sample is directly proportional to the plasma iodine level, which falls as the agent is removed by the kidneys.

A radionuclide of *iothalamate* is injected subcutaneously. Timed samples of urine and plasma are assayed to measure the rate at which radioactivity falls in the plasma and rises in the urine as the radionuclide is removed by the kidneys.

In both tests, the clearance is calculated from the rate of change in measured radiation from one timed sample to the next. These methods offer highly accurate reflections of GFR, but their use is still limited.

---

[5]Source: D. W. Cockroft and M. H. Gault, "Prediction of creatinine clearance from serum creatinine," *Nephron* (1976), 16:31–41.

VOLUME CONSIDERATIONS RELATED TO GFR

The GFR is the volume of fluid that gets filtered each minute. Obviously, the total volume of plasma passing through the kidneys each minute is this filtered volume *plus* the volume of the plasma that does *not* get filtered at the glomerulus.

To find the *total renal plasma flow,* we calculate the clearance of a substance that is totally secreted—transported *from* the unfiltered blood in the peritubular capillaries *into* the tubular fluid in the nephron (look again at Figure 5.8). The best substance for this purpose is a chemical called para-aminohippuric acid (PAH). Because PAH that does *not* get filtered at the glomerulus ends up being excreted anyway, its clearance is approximately equal to the rate at which plasma enters the kidney.

Here is the clearance formula with typical values for the urinary and plasma concentrations (in brackets) of PAH:

$$\text{Clearance}_{PAH} = \frac{[\text{PAH}]_{urine} \cdot \text{urine formation rate}}{[\text{PAH}]_{serum}}$$

$$= \frac{15 \frac{mg}{mL} \cdot 1 \frac{mL}{min}}{0.025 \frac{mg}{mL}} = 600 \text{ mL} \cdot \text{min}^{-1}$$

Note that there is no conversion constant in this case, because both concentrations are shown in the same units, mg/mL, and the urine formation rate is shown in mL/min. Also note that the plasma level is extremely low, as we would expect; a substance that is rapidly secreted will rapidly disappear from the circulation.

About 90% of the PAH in the unfiltered blood in the peritubular capillaries is secreted into the tubule during a single pass through the kidney. Therefore:

$$\text{Total renal plasma flow} \approx \frac{600 \text{ mL} \cdot \text{min}^{-1}}{0.9} \approx 667 \text{ mL} \cdot \text{min}^{-1}$$

The *hematocrit* is the percentage of blood volume represented by blood cells; plasma represents the remaining volume. Then with the hematocrit expressed as a decimal:

$$\text{Total renal blood flow} = \frac{\text{renal plasma flow}}{1 - \text{hematocrit}}$$

Using the renal plasma flow previously given and a typical hematocrit of 40%:

$$\text{Total renal blood flow} = \frac{667 \text{ mL} \cdot \text{min}^{-1}}{0.6} \approx 1112 \text{ mL} \cdot \text{min}^{-1}$$

In other words, about one-fifth of the body's total blood supply (which is about 5 to 6 liters in a normal adult) is cycled through the kidneys every minute.

Knowing the GFR and the total renal plasma flow, the *filtration fraction* is simply their ratio. At the same values we have been using:

$$\text{Filtration fraction} = \frac{\text{GFR}}{\text{renal plasma flow}} = \frac{120 \text{ mL} \cdot \text{min}^{-1}}{667 \text{ mL} \cdot \text{min}^{-1}} \approx 0.18 = 18\%$$

Recall the two pathways mentioned near the beginning of this chapter: fluid that gets filtered and fluid that remains in the peritubular capillaries. Of the total volume of plasma reaching the glomerulus, about 18% gets filtered into the nephron tubule, while the other 72% remains in the unfiltered blood that enters the peritubular capillaries. The filtration fraction is typically close to one-fifth.

With reference to the four-fifths that does *not* get filtered, the flow rate of plasma into the peritubular capillaries would be the difference between the rate at which plasma enters the kidney and the rate at which it is filtered. Again, using the same values:

$$\text{Nonfiltration rate} = \text{total renal plasma flow} - \text{GFR}$$

$$= 667 - 120 = 547 \text{ mL} \cdot \text{min}^{-1}$$

Finally, let's compare a typical urine production rate with a typical GFR:

$$\frac{\text{Urine volume} \cdot \text{minute}^{-1}}{\text{Filtered volume} \cdot \text{minute}^{-1}} = \frac{1 \text{ mL}}{120 \text{ mL}} \approx 0.008 = 0.8\%$$

We see the same results using rates per day. A typical 24-hour urine volume might be 1.3 liters. Converting GFR to volume filtered per day and then taking the ratio:

$$120 \frac{\text{mL}}{\text{min}} \cdot \frac{1440 \text{ min}}{\text{day}} = 172{,}800 \frac{\text{mL}}{\text{day}} \approx 173 \text{ L} \cdot \text{day}^{-1}$$

$$\frac{\text{Urine volume} \cdot \text{day}^{-1}}{\text{Filtered volume} \cdot \text{day}^{-1}} \approx \frac{1.3 \text{ L}}{173 \text{ L}} \approx 0.008 = 0.8\%$$

In other words, about 99% of the water in the glomerular filtrate entering the tubule is returned to the circulation by the time the tubular fluid enters the ureters as urine. Roughly two-thirds of this amount passively follows the reabsorption of osmotic solutes at the proximal convolution. More water follows the reabsorption of solutes at the distal convolution. A final amount of water is extracted at the collecting duct, as regulated by the ADH mechanism described earlier.

TRANSPORT MAXIMUM

There is a limit to the rate at which any substance can be transported from one physiologic location to another; this limit is called the *tranport maximum* (Tm). The Tm for glucose is the maximum rate at which it can be reabsorbed from the nephron tubule into the peritubular capillaries. Since glucose does not normally appear in the urine at all, the rate at which it enters the tubular fluid must be below the Tm.

For any substance in the blood, the "tubular load" (the rate of delivery into the tubular fluid) is the product of its serum or plasma concentration and the GFR. For glucose, a typical concentration in the serum might be 90 mg/dL = 0.9 mg/mL; then at a typical GFR of 120 mL/min:

$$\text{Tubular load} = 0.9\,\frac{\text{mg}}{\text{mL}} \cdot 120\,\frac{\text{mL}}{\text{min}} = 108\ \text{mg} \cdot \text{min}^{-1}$$

This rate of delivery happens to be well below the Tm for glucose; that is, glucose can be reabsorbed from the tubular fluid far faster than it normally enters the glomerular filtrate. Therefore, total reabsorption is easily achieved, and virtually no glucose appears in the urine.

However, at the same GFR, if the serum glucose concentration is much higher (a condition called hyperglycemia), the rate of delivery into the tubular fluid is also much higher. If the rate of delivery is faster than the rate of reabsorption, glucose starts spilling over into the urine—a condition called *glycosuria*.

Clinical observations reveal that the "renal threshold" for glucose—that is, the serum concentration at which glycosuria begins—is usually about 200 mg/dL. Thereafter, as the serum level rises still higher, the rate of appearance in the urine also rises, in a simple linear relationship (Figure 5.11). The graph represents the concentration of glucose in urine as a function of its concentration in serum. Note that the line begins as a gradually rising curve but then continues as a straight line. Theoretically, the renal threshold should be the serum concentration at which the tubular load just surpasses the reabsorption Tm. However, the system does not work with such me

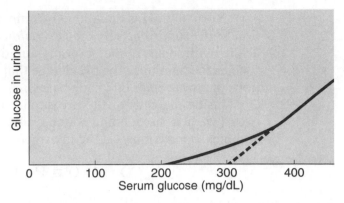

Figure 5.11 The renal threshold for glucose; glycosuria typically appears when serum glucose level exceeds about 200 mg/dL.

chanical exactness, and the actual renal threshold is somewhat below the theoretical "ideal" threshold.

Look again at Figure 5.11. The ideal threshold is found by extending the straight-line segment of the graph back down to the horizontal axis, as shown by the dashed line; the line crosses the axis at approximately 300 mg/dL = 3 mg/mL. Therefore, because the Tm is equal to the tubular load at that ideal threshold:

$$\text{Tm}_{\text{glucose}} = 3\,\frac{\text{mg}}{\text{mL}} \cdot 120\,\frac{\text{mL}}{\text{min}} = 360\,\text{mg} \cdot \text{min}^{-1}$$

Indeed, the Tm for reabsorption of glucose is, on average, about 360 mg/min in men and slightly lower in women.

More often, Tm refers to reabsorption; however, the same principles apply to secretion from the unfiltered blood in the peritubular capillaries into the tubular fluid in the nephron. The rate at which a substance enters the peritubular capillaries is the product of its plasma concentration and the nonfiltration flow rate:

Peritubular capillary load = [substance]$_{\text{plasma}}$ · (renal plasma flow − GFR)

Compare this rate of entry into the peritubular vessels with the Tm for secretion of the substance. With a relatively high Tm, most of the substance will be secreted into the tubule for excretion on one pass through the kidney, and little of it will return to the general circulation (as we see with PAH); with a relatively low Tm, the substance will begin building up in the blood.

RENAL CONTROL OF pH

The processes of tubular reabsorption and secretion are integral to the kidney's contribution to acid-base balance in the body. Acid-base status is measured on a logarithmic scale called pH. The normal pH range for arterial blood is 7.35–7.45; values below 7.35 indicate a high concentration of hydrogen ion (acidity), while values above 7.45 indicate a low concentration of hydrogen ion (alkalinity). These concepts are discussed in detail in Chapter 6.

Normally, pH is maintained within a narrow range. The renal contribution to pH control involves the *excretion of excess acids* (inorganic acids and acidic by-products of metabolism), and the *preservation of buffers* (chemical species that minimize the pH change that would otherwise result from the actions of strong acids or strong bases).

Inside the cells of the proximal tubular convolution, the reaction between carbon dioxide and water results in the formation of hydrogen ions and bicarbonate ions, via the intermediary compound, carbonic acid:

$$CO_2 + H_2O \rightleftarrows H_2CO_3 \rightleftarrows H^+ + HCO_3{}^-$$

The hydrogen ion represents acid to be eliminated; the bicarbonate ion represents buffer to be preserved. As sodium ion is reabsorbed, hydrogen ion is secreted, and there is no net change in electrical charge. The sodium cation and the bicarbonate anion then enter the peritubular blood as the dissociated form of the electrically neutral compound, sodium bicarbonate (Figure 5.12). This process is the key—we will return to it several times.

As more and more hydrogen ion is secreted, its concentration in the tubular fluid keeps rising and the pH in the tubular fluid keeps falling. However, as the hydrogen ion concentration rises, it becomes more difficult to secrete additional hydrogen ion. At some high concentration marked by a very low pH, no additional hydrogen ion could be secreted. That limit would soon be reached were it not for the action of buffers working within the urinary system.

One such buffer is bicarbonate. Bicarbonate is a normal constituent of plasma, so a certain amount gets filtered. The presence of bicarbonate in the tubular fluid buffers the effects of hydrogen ions being secreted into the tubular fluid. The reaction of hydrogen ion and bicarbonate ion in the tubular fluid produces carbonic acid, which reenters the tubular cells; thereafter, events proceed as shown in Figure 5.12, with the recovery of sodium bicarbonate into the circulation.

Another buffer is *dibasic sodium phosphate*. Its chemical reaction with a secreted hydrogen ion is as follows:

$$Na_2HPO_4 + H^+ \rightleftarrows NaH_2PO_4 + Na^+$$

One of the sodium ions in the compound is replaced by a hydrogen ion (forming monobasic sodium phosphate); the freed sodium ion is then reabsorbed

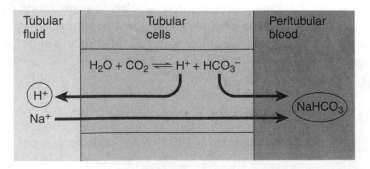

Figure 5.12 Hydrogen ion is secreted into tubule as sodium and bicarbonate are reabsorbed.

in exchange for the secretion of another hydrogen ion. Again, events then proceed as shown in Figure 5.12.

A third buffer is *ammonia* (NH_3), a highly toxic compound derived from protein metabolism. Ammonia in the unfiltered plasma readily crosses the tubular cells and enters the tubular fluid, where it combines with a secreted hydrogen ion to form an ammonium ion. Ammonium ion does *not* readily cross cell membrane barriers; it stays in the tubular fluid, accompanied by any available anion (such as chloride). The overall chemical process is as follows:

$$NH_3 + H^+ \rightleftarrows NH_4^+$$

$$NH_4^+ + NaCl \rightleftarrows NH_4Cl + Na^+$$

Once more, the sodium ion is reabsorbed in exchange for the secretion of another hydrogen ion; and again, events proceed as shown in Figure 5.12.

REMEMBER: Hydrogen ion represents excess acid to be excreted via the urine. Bicarbonate ion represents buffer to be preserved. Secretion of hydrogen ion into the tubular fluid occurs as sodium ion and bicarbonate ion are reabsorbed back into the circulation. These processes maintain the body's supply of bicarbonate.

Azotemia and Uremia

Two substances in the serum are of special significance with respect to renal function—blood urea nitrogen (BUN) and creatinine. The normal serum concentrations are about 20 mg/dL for BUN, and 1 mg/dL for creatinine. Thus, the normal BUN/creatinine ratio is about 20.

An abnormal elevation in BUN is called *azotemia* (the prefix *azo-* denotes nitrogen). If the creatinine level remains normal in a patient with azotemia, the BUN/creatinine ratio is high. Azotemia can result from a wide range of conditions, some of which are unrelated to renal function; for example, dehydration will cause a marked rise in BUN. Because this type of systemic condition is already present as blood reaches the kidney, the resulting BUN elevation is called *prerenal azotemia.* However, in severe and prolonged dehydration and hypovolemia, the kidney itself may be damaged by inadequate circulation.

By contrast, an inability to void the urine produced by the kidney is called *postrenal uropathy.* The backup of fluid in the bladder and ureters can interfere with renal function and may eventually cause structural damage in the kidney.

Chronic renal failure may result from an intrinsic kidney disease or the cumulative effects of severe and prolonged prerenal or postrenal problems. Whatever the cause, the effect is a reduction in all renal functions—glomerular filtration, reabsorption and secretion, pH control, and volume adjustment. The hallmark of progressive dysfunction is a rising serum creatinine concentration.

The term *uremia* refers to a generalized toxic state caused by the accumulation of creatinine and all the acidic metabolic waste products that the failing kidneys are unable to eliminate.

Even with the actions of the urinary buffers, the hydrogen ion concentration is much higher in urine than in serum; a typical urinary pH might be about 6.0, which represents a hydrogen ion concentration 25 times as high as the concentration in arterial serum and glomerular filtrate, where the pH is normally about 7.4. *Titratable acidity* refers to the difference between the pH in urine and arterial serum; it is measured as the amount of alkaline chemical it would take to titrate a unit volume of urine to pH 7.4.

QUIZ

1. A patient has a glomerular filtration rate (GFR) of 96 mL/min, a serum glucose concentration of 135 mg/dL, and a normal transport maximum (Tm) for glucose reabsorption, 360 mg/min. Would you expect glycosuria in this patient?

2. Hypothetically, suppose that two substances, R and S, are normally at the same concentration in the plasma. Further suppose that substance R undergoes tubular reabsorption and substance S undergoes tubular secretion; and that the Tm for reabsorption of R is the same as the Tm for secretion of S. If both substances are equally elevated in plasma, compare the likelihood that R will spill over into the urine versus the likelihood that S will build up in the blood.

3. In Patient A, the creatinine concentration is 1.3 mg/dL in serum, and 0.8 mg/mL in a 24-hour urinary volume of 0.9 liter. In Patient B, serum creatinine is 1.2 mg/dL and a 24-hour urine contains a total of 1.4 grams of creatinine. Compare the creatinine clearance rates in these patients. In which patient does the creatinine clearance rate provide a better estimate of the true GFR?

4. If PAH clearance is measured at 594 mL/min, determine the total renal plasma flow and the filtration fraction in a patient with a GFR of 125 mL/min. (Assume 90% secretion in one pass through the kidneys.) If serum glucose is 90 mg/dL, how fast does glucose enter the peritubular capillaries? If the hematocrit is 37%, what is the total renal blood flow? If the patient's total blood volume is 5.40 liters, what percentage of the blood goes through the kidneys each minute?

5. A 46-year-old man is 15% overweight at 200 pounds; his serum creatinine level is 1.6 mg/dL, and the urine creatinine level is 1.1 mg/mL in a 24-hour urine volume of 1.3 liters. Compare his actual creatinine clearance to the estimated clearance using the Cockroft-Gault formula.

6. Suppose that glycosuria appears in a diabetic patient when the serum glucose level passes 220 mg/dL. At this glucose level, what is the tubular load of glucose if the patient's GFR is 105 mL/min? Is this rate of delivery into the tubular fluid higher than, lower than, or equal to the patient's transport maximum for glucose reabsorption?

7. In a healthy person with normal renal function, which of the following ions would be mainly excreted and which would be mainly recovered into general circulation?

   Ammonium ions   Hydrogen ions
   Bicarbonate ions   Sodium ions

8. If an otherwise healthy person loses one kidney due to trauma, what is likely to happen to the arterial pressure in the remaining kidney. How will this change in pressure affect GFR?

9. In an otherwise healthy person who is becoming dehydrated, will ADH activity initially increase, decrease, or remain constant? Will urine specific gravity rise, fall or remain constant?

10. If a poorly controlled diabetic patient has a serum glucose level of 550 mg/dL and a GFR of 110 mL/min, what is the rate at which glucose will spill over into the urine? (Assume that the Tm for glucose reabsorption is 360 mg/min.)

Answers and explanations at end of book.

chapter 6

Acid-Base Balance

Chapter Outline

Acids and Bases: Fundamental Concepts
The Dissociation Constant
The Meaning of pH
Intervals on the pH Scale
The Henderson-Hasselbalch Equation
Buffers
Acidemia and Acidosis, Alkalemia and Alkalosis
Primary and Compensatory Mechanisms
The Major Categories of Acid-Base Disturbance
Diagnostic Tools

Mathematical Tools Used in This Chapter

Negative exponents (Appendix C)
Dimensional units with negative exponents (Appendix D)
Logarithms and logarithmic scales (Appendix F)
Order of magnitude (Appendix G)
Quadratic equations (Appendix H)

Homeostasis is the overall process of maintaining stability in the body's internal physical and chemical systems. These processes involve rapid correction of disturbances that may arise, as well as instant-by-instant adjustments to prevent gross disturbances from arising.

A simple example is the response to exercise. At rest, a healthy adult might show a heart rate of 72 beats per minute and a respiratory rate of 15 breaths per minute; however, those rates may double or more during physical exertion. Increased muscular activity requires larger supplies of oxygen and produces larger quantities of carbon dioxide (a waste product of the oxidative process). To supply the muscles with the extra oxygen needed and carry away the extra carbon dioxide produced, the heart and lungs must work faster.

This process is homeostasis—despite the increased activity, the body maintains its normal internal state. If the heart rate and respiratory rate did not increase during physical exertion, body chemistry would be significantly altered by the resulting deficit in oxygen and accumulation of carbon dioxide.

The amount of carbon dioxide in the blood has an immediate and direct effect on the body's *acid-base balance,* a key aspect of the internal chemical state. We measure acid-base balance on a mathematical scale called pH.

ACIDS AND BASES: FUNDAMENTAL CONCEPTS

Acids and bases are defined in terms of hydrogen ions. An electrically neutral hydrogen *atom* has a single proton as its nucleus and a single electron. A hydrogen *ion* is formed when the electron is removed, leaving just the proton. An *acid* is a compound that can release hydrogen ions; thus, acids are called "proton donors." A *base* is a "proton receiver" (an ion or molecule that can take up a free proton); the term also refers to a compound that can release a proton receiver.

Let's look first at acids. In water solution, certain types of electrically neutral molecules split into positively charged ions (cations) and negatively charged ions (anions). If the cation thus formed is hydrogen (H^+), the compound that underwent this process of ionic dissociation is an acid.[1] If the cation is *not* hydrogen, the compound is *not* an acid. For example, sodium chloride (table salt) in water dissociates into ions of sodium and chlorine:

$$NaCl \rightarrow Na^+ + Cl^-$$

Because the cation is sodium rather than hydrogen, the compound is not an acid. However, with hydrogen chloride:

$$HCl \rightarrow H^+ + Cl^-$$

Here, the cation *is* hydrogen, and so the compound *is* an acid (hydrochloric acid).

Note that the release of a hydrogen ion leaves the remainder of the acid molecule as an anion. With hydrochloric acid, as we just saw, the anion is chloride. With other acids, other types of anion are released; for example, nitric acid (HNO_3) dissociates into hydrogen ions and nitrate anions (NO_3^-). Then

---

[1]In solution, free protons released by the dissociation of acid molecules link up with water molecules to form *hydronium* ions: $H^+ + H_2O \rightarrow H_3O^+$. For our purposes, we can refer to hydrogen ions or hydronium ions interchangeably.

in general terms, acids can be symbolized as HA, and their dissociation into hydrogen ion and some type of anion can be symbolized as:

$$HA \rightarrow H^+ + A^-$$

So far, we have shown what happens in *strong* acids—dissociation proceeds to completion, with none of the original acid left intact. In *weak* acids, dissociation rapidly reaches a point of equilibrium, with the nondissociated acid molecules constantly splitting into ions and ions recombining to form intact acid molecules:

$$HA \rightleftarrows H^+ + A^-$$

At equilibrium, the forward and reverse processes are taking place at the same rate, so that the concentrations and proportions of intact molecules and ions present at that time will remain constant thereafter.

The exact point of equilibrium varies with the extent of ionic dissociation. With weak acids, very little dissociation takes place; most of the system remains in the form of the intact acid, and the ionic concentrations are low. However, with stronger acids, a correspondingly greater degree of dissociation occurs; less intact acid remains, and the ionic concentrations are higher (Figure 6.1).

Bases are defined as proton receivers. An example is the *hydroxide* group (OH^-), which combines with hydrogen ion to form water:[2]

$$H^+ + OH^- \rightarrow H_2O$$

Hydroxide is part of numerous compounds, such as sodium hydroxide (NaOH); when such compounds dissociate in solution, they release hydroxide and a cation. Thus, the other meaning of *base* is a compound that dissociates to release a proton receiver—whether it is hydroxide or any other anion or radical that takes up protons. As with acids, compounds that dissociate to release a proton receiver are described as strong or weak bases, depending on the extent of ionic dissociation at equilibrium.

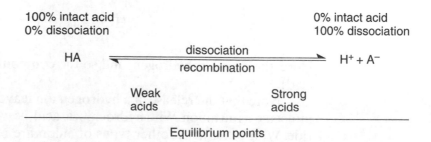

Figure 6.1 At equilibrium, strong acids have undergone extensive dissociation; weak acids have undergone only limited dissociation.

---

[2]Technically, $H_3O^+ + OH^- \rightarrow 2H_2O$.

When Is an Acid Also a Base?

Theoretically, certain substances can *either* release *or* take up protons. Consider the compound $NaHCO_3$ (sodium bicarbonate; also called "bicarbonate of soda" or "baking soda"). In solution, this compound splits to form sodium cations and bicarbonate anions:

$$NaHCO_3 \rightarrow Na^+ + HCO_3^-$$

The monovalent bicarbonate anion can take up a hydrogen ion to form a molecule of a weak acid called *carbonic acid:*

$$HCO_3^- + H^+ \rightarrow H_2CO_3$$

Thus, bicarbonate is a base. (In fact, this process is just the reverse direction of the carbonic acid dissociation equilibrium; that is, bicarbonate is the conjugate base of carbonic acid.) Alternatively, a bicarbonate ion itself could split, releasing a hydrogen ion and leaving the remainder of the group as a divalent anion called *carbonate:*

$$HCO_3^- \rightarrow H^+ + CO_3^{2-}$$

In releasing a proton, this bicarbonate ion is acting as an acid. However, the number of bicarbonate ions taking up protons is vastly greater than the number releasing protons, so we classify bicarbonate as a base. While it is theoretically possible for an individual bicarbonate ion to function as either an acid or a base at a given instant, the net action of all the bicarbonate ions present in a system is that of a base.

Returning to weak acids, equilibrium is the point at which the reverse process (recombination) is taking place at the same rate as the forward process (dissociation). Here is that reverse process by itself:

$$HA \leftarrow A^- + H^+$$

In other words, the anion can take up the hydrogen ion; therefore, A^- may be considered a base. Such an anion, formed as a molecule of acid dissociates to release a hydrogen ion, is called the *conjugate base* of the acid.

THE DISSOCIATION CONSTANT

When dissociation does not go to completion, the point of equilibrium is defined by a mathematical constant. Consider the hypothetical compound Q_2X_3:

$$Q_2X_3 \rightleftarrows 2Q^{3+} + 3X^{2-}$$

This system has three components or "chemical species"—the intact compound, the cation, and the anion. The dissociation constant K is defined as follows:

$$K = \frac{[Q^{3+}]^2 \cdot [X^{2-}]^3}{[Q_2X_3]}$$

The square brackets denote concentrations at equilibrium. Notice that the *coefficients* of the ions on the right side of the dissociation equation become *exponents* in the numerator of the equilibrium equation. As each intact molecule dissociates, *two* Q ions are released, so the concentration of this cation is *squared;* and *three* X ions are released, so the concentration of this anion is *cubed.* With concentrations thus raised to the appropriate powers, K is the ratio obtained as the product of the ion concentrations divided by the concentration of the compound.[3]

For strong acids and bases, dissociation is so extensive that the ion concentrations are much higher than the concentration of intact compound. With the product of high values in the numerator of the equilibrium equation and a low value in the denominator, K would be huge—if we bothered to calculate it (and for strong acids and bases, we generally do not).

In fact, dissociation constants apply mainly to weak acids and weak bases. With lower ion concentrations and a higher concentration of intact compound, the equilibrium equation yields a useful value for K.

Suppose that the introduction of 12 mmol of weak acid HA into 1 liter of water yields the following concentrations at equilibrium (in mmol/L):

$$HA \rightleftarrows H^+ + A^-$$
$$10 \qquad 2 \qquad 2$$

If we still have 10 mmol of intact acid, 2 mmol dissociated; each molecule that dissociates yields one hydrogen ion and one conjugate-base anion, so the dissociation of 2 mmol of HA yields 2 mmol of H^+ and 2 mmol of A^-, as shown. (These simple numbers are arbitrarily selected to illustrate the general concept; they do not represent any specific chemical process.)

Now we can calculate the dissociation constant:

$$K = \frac{[H^+] \cdot [A^-]}{[HA]} = \frac{2 \cdot 2}{10} = \frac{4}{10} = 0.4$$

Knowing the value of K, we can *predict* the species concentrations if a different amount of acid is used. Set up an equilibrium equation with K given and the unknown concentrations expressed in terms of x, the number of millimoles of acid that will dissociate. Starting with 6 mmol of acid, the dissociation of x mmol yields x mmol of hydrogen ion and x mmol of conjugate-base anion; the amount of intact compound left is $6 - x$. Plugging these values into the equilibrium equation:

---

[3]In any equilibrium process, K is obtained as the product of the concentrations of right-side species divided by the product of the concentrations of left-side species, with each concentration raised to a power equal to that species' coefficient in the equilibrium equation. K is called a dissociation constant for processes involving dissociation to equilibrium, or an equilibrium constant for other processes. By either name, the method of calculating K is the same.

$$K = \frac{x \cdot x}{6 - x} = 0.4$$

$$x^2 = 2.4 - 0.4x$$

From this quadratic equation,[4] $x \approx 1.36$. To check:

$$K = \frac{1.36 \cdot 1.36}{6 - 1.36} \approx \frac{1.85}{4.64} \approx 0.4$$

 Do not confuse the dissociation constant with the percentage of dissociation that occurs in a system. For any given compound, the dissociation constant is always the same; but the proportions of the species vary with conditions in the system, as we see in the previous example. At 12 mmol/L, 2 mmol (about 16.7%) dissociates; at 6 mmol/L, 1.36 mmol (about 22.7%) dissociates. The percentage of dissociation depends on such variables as the amount of compound introduced (as we have just seen), the temperature and volume of the solution, and the presence or absence of other substances in the system; of course, the dissociation constant itself is also a factor.

If any chemical species in a system is increased or decreased, the point of equilibrium (that is, the percentage of dissociation) shifts. The shift is toward the opposite side to that of a species that has been increased, or toward the same side as that of a species that has been decreased (Figure 6.2).

REMEMBER: At equilibrium, the percentage of dissociation (reflected in the proportions of intact compound and ions) can be affected by variables in the system. In contrast, the dissociation constant is an intrinsic and unchanging characteristic of the compound itself.

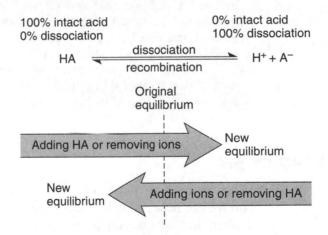

Figure 6.2 Adding or removing components shifts the equilibrium point, but the dissociation constant is always the same.

---

[4]If dissociation yields ion coefficients greater than 1, we end up with a higher-degree equation, which is far beyond the scope of entry-level clinical science.

Carbonic Acid: A Transitional Equilibrium

Carbon dioxide is one of the gases carried in blood. Hydration (combination with water) of dissolved carbon dioxide sets up an equilibrium with carbonic acid, which plays a key role in acid-base balance:

$$H_2O + CO_2 \rightleftarrows H_2CO_3$$

In turn, dissociation of this weak acid yields hydrogen ion and the conjugate base, bicarbonate ion:

$$H_2CO_3 \rightleftarrows H^+ + HCO_3^-$$

Therefore, carbonic acid can be viewed as a transitional stage between the hydration of dissolved carbon dioxide on one side, and the dissociation into hydrogen and bicarbonate ions on the other side:

$$H_2O + CO_2 \leftrightarrows H_2CO_3 \leftrightarrows H^+ + HCO_3^-$$

"Hydration   "Dissociation
equilibrium"   equilibrium"

Increasing one of the chemical species in a system pushes the equilibrium toward the opposite side. Thus, an increase in carbon dioxide (for example, during physical exertion) pushes the "hydration equilibrium" toward the formation of additional carbonic acid; in turn, the increase in carbonic acid pushes the "dissociation equilibrium" toward an increase in hydrogen and bicarbonate ions. *Thus, a rise in carbon dioxide levels in the blood causes an increase in the hydrogen ion concentration.*

This acid-base disturbance triggers an immediate increase in the heart rate and respiratory rate, resulting in more rapid removal of carbon dioxide from the blood. Removal of carbon dioxide pulls the hydration equilibrium away from carbonic acid, and a lower level of carbonic acid pulls the dissociation equilibrium away from hydrogen and bicarbonate ions, thereby restoring normal acid-base balance. *Thus, a fall in carbon dioxide levels causes a decrease in the hydrogen ion concentration.*

In short, the hydrogen ion concentration varies directly with the carbon dioxide level; other things being equal, a high carbon dioxide level creates a high concentration of hydrogen ion, and vice versa.

THE MEANING OF pH

We have seen the central role of the hydrogen ion in acid-base phenomena, and we are now ready to measure its concentration. The concentration of hydrogen ion in arterial blood is very low—typically, about 40 nanomoles ($4 \cdot 10^{-8}$ mole) per liter. Rather than handle such a minuscule quantity, we look at its common logarithm:

$$\log [H^+] = \log 0.00000004 \approx -7.40$$

If a logarithm is below 0, the antilogarithm is between 0 and 1 (in this case, the hydrogen ion concentration is much closer to 0). To avoid the negative value, multiply the equation by −1, thereby reversing all the signs;

$$-\log [H^+] = -\log 0.00000004 \approx 7.40$$

Now we have a manageable number—the negative logarithm of a normal concentration of hydrogen ion in arterial blood is approximately 7.40. Here is another way to see how this result is derived. A logarithm is an exponent, and a negative exponent indicates a reciprocal. Therefore:

$$-\log[H^+] = \log[H^+]^{-1} = \log\left(\frac{1}{[H^+]}\right)$$

$$= \log\left(\frac{1}{0.00000004}\right) = \log 25{,}000{,}000 \approx 7.40$$

We symbolize "negative logarithm of . . . " by lowercase p. Then for the negative logarithm of the hydrogen ion concentration:

$$-\log [H^+] = p[H^+] = pH$$

And so we arrive at this definition—*pH is the negative logarithm of the hydrogen ion concentration.* Alternatively, pH is the logarithm of the reciprocal of the hydrogen ion concentration.[5]

Now let's see how changes in the hydrogen ion concentration show up on the pH scale. Suppose the hydrogen ion concentration increases to 50 nanomoles ($5 \cdot 10^{-8}$ mole) per liter. Going from 0.00000004 to 0.00000005 may not seem impressive, but the difference is more obvious on the pH scale:

$$pH = \log\left(\frac{1}{0.00000005}\right) = \log 20{,}000{,}000 \approx 7.30$$

Note that the pH value fell as the hydrogen ion concentration increased. Conversely, pH rises as the hydrogen ion concentration decreases.

---

pH Units?

pH is a logarithm—a pure number, with no dimensional units (a logarithm is an exponent, and an exponent is just a number). However, in some reference listings of normal laboratory values, the pH range is given in units called "pH units." This terminology is arbitrary, because the "pH unit" is simply the unit-value on the number line.

---

 [5]Reversing the sign of "log x" yields the "negative logarithm" of x. Do not confuse "negative logarithm" with a logarithm whose value happens to be a negative number (which occurs when $1 > x > 0$). If the hydrogen ion concentration is 0.00000004 mol/L, the logarithm is a negative number, -7.40; then the negative logarithm is 7.40. The logarithm of any positive number is the negative logarithm of the reciprocal of that number; that is, $\log x = -\log 1/x$, for any real number $x > 0$ (there is no logarithm for $x \le 0$).

For example, suppose the concentration decreases to 30 nanomoles ($3 \cdot 10^{-8}$ mole) per liter:

$$pH = \log\left(\frac{1}{0.00000003}\right) \approx \log 33{,}333{,}333 \approx 7.52$$

REMEMBER: pH varies *inversely* with the hydrogen ion concentration—a lower pH signifies a higher concentration, and a higher pH signifies a lower concentration.

Since pH is a logarithm, a difference of 1.00 on the pH scale represents a tenfold difference in the hydrogen ion concentration. For example, pH 6.40 represents a hydrogen ion concentration ten times the normal concentration in arterial blood (pH 7.40); conversely, pH 8.40 represents a hydrogen ion concentration one-tenth of normal.

Such severe abnormalities are incompatible with life. The hydrogen ion concentration in arterial blood[6] is tightly regulated, which means that pH is maintained within a narrow range; the usual reference range for a normal pH in arterial blood is 7.35–7.45.

INTERVALS ON THE pH SCALE

Consider the ratios 10/2 and 20/4. Both quotients are 5, but the magnitude of the difference between numerator and denominator is larger with the higher numbers ($10 - 2 = 8$ while $20 - 4 = 16$). Similarly, any constant interval on the pH scale always represents the same *ratio* between hydrogen ion concentrations; but the *difference* in concentration increases at higher concentrations. For example, a pH interval of 0.3 represents a ratio of about 2; thus, the hydrogen ion concentration at pH 7.1 is twice the concentration at pH 7.4, which is twice the concentration at pH 7.7. Yet while both ratios are the same, as indicated by the same 0.3 interval on the pH scale, the difference in hydrogen ion concentration between pH 7.1 and 7.4 is greater than that between pH 7.4 and 7.7 (Figure 6.3).

In the clinical setting, changes or differences in arterial blood pH are often measured in intervals of 0.1 above and below the normal value (7.4):

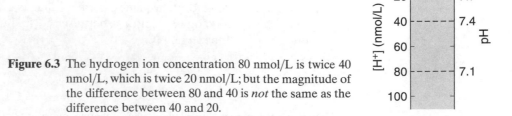

Figure 6.3 The hydrogen ion concentration 80 nmol/L is twice 40 nmol/L, which is twice 20 nmol/L; but the magnitude of the difference between 80 and 40 is *not* the same as the difference between 40 and 20.

[6]In venous blood, the greater concentration of dissolved carbon dioxide shifts the equilibrium toward hydrogen ions, as explained in the box, "Carbonic Acid." Therefore, pH is typically lower in venous blood than in arterial blood. In addition, pH in venous blood is more variable and therefore less clinically reliable than its measurement in arterial blood.

pH	$[H^+]$	Magnitude of Change in $[H^+]$
7.0	100 nmol/L	—
7.1	79	$100 - 79 = 21$ nmol/L
7.2	63	$79 - 63 = 16$
7.3	50	$63 - 50 = 13$
7.4	40	$50 - 40 = 10$
7.5	32	$40 - 32 = 8$
7.6	25	$32 - 25 = 7$
7.7	20	$25 - 20 = 5$
7.8	16	$20 - 16 = 4$

Reading down, the pH rises in uniform intervals of 0.1, so the concentration ratios at consecutive steps are all the same; $100/79 \approx 79/63 \approx 63/50$, etc. The quotient is always about 1.26, but the magnitude of the difference diminishes with each step.

THE HENDERSON-HASSELBALCH EQUATION

To see how pH relates to acid-base homeostasis, we must return to carbonic acid and its dissociation constant:

$$H_2CO_3 \leftrightarrows H^+ + HCO_3^-$$

$$K = \frac{[H^+] \cdot [HCO_3^-]}{[H_2CO_3]}$$

Since we are interested in the hydrogen ion concentration, solve for $[H^+]$ by rearranging the equilibrium equation:

$$[H^+] = \frac{K \cdot [H_2CO_3]}{[HCO_3^-]}$$

Place the preceding equation in logarithmic form and then multiply both sides of the logarithmic equation by -1 to reverse the signs:

$$\log [H^+] = \log K + \log [H_2CO_3] - \log [HCO_3^-]$$

$$-\log [H^+] = -\log K - \log [H_2CO_3] + \log [HCO_3^-]$$

The symbol p replaces "negative logarithm of . . ." in both $-\log [H^+]$ and $-\log K$:

$$pH = pK - \log [H_2CO_3] + \log [HCO_3^-]$$

To set up the final step, reverse the order of the last two terms:

$$pH = pK + \log [HCO_3^-] - \log [H_2CO_3]$$

Finally, since $\log x - \log y = \log (x/y)$:

$$pH = pK + \log\left(\frac{[HCO_3^-]}{[H_2CO_3]}\right)$$

This equation is the *Henderson-Hasselbalch equation*—the key to understanding acid-base balance in the human body.[7] Since K (the dissociation constant for carbonic acid) is a constant, pK (the negative logarithm of K) is also a constant.[8] Then the only variables in the equation are the concentrations of bicarbonate ion and carbonic acid; and as the equation shows, what counts is the *ratio* of those concentrations. To obtain pH, add the logarithm of that ratio to the constant, pK. In other words, pH depends largely on just two variables.[9]

However, the equation is not useful in this form. There is no practical way to measure the concentration of carbonic acid. Recall that this weak acid is poised between dissolved carbon dioxide in the blood, and hydrogen and bicarbonate ions; and what really drives the system is the level of dissolved carbon dioxide in the blood (see the box, "Carbonic Acid"). When the carbon dioxide level rises, the equilibrium of the system is pushed rightward, resulting in higher concentrations of hydrogen and bicarbonate ions (Figure 6.4). Conversely, when the carbon dioxide level falls, the equilibrium is pulled leftward, resulting in lower ion concentrations (Figure 6.5).

Fortunately, we *can* measure carbon dioxide. An arterial blood gas report includes a measurement called "partial pressure of carbon dioxide" or "carbon dioxide tension" (symbolized generally as P_{CO_2}, and specifically in arterial blood as P_aCO_2; see Chapter 4).[10] Partial pressure reflects the amount of dissolved gas in the plasma; it is a relatively small portion of the total amount of carbon dioxide in the blood, but it is the portion that directly affects pH. Therefore, we might be tempted to try replacing $[H_2CO_3]$ with P_aCO_2

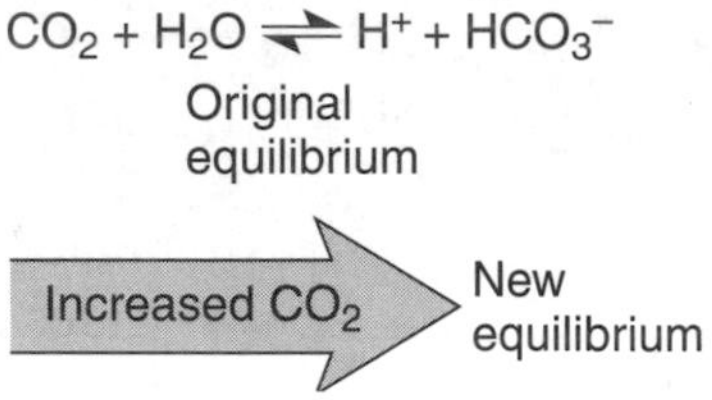

$$CO_2 + H_2O \rightleftharpoons H^+ + HCO_3^-$$

Original
equilibrium

Increased CO_2

New
equilibrium

Figure 6.4 Increasing a left-side species pushes the equilibrium rightward.

---

[7]Lawrence J. Henderson (1879–1942) was an American biochemist. Karl Hasselbalch was a Danish biochemist and physician.

[8]The chemical symbol for the negative logarithm of the dissociation constant of a weak acid is pK_a; of a weak base, pK_b; pK is generic for either type.

[9]As mentioned later in this chapter (under "Buffers"), other systems in the body can affect the hydrogen ion concentration—in particular, blood proteins. However, the carbonic acid system provides exceptionally fine adjustment and control.

[10]The usual symbol for partial pressure is uppercase P, as in P_{CO_2}. However, partial pressures are often written as "pCO_2" and "pO_2." Do not confuse this use of lowercase p (to denote partial pressure) with its meaning in pH and pK (negative logarithm).

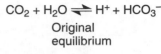

$$CO_2 + H_2O \rightleftharpoons H^+ + HCO_3^-$$
Original
equilibrium

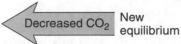
Decreased CO_2

New
equilibrium

Figure 6.5 Decreasing a left-side species
pulls the equilibrium leftward.

as the denominator of the ratio in the Henderson-Hasselbalch equation. However, we cannot have a ratio whose numerator is an ion concentration in milliequivalents per liter (meq/L) and whose denominator is a partial pressure in millimeters of mercury (mm Hg); we must convert carbon dioxide tension to the corresponding molar concentration.

A typical normal value for P_aco_2 is about 40 mm Hg. This partial pressure corresponds to a molar concentration of approximately 1.2 mmol/L. Therefore, the conversion factor (or "solubility coefficient") is:

$$\frac{1.2 \text{ mmol} \cdot L^{-1}}{40 \text{ mm Hg}} = \frac{0.03 \text{ mmol} \cdot L^{-1}}{\text{mm Hg}}$$

Then to obtain the molar concentration of dissolved carbon dioxide, multiply the carbon dioxide tension (given in mm Hg on the arterial blood gas report) by 0.03. For example, if the carbon dioxide tension is 54 mm Hg:

$$54 \text{ mm Hg} \cdot \frac{0.03 \text{ mmol} \cdot L^{-1}}{\text{mm Hg}} \approx 1.6 \text{ mmol} \cdot L^{-1}$$

One last problem: The dissociation constant for carbonic acid does not apply to living systems; it is a theoretical concept that applies only to idealized, ultra-dilute solutions. In its place, we need a constant that defines the real-life equilibrium between dissolved carbon dioxide and the hydrogen ion concentration. The value of this ionization constant is approximately $8 \cdot 10^{-7}$, and the negative logarithm of this constant—pK in the revised equation—is 6.10.

At last, we have a useful formula[11] relating pH to the measured concentration of dissolved carbon dioxide in the blood:

$$pH = 6.10 + \log\left(\frac{[HCO_3^-]}{P_aco_2 \cdot 0.03}\right)$$

Within the parentheses, all the dimensional units cancel out, leaving just a pure numeric ratio. If the numeric values for the bicarbonate ion concentration and the carbon dioxide tension are represented by B and C, respectively:

---

[11]We do not need the Henderson-Hasselbalch equation to calculate pH, since the laboratory can measure the hydrogen ion concentration directly, using a device called the "pH electrode." The value of the equation is that it shows how bicarbonate and carbon dioxide determine pH.

$$\frac{B \text{ mmol} \cdot \text{L}^{-1}}{C \text{ mm Hg} \cdot \dfrac{0.03 \text{ mmol} \cdot \text{L}^{-1}}{\text{mm Hg}}} = \frac{B}{C \cdot 0.03}$$

Let's try out this modified Henderson-Hasselbalch equation, plugging in normal values for the bicarbonate ion concentration[12] (24 mmol/L) and the carbon dioxide tension (40 mm Hg):

$$pH = 6.10 + \log\left(\frac{24}{40 \cdot 0.03}\right) = 6.10 + \log\left(\frac{24}{1.2}\right) = 6.10 + \log 20$$

$$\approx 6.10 + 1.30 = 7.40$$

Bicarbonate and carbon dioxide are both normal, so the predicted pH is normal. Note that when bicarbonate and carbon dioxide levels (both in mmol/L) are both normal, the ratio of their concentrations is about 20. The logarithm of 20 is approximately 1.30; and when this value is added to the pK, the predicted pH is normal (6.10 + 1.30 = 7.40).

If bicarbonate is normal but carbon dioxide is high (for example, 56 mm Hg, as might be seen in a patient in acute respiratory failure):

$$pH = 6.10 + \log\left(\frac{24}{56 \cdot 0.03}\right) \approx 6.10 + \log\left(\frac{24}{1.7}\right) \approx 6.10 + \log 14.1$$

$$\approx 6.10 + 1.15 = 7.25$$

A high level of carbon dioxide pushes the carbonic acid system equilibrium rightward (toward ionic dissociation), resulting in a higher concentration of hydrogen ion; thus, we see a lower pH.

If carbon dioxide is normal but bicarbonate is low (20 mmol/L, as might be seen in a patient with certain types of renal dysfunction):

$$pH = 6.10 + \log\left(\frac{20}{40 \cdot 0.03}\right) = 6.10 + \log\left(\frac{20}{1.2}\right) \approx 6.10 + \log 16.7$$

$$\approx 6.10 + 1.22 = 7.32$$

In both of these examples, the bicarbonate/CO_2 ratio is substantially below 20; therefore, the logarithm of the ratio is substantially less than 1.30. Adding that low value to the constant, 6.10, yields a low pH.

---

[12]Carbonic acid dissociation yields equal amounts of hydrogen ion and bicarbonate, but the concentration of bicarbonate in body fluids is vastly greater; its main source is sodium bicarbonate, a salt kept in the body by the kidney; see Chapter 5.

If bicarbonate is normal but carbon dioxide is low (32 mm Hg, as might be seen in a patient who is hyperventilating):

$$pH = 6.10 + \log\left(\frac{24}{32 \cdot 0.03}\right) \approx 6.10 + \log\left(\frac{24}{0.96}\right) \approx 6.10 + \log 25$$

$$\approx 6.10 + 1.40 = 7.50$$

A low level of carbon dioxide pulls the carbonic acid system equilibrium leftward (away from ionic dissociation), resulting in a lower concentration of hydrogen ions; thus, we see a higher pH.

If carbon dioxide is normal but bicarbonate is high (28 mmol/L, as might be seen in a patient who has consumed a large quantity of antacids):

$$pH = 6.10 + \log\left(\frac{28}{40 \cdot 0.03}\right) = 6.10 + \log\left(\frac{28}{1.2}\right) \approx 6.10 + \log 23.3$$

$$\approx 6.10 + 1.37 = 7.47$$

In the last two examples, the bicarbonate/CO_2 ratio is substantially above 20; the logarithm of the ratio is substantially more than 1.30. Adding that high value to 6.10 yields a high pH.

If the bicarbonate concentration were equal to the adjusted carbon dioxide tension (that is, if $[HCO_3^-] = P_aco_2 \cdot 0.03$), their ratio would be 1; and because the logarithm of 1 is 0:

$$pH = pK + logarithm \ of \ ratio$$

$$pH = 6.10 + \log 1$$

$$pH = 6.10 + 0 = 6.10$$

That is, pH = pK if the bicarbonate/CO_2 ratio = 1. There is a clinically relevant application of this concept. Thus far, we have applied the Henderson-Hasselbalch equation only to the carbonic acid system, but it can be applied to any weak acid. Many drugs are weak acids or weak bases; each has its own dissociation constant, the negative logarithm of which is its pK. For different drugs, the pK may be close to the pH in body fluids, or considerably higher or lower. As shown in Chapter 7, the difference between pH and pK determines the percentage of drug ionization in body fluids. Here, just note that if a drug is a weak acid (HA) with a pK of 7.4:

$$pH = pK + \log\left(\frac{[A^-]}{[HA]}\right)$$

$$7.40 = 7.4 + \log\left(\frac{[ionized]}{[nonionized]}\right)$$

Acidity, Alkalinity, and Neutrality

Under controlled conditions, pure water dissociates slightly, releasing equal amounts of hydrogen ion and hydroxide ion; the free protons link up with water molecules to form hydronium ions. As a result of this process, $[H_3O^+] = 10^{-7}$ mol/L and pH = 7.0. Note that pOH (the negative logarithm of the hydroxide ion concentration) is the same, which is why pH 7.0 is defined as neutral. On the pH scale, values below 7.0 are in the *acid* range, and values above 7.0 are in the *alkaline* range.

pH	$[H_3O^+]$ in mol/L		$[OH^-]$	pOH
0	$10^0 = 1$	Maximum acidity	10^{-14}	14
1	10^{-1}		10^{-13}	13
2	10^{-2}		10^{-12}	12
3	10^{-3}	ACID	10^{-11}	11
4	10^{-4}	RANGE	10^{-10}	10
5	10^{-5}		10^{-9}	9
6	10^{-6}		10^{-8}	8
7	10^{-7}	Neutrality	10^{-7}	7
8	10^{-8}		10^{-6}	6
9	10^{-9}		10^{-5}	5
10	10^{-10}	ALKALINE	10^{-4}	4
11	10^{-11}	RANGE	10^{-3}	3
12	10^{-12}		10^{-2}	2
13	10^{-13}		10^{-1}	1
14	10^{-14}	Maximum alkalinity	$10^0 = 1$	0

In theory, pH < 0 would represent $[H_3O^+] > 1$ mol/L, and pH > 14 would represent $[H_3O^+] < 10^{-14}$ mol/L. In reality, such conditions do not exist.

While pH 7.0 is chemically neutral, such a pH value in arterial blood would be severely abnormal, representing a hydrogen ion concentration two-and-a-half times higher than normal (usually close to pH 7.4).

Obviously, if $x = x + y$, then $y = 0$. Therefore:

$$\log\left(\frac{[\text{ionized}]}{[\text{nonionized}]}\right) \approx 0$$

$$\frac{[\text{ionized}]}{[\text{nonionized}]} = \text{antilog } 0 = 1$$

$$[\text{ionized}] = [\text{nonionized}]$$

REMEMBER: pK, the negative logarithm of a drug's dissociation constant, indicates the body-fluid pH at which the concentrations of ionized and nonionized drug would be equal—that is, the pH at which there would be 50% ionic dissociation at equilibrium.

BUFFERS

Introducing a strong acid or base into a system causes a dramatic change in pH. For example, if we introduce just 1 millimole of sulfuric acid into 1 liter of water:

$$H_2SO_4 \rightarrow 2H^+ + SO_4^{2-}$$

The hydrogen ions link up with water molecules to form hydronium ions. Since dissociation is almost complete and each molecule that dissociates releases two hydrogen ions and one sulfate group, the hydronium ion concentration would be about 2 mmol/L. The pH is computed as follows:

$$[H_3O^+] = 2\,\frac{mmol}{L} = 0.002\ mol \cdot L^{-1}$$

$$\log 0.002 \approx -2.7$$

$$pH = 2.7$$

Hydronium from the water itself is ignored, because its relative contribution to the total is negligible: $0.002 + 0.0000001 \approx 0.002$. If the pH in the water was around 7.0 prior to the introduction of the strong acid, the increase in the hydronium ion concentration exceeds four orders of magnitude.

The body does not tolerate sudden or severe changes in pH. Keeping the hydrogen ion concentration in body fluids stable is the task of a *buffer*— a chemical system that minimizes pH changes that would otherwise occur due to the action of strong acids or bases. The mechanism is based on the actions of a *buffer pair,* which consists of a weak acid and its own conjugate base—that is, HA and A^-. Because a weak acid, by definition, undergoes only limited dissociation, the main source of the conjugate base is not the weak acid itself, but a *salt* of the weak acid. A salt is a product formed by the reaction of an acid and a base, in which the cation released by the base replaces the hydrogen ion from the acid. For example, the sodium cation from the base sodium hydroxide replaces the hydrogen ion from hydrochloric acid to form table salt (and water):

$$HCl + NaOH \rightarrow NaCl + H_2O$$

An important buffer pair in the human body is carbonic acid (H_2CO_3) and its conjugate base, bicarbonate ion (HCO_3^-), which comes from the salt sodium bicarbonate ($NaHCO_3$).

In pure water (as we saw earlier), hydrogen ions, released by dissociation of a strong acid, link up with water molecules to form hydronium ions; the result is a sharp increase in the hydronium ion concentration and a sharp drop in pH. In a buffered system, the pH change is more limited. Instead of forming hydronium ions by linking up with water, hydrogen ions from the strong acid form molecules of the weak acid by combining with the conjugate base anions. The normal concentration of hydrogen ion is maintained, and there is no significant drop in pH.

Let's see what happens with hydrochloric acid in the presence of sodium bicarbonate. Dissociation of this strong acid releases hydrogen ions, which link up with water to form hydronium ions:

$$HCl \rightarrow H^+ + Cl^-$$

$$H^+ + H_2O \rightarrow H_3O^+$$

The salt of the weak acid also exhibits almost complete dissociation:

$$NaHCO_3 \rightarrow Na^+ + HCO_3^-$$

Here is the reaction involving all the species present:

$$H_3O^+ + HCO_3^- + Na^+ + Cl^- \rightarrow H_2CO_3 + H_2O + NaCl$$

The critical process is the reaction of the extra hydrogen ions (from the strong acid) and bicarbonate ions (from the salt of the weak acid), which combine to form additional carbonic acid. The net result can be summarized in one step by showing only those species that are significantly affected:

$$H_3O^+ + HCO_3^- \rightarrow H_2CO_3 + H_2O$$

What we are really seeing is just the reverse direction of the carbonic acid dissociation equilibrium:

$$H_2CO_3 \leftarrow H^+ + HCO_3^-$$

This process illustrates the principle that the equilibrium of a system is pushed toward the side opposite to that of a species whose concentration is increased. Look at the last equation. The equilibrium shifts toward the formation of carbonic acid due to the increased concentration of hydrogen ion, which was caused by the introduction of the strong acid.

This buffer also limits the pH change brought about by a strong base, such as sodium hydroxide. In pure water, hydroxide anions, released by dissociation of the base, react with hydronium ions to form additional water molecules; the result is a sharp decrease in the hydronium ion concentration and a sharp rise in pH. In a buffered system, this effect is at least partially offset by the dissociation of some intact carbonic acid, releasing hydrogen ions and bicarbonate ions. The normal concentration of hydrogen ion is maintained, and there is no significant rise in pH.

Taking it step by step, dissociation of the strong base releases hydroxide anions, which react with hydronium ions to form water molecules:

$$NaOH \rightarrow Na^+ + OH^-$$

$$OH^- + H_3O^+ \rightarrow 2H_2O$$

This time, the overall reaction involves intact carbonic acid:

$$Na^+ + H_2CO_3 + H_2O \rightarrow NaHCO_3 + H_3O^+$$

The net result:

$$OH^- + H_2CO_3 \rightarrow HCO_3^- + H_2O$$

This process illustrates the principle that the equilibrium of a system is pulled toward the same side as that of a species whose concentration is decreased. Again, look at the carbonic acid dissocation equilibrium. This time, the equilibrium shifts toward the release of ions by the dissociation of carbonic acid due to the decreased concentration of hydrogen ion, which was caused by the introduction of the strong base.

Other Buffers

The purpose of a buffer system is to prevent sudden and severe changes in the hydrogen ion concentration. The carbonic acid-bicarbonate system is one of the primary buffers in the body. Dissolved carbon dioxide in the blood guarantees a source of carbonic acid (the weak acid in the system). Sodium bicarbonate provides bicarbonate ion (the conjugate base). This buffer system is active in blood as well as in the interstitial fluids.

Other systems also serve as buffers. Plasma proteins have a *carboxyl group* ($-COOH$), which can release or take up hydrogen ions:

$$-COOH \rightleftarrows -COO^- + H^+$$

An increase in the concentration of hydrogen ion in the blood pushes this equilibrium leftward; a decrease in hydrogen ion concentration pulls the equilibrium rightward. In short, plasma proteins act as a reservoir, taking up hydrogen ions through recombination when there is an excess of hydrogen ion in the blood, and releasing ions through dissociation when there is a deficit. Similarly, inside red blood cells, part of the hemoglobin molecule releases or takes up hydrogen ions as needed to maintain a stable concentration.

When the extracellular pH is low, excess hydrogen ions enter the cells (in exchange for potassium ions leaving the cells to preserve electrical balance). Inside the cells, buffering is provided by a protein system similar to that of plasma proteins.

Elimination of excess metabolic acids via the urine depends on a buffer system in the kidneys. In addition to bicarbonate, there is a phosphate buffer and an ammonium buffer (the urinary buffers are discussed in more detail in Chapter 5).

ACIDEMIA AND ACIDOSIS, ALKALEMIA AND ALKALOSIS

The principles of acid-base physiology apply directly to the clinical assessment of disturbances in acid-base homeostasis. First, however, we must define some terms that are often used in a generic and interchangeable manner.

Acidemia refers to a high concentration of hydrogen ion in the blood, indicated by a pH below 7.35—but the amounts of bicarbonate ion and other bases (alkaline species) are not defined. Acidemia can occur if the bicarbonate level is low, or if the bicarbonate level is normal but the carbon dioxide tension is high. *Acidosis* refers to an absolute deficit in alkaline species or a relative deficit caused by an increase in acid species—but pH is not defined. If the alkaline deficit is matched by a proportional decrease in carbon dioxide tension, pH remains normal; but without a fully compensatory change in carbon dioxide, pH is low.

Conversely, *alkalemia* refers to a low concentration of hydrogen ion, indicated by a pH above 7.45. Alkalemia can occur if the bicarbonate level is high, or if the bicarbonate level is normal but the carbon dioxide tension is low. *Alkalosis* refers to an absolute excess in alkaline species or a relative excess caused by loss of normal acid species. If the alkaline excess is matched by a proportional increase in carbon dioxide tension, pH remains normal; but without a fully compensatory change in carbon dioxide, pH is high.

REMEMBER: *Acidemia* (pH < 7.35) is caused by uncompensated or incompletely compensated *acidosis* (absolute or relative alkaline deficit). *Alkalemia* (pH > 7.45) is caused by uncompensated or incompletely compensated *alkalosis* (absolute or relative alkaline excess).

PRIMARY AND COMPENSATORY MECHANISMS

The kidney is responsible for maintaining an adequate supply of bicarbonate in the body (see Chapter 5); however, this system does not work fast enough to make continual fine adjustments in the bicarbonate level in arterial blood. In contrast, the respiratory system *can* make continual adjustments in the carbon dioxide tension, through control of the depth and rate of respiration (see Chapter 4). The renal contribution to pH (the numerator of the bicarbonate/ CO_2 ratio) is powerful but slow, while the respiratory contribution (the denominator) is fast enough to maintain instant-by-instant homeostasis in the hydrogen ion concentration in arterial blood.

As shown in the box, "pH and the Ratio of the Variables," acidemia is associated with a *low* bicarbonate/CO_2 ratio; and a low ratio could be due to either a *low* bicarbonate level or a *high* carbon dioxide tension (hypercapnia). Therefore, if either of these conditions is present in a patient with acidemia, it is likely to be the *primary cause* of the pH abnormality.

In a patient with acidemia due to low bicarbonate, a *low* carbon dioxide tension is probably a *compensatory response*—eliminating carbon dioxide through hyperventilation, to hold the bicarbonate/CO_2 ratio up as close as possible to 20. Likewise, in a patient with acidemia due to chronic hypercapnia, a *high* bicarbonate level could be a compensatory response.

Conversely, alkalemia is associated with a *high* bicarbonate/CO_2 ratio; and a high ratio could be due to either a *high* bicarbonate level or a *low* car-

pH and the Ratio of the Variables

As shown by the Henderson-Hasselbalch equation, pH depends on the ratio of bicarbonate to carbon dioxide. Different combinations of values for these two variables can yield a normal, high, or low ratio. For example:

Normal Ratio (20), Normal pH (7.35–7.45)

- Both variables normal.
- Both variables high or both low, but in normal proportion to each other.

Low Ratio (< 20), Low pH (< 7.35)

- Bicarbonate normal, carbon dioxide high.
- Bicarbonate low, carbon dioxide normal or slightly low.

High Ratio (> 20), High pH (> 7.45)

- Bicarbonate normal, carbon dioxide low.
- Bicarbonate high, carbon dioxide normal or slightly high.

bon dioxide tension (hypocapnia). Therefore, if either of these conditions is present in a patient with alkalemia, it is likely to be the primary cause of the pH abnormality.

In a patient with alkalemia due to high bicarbonate, hypercapnia could be a compensatory response—retaining carbon dioxide through hypoventilation, to hold the bicarbonate/CO_2 ratio down as close as possible to 20. Likewise, in a patient with alkalemia due to chronic hypocapnia, a low bicarbonate level could be a compensatory response (although this last scenario would be unusual).

REMEMBER: pH varies directly with bicarbonate and inversely with carbon dioxide. As bicarbonate rises, pH rises, and vice versa. As carbon dioxide rises, pH falls, and vice versa.

THE MAJOR CATEGORIES OF ACID-BASE DISTURBANCE

Changes in the carbon dioxide tension are called *respiratory;* changes in the bicarbonate ion concentration are called *metabolic.* Thus, acidemia due to hypercapnia is *respiratory acidosis;* alkalemia due to hypocapnia is *respiratory alkalosis.* Acidemia due to bicarbonate deficit is *metabolic acidosis;* alkalemia due to bicarbonate excess is *metabolic alkalosis.*

Compensatory respiratory responses occur quickly in any case of primary metabolic disturbance; in contrast, compensatory metabolic responses occur slowly and are therefore seen only when a primary respiratory disturbance is chronic or long-standing.

REMEMBER: Maintaining a stable and normal concentration of hydrogen ion depends on renal and respiratory mechanisms. For the sake of chemical homeostasis, primary disturbances in one system tend to trigger compensatory responses in the other. Respiratory compensation in

Causes and Responses in Acid-Base Disturbances

Respiratory Acidosis

- Primary cause: Increased carbon dioxide (hypercapnia)
- Compensatory response: Increased bicarbonate (in chronic condition)

Metabolic Acidosis

- Primary cause: Bicarbonate deficit
- Compensatory response: Hyperventilation to reduce carbon dioxide

Respiratory Alkalosis

- Primary cause: Decreased carbon dioxide (hypocapnia)
- Compensatory response: Decreased bicarbonate (rare)

Metabolic Alkalosis

- Primary cause: Bicarbonate excess
- Compensatory response: Hypoventilation to raise carbon dioxide

primary metabolic disturbance is fast; metabolic compensation in primary respiratory disturbance is slow and therefore occurs only in chronic conditions.

DIAGNOSTIC TOOLS

An acid-base disturbance may be suspected on the basis of clinical evidence, but the precise diagnosis depends the results of an *arterial blood gas* (ABG) analysis.

The first step in interpreting the ABG report is to see whether the pH is low (acidemia) or high (alkalemia). Next, identify the primary cause of the disturbance—the bicarbonate or carbon dioxide abnormality that would produce that type of pH abnormality. Low bicarbonate or high carbon dioxide is consistent with acidemia; high bicarbonate or low carbon dioxide is consistent with alkalemia. Finally, look for compensatory responses—changes in bicarbonate or carbon dioxide that are *not* consistent with the observed pH abnormality, but would tend to draw the bicarbonate/CO_2 ratio back toward 20 and the pH back toward 7.4. Note that the pH is normal in cases of fully compensated acidosis or alkalosis.

In some cases, no compensation occurs (for example, with acute changes in carbon dioxide tension). In other cases, combined primary disturbances are present (respiratory and metabolic factors both contributing to acidosis, or, less commonly, both contributing to alkalosis); the pH change can be especially severe in combined primary disturbances.

Another aid in diagnosis is an acid-base nomogram. There are many different acid-base nomograms in clinical use, but the general design involves plotting the pH, bicarbonate level, and carbon dioxide tension on their respective scales and finding the place where the lines cross. The nomogram is like a map, with different regions corresponding to acute and chronic acid-base disturbances; the region in which the lines cross identifies the condition with 95% accuracy.

Arterial Blood Gas Analysis: Normal Values

- pH: 7.35–7.45
  Negative logarithm of the hydrogen ion concentration; lower values are increasingly acidic, higher values are increasingly alkalinic.

- Oxygen tension (P_aO_2): 75–100 mm Hg
  Reflects only dissolved oxygen (99% of oxygen in arterial blood is bound to hemoglobin). Oxygen tension is not a determinant of pH, but it may be useful in diagnosis; for example, low P_aO_2 is consistent with respiratory acidosis due to chronic obstructive lung disease.

- Carbon dioxide tension (P_aCO_2): 35–45 mm Hg
  Reflects only dissolved CO_2. *Total CO_2* (or CO_2 content, mmol/L) is the combined molar concentration of all forms of CO_2 in the blood (mainly bicarbonate and compounds formed with protein amino groups).

- Bicarbonate ion concentration (HCO_3^-): 22–26 meq/L
  Reflects bicarbonate from all sources. *Standard bicarbonate* is the plasma concentration when the sample is taken from fully oxygenated whole blood and equilibrated with CO_2 at 40 mm Hg.

- Base excess (BE): 0 meq/L.
  The amount of acid or base needed to titrate 1 L of blood to pH 7.4. Titration is with acid in alkalosis (positive BE); with base in acidosis (negative BE = base deficit).

Heavy Breathing

A 38-year-old male with a history of Type-1 diabetes mellitus is brought into the emergency department in a state of coma. The patient is breathing deeply and rapidly, and an acetone odor is evident on his breath. Urine obtained by catheter is positive for sugar and ketones.

From the arterial blood gas analysis:

- pH: 7.27
- P_aCO_2: 31 mm Hg
- HCO_3^-: 16 meq/L

The pH is low, so the condition is acidemia. Bicarbonate and carbon dioxide are both low. A low bicarbonate is consistent with acidemia, so the condition is metabolic acidosis. The low carbon dioxide level is a compensatory mechanism—the patient is hyperventilating to eliminate carbon dioxide, because a lower carbon dioxide tension pulls the carbonic acid system equilibrium away from hydrogen ion formation.

(This patient is in *diabetic ketoacidosis*, a classic example of acute metabolic acidosis with a partial compensatory respiratory response.)

QUIZ

1. If a patient's bicarbonate concentration is 28 meq/L, and the carbon dioxide tension is 48 mm Hg, what is the pH? What type of condition is present?

2. If a patient's pH is 7.52 and the bicarbonate level is 25 meq/L, what is the carbon dioxide tension? What type of condition is present?

3. If a patient's pH is 7.37 and the carbon dioxide tension is 33 mm Hg, what is the bicarbonate level? What type of condition is present?

4. If a patient's pH, bicarbonate, and carbon dioxide are all lower than normal, what type of condition is present?

5. If a patient's pH is low, bicarbonate is normal, and carbon dioxide is high, what type of condition is present?

6. If a patient's bicarbonate concentration is 20 meq/L and the carbon dioxide tension is 46 mm Hg, what is the bicarbonate/CO_2 ratio? Without further calculation, predict whether the pH will be high, low, or normal.

7. Introduction of 24 mmol of weak acid HA into three liters of water results in the following concentrations at equilibrium: [HA] = 7 mmol/L, [H^+] = 1 mmol/L, [A^-] = 1 mmol/L. What is the dissociation constant? What is the pK?

8. Suppose we have the same amount of the same weak acid (from Question 7) in two liters of water. Predict the equilibrium concentrations of the ions and the intact compound.

9. If 0.5 mmol of nitric acid (HNO_3, a strong acid) is introduced into two liters of pure water, what is the new pH?

10. If carbonic acid and sodium bicarbonate had been present in the water before the strong acid (from Question 9) was introduced, would the new pH be higher or lower than the result seen in pure water?

Answers and explanations at end of book.

Pharmacokinetics

Chapter Outline

Mathematical Tools Used in This Chapter

Pharmacokinetics is the scientific description of what happens to a systemically acting drug in the body over the course of time—its *absorption, distribution, metabolism,* and *excretion.* Clinical pharmacokinetics is the application of these concepts to individual patients—monitoring and adjusting the concentration of a drug in the body so as to maximize effectiveness and minimize toxicity.

THE FATE OF A SYSTEMIC DRUG

A drug may enter the body by any of several routes—oral, sublingual, intranasal, inhaled, transdermal, subcutaneous, intramuscular, or intravenous.[1]

---

[1]By its literal meaning, the term *parenteral* ("alongside rather than inside the enteral system") is descriptive of all the non-oral routes. In routine clinical usage, however, *parenteral* more often refers exclusively to the intravenous and intramuscular routes.

Whatever the route of administration, the drug must get to the circulatory system if it is to act systemically. With intravenous administration, the drug goes directly into the bloodstream; by any other route, the drug must be *absorbed* into circulation.

In the plasma, the drug may become partially bound to plasma proteins, which carry it in the blood as it is *distributed* throughout the body. For drugs that act on specific organs or tissues of the body (as distinct from antimicrobials, which act on pathogenic organisms), the unbound or "free" portion moves from the plasma to the interstitial fluids to reach the target organ or tissue; there, it binds to receptor sites on the cell membranes to produce its intended therapeutic effect. The shift from the intravascular to the extravascular space enlarges the volume of the drug's distribution in the body and creates an equilibration between the drug levels in plasma and the interstitial fluid.

Starting from the moment the drug enters the bloodstream and continuing for as long as any of it remains in circulation, the body works to rid itself of this foreign substance. *Elimination* involves two processes: *metabolism* (chemical alteration and breakdown of the drug as blood passes through the liver and lungs) and *excretion* (removal from the body). Excretion takes place largely through *clearance,* which is extraction of drug molecules and metabolic breakdown products from the plasma as blood passes through the liver and kidneys. Because of the equilibrium between plasma and tissue levels of the drug, the decline in plasma levels due to clearance is accompanied by a corresponding decline in tissue levels.

PLASMA CONCENTRATION

Most often, we measure drugs in the body by their concentration in the plasma. Rising and falling plasma levels reflect the relative rates of absorption and removal. Initially after administration, the rate of absorption exceeds the rate of removal, and the plasma concentration rises. If no additional doses are administered, absorption peaks and declines, and the concentration begins to fall.

During ongoing therapy, a drug will start to accumulate in the body. At a proper maintenance dosage, the amounts being absorbed into the bloodstream are just enough to balance the amounts being eliminated.[2] With continuous intravenous infusion, the drug concentration in the plasma can be held steady. With oral administration, the concentration fluctuates, rising to a peak (the highest level) after each dose and then gradually falling to a trough (the lowest level) before the next dose is given. Thus, the elapsed time between the last dose and the collection of the blood sample can significantly affect the measured concentration.

SOLUBILITY CHARACTERISTICS OF A DRUG

Drugs dissolved in the body fluids may become partially ionized. In general, ionized molecules are more *hydrophilic* (water-soluble), and nonionized, electrically neutral molecules are more *lipophilic* (fat-soluble).

---

[2]A *dose* is the amount of drug administered at any one time. If the patient takes two 500-mg tablets of a drug three times a day, each of the three daily doses is 1 gram. The use of *dose* as a verb (as in, "The patient was last dosed at 8 p.m.") is to be avoided. *Dosage* is not a synonym for dose; it is the recommended regimen for a specific condition in a specific patient: "(Name of drug), 100 mg by mouth, twice daily for ten days."

Many drugs are chemically classified as weak acids or weak bases—they undergo partial dissociation to a point of equilibrium between the amounts of ionized and nonionized drug. The portion that is ionized in solution is more hydrophilic; the portion that remains nondissociated is more lipophilic. The point of equilibrium depends on the pH in body fluids and the drug's pK (the negative logarithm of the dissociation constant).[3] pK is equal to the pH at which half the molecules become ionized and half remain nondissociated and nonionized, as shown by the Henderson-Hasselbalch equation (see Chapter 6):

$$pH = pK + \log\left(\frac{[\text{ionized}]}{[\text{nonionized}]}\right)$$

If half of the drug is ionized, the ionized and nonionized concentrations [shown in brackets] are equal and their ratio is 1; and because $\log 1 = 0$, $pH = pK + 0 = pK$.

A rearrangement shows that the logarithm of the ratio is given by the difference between pH and pK:

$$\log\left(\frac{[\text{ionized}]}{[\text{nonionized}]}\right) = pH - pK \quad or \quad \log\left(\frac{[\text{nonionized}]}{[\text{ionized}]}\right) = pK - pH$$

We'll use the latter version. If $pK - pH$ gives the logarithm of the ratio of nonionized to ionized drug concentrations, the *antilogarithm* of $(pK - pH)$ gives the ratio itself. For example, if a drug's pK is 7.1 and pH is 7.4, the logarithm of the concentration ratio is $7.1 - 7.4 = -0.3$; the antilogarithm of -0.3 is about 0.5, which means that there is twice as much ionized drug as nonionized drug.

---

pK, pH, and Drug Ionization

If $pK < pH$, then $pK - pH < 0$; thus, more than 50% is ionized, because

$$\text{if } \log\left(\frac{[\text{nonionized}]}{[\text{ionized}]}\right) < 0, \text{ then } \frac{[\text{nonionized}]}{[\text{ionized}]} < 1$$

If $pK > pH$, then $pK - pH > 0$; thus, less than 50% is ionized, because

$$\text{if } \log\left(\frac{[\text{nonionized}]}{[\text{ionized}]}\right) > 0, \text{ then } \frac{[\text{nonionized}]}{[\text{ionized}]} > 1$$

If $pK = pH$, then $pK - pH = 0$; thus, 50% is ionized, because

$$\text{if } \log\left(\frac{[\text{nonionized}]}{[\text{ionized}]}\right) = 0, \text{ then } \frac{[\text{nonionized}]}{[\text{ionized}]} = 1$$

---

[3]The chemical symbol for the negative logarithm of the dissociation constant of a weak acid is pK_a; of a weak base, pK_b; pK is generic for either type.

ABSORPTION INTO SYSTEMIC CIRCULATION

Absorption into the bloodstream depends on a drug's ability to cross a series of biological barriers; in turn, this ability depends on the size and solubility of the molecule. To enter the circulation and reach its site of action, a drug must be small enough to pass through the capillary pores. To be carried as a dissolved substance in plasma, it must have some degree of solubility in water. Outside the circulatory system, to cross cell membranes in the tissues, a drug must also have some solubility in lipids.

With an orally administered agent, the drug molecule must remain intact as it moves through the stomach to reach the intestinal villi. It passes through the mucosa to enter the capillaries of the portal system, which carries blood from the alimentary canal directly to the liver. The liver extracts some portion of the drug from blood passing through ("first-pass effect"). Only the amount of drug that gets past the liver is absorbed into circulation.

BIOAVAILABILITY AND "AREA UNDER THE CURVE"

Bioavailability is the percentage or fraction (hence, its symbol, F) of the administered dose that ultimately reaches the systemic circulation:

$$F = \frac{\text{amount of drug reaching systemic circulation}}{\text{amount of drug administered}}$$

Most often, we speak of bioavailability with respect to oral drugs, and not at all with respect to intravenous drugs (by definition, the bioavailability of an intravenously administered agent is 100%). Bioavailability reflects the extent but not the rate of absorption; it is the portion of a dose that ultimately reaches systemic circulation, regardless of how long that process may take.[4]

---

[4]Some oral agents are formulated in compounds that are not pure drug. For example, when the antibiotic carbenicillin is to be used orally, it must be given as an acid-stable salt (carbenicillin indanyl sodium) so that it can resist breakdown in the stomach. A 500-mg tablet of this salt delivers 382 mg of active drug.

Measuring a drug's concentration in plasma at a single point in time is a routine laboratory procedure. Measuring the total amount of drug that enters the circulation and is present in the plasma over an interval of time involves a mathematical concept called *area under the curve* (AUC).

Suppose we administer a single dose of a drug and then take repeated measurements of its concentration in plasma. On a graph, we can record the plasma concentration (on the *y*-axis) at each point in time (on the *x*-axis). This set of points can then be assembled to form a continuous curved line that rises from the *x*-axis and then falls back over a certain period of time (Figure 7.1).

AUC is the area enclosed by this concentration-over-time curve and the horizontal axis. Just as the area of a rectangle is obtained at the product of its height and width, AUC is obtained as the product of concentration and time. With plasma concentration measured in mg/mL and time in hours, the dimensional units for AUC would be mg · mL^{-1} · hr.

AUC can be estimated by dividing the area into narrow vertical slices of uniform width representing a short time interval. If the width is small enough, we can treat the short curved segment at the top of each slice as a straight line. The result is a set of trapezoids. The area of each trapezoidal slice is the product of its horizontal width and the average of its two vertical heights (Figure 7.2). The sum of these trapezoidal areas approximates the true value for AUC.

Now let's perform two simple mathematical operations. First, multiply the numeric value for AUC by the total number of milliliters of plasma, representing the total plasma volume (as measured by the methods discussed in Chapter 2). Next, divide by the number of hours during which the drug was detectable in the plasma, representing the total time in circulation. Looking at these two operations in terms of dimensional units:

$$\frac{\text{mg} \cdot \text{hr}}{\text{mL}} \cdot \text{mL} \cdot \text{hr}^{-1} = \text{mg}$$

Note that the units for volume and time cancel out, leaving just weight in milligrams. The value obtained is the total amount of drug that was present in

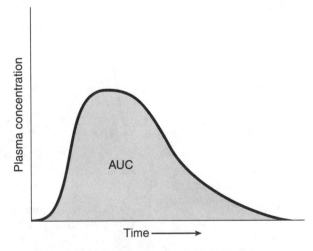

Figure 7.1 Area under the curve (AUC) for plasma concentration over time.

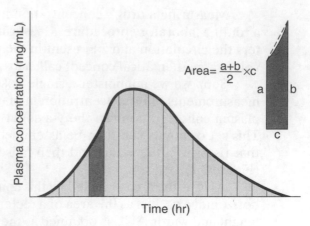

Figure 7.2 AUC can be estimated by dividing the area into narrow vertical slices with straight-line tops. Each slice is a trapezoid whose area is easily obtained.

the plasma throughout the entire time in circulation—that is, the total amount absorbed into circulation. Thus, we have mathematically derived the numerator of the bioavailability fraction. *The ratio between this absorbed amount and the total amount administered in a single dose is the drug's bioavailability.*

Bioavailability and AUC are related concepts—the amount of drug absorbed into circulation from a single dose defines the amount present in circulation over time.[5] However, whereas bioavailability is a direct function of absorption, the shape of the curve representing the drug's plasma concentration over time is a function not only of its absorption but also of its distribution, metabolism, and excretion.

Different formulations of the same drug may differ in their bioavailability. For the cardiac drug digoxin in capsule form, F = 1.00 (100% bioavailability); in elixir form, F = 0.77; and in tablet form, F = 0.70. Different formulations of the same drug are *bioequivalent* if equal doses yield the same AUC; a difference in AUC implies that the formulations differ in bioavailability (Figure 7.3).

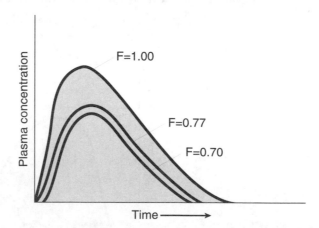

Figure 7.3 Different formulations of a drug can have different bioavailability, as seen in different AUCs.

---

[5]A drug's bioavailability is determined by collecting AUC data from a number of test subjects.

DISTRIBUTION THROUGHOUT THE BODY

Following absorption, a drug is distributed throughout the body. The simplest way to look at drug distribution is the *one-compartment model*—the entire body is regarded as a single compartment throughout which a drug is distributed instantaneously following administration (Figure 7.4). Intravenous drugs come closest to illustrating this model, with rapid distribution throughout the systemic circulation and then throughout all the extracellular fluid, both intravascular (plasma) and extravascular (interstitial fluid).

In reality, systemic distribution is not instantaneous. Typically, there is a relatively fast distribution to the highly perfused central organs, followed by a slower distribution to the peripheral tissues (Figure 7.5). In this *two-compartment model,* the drug first reaches the central compartment (the circulatory system and the main internal organs). From there, it goes to the peripheral compartment (skin, muscle, fat), which serves as a drug reservoir, maintaining an equilibrium with the central compartment[6] (Figure 7.6).

Plasma concentrations reflect the central compartment. A rapid initial decline after intravenous administration corresponds to the outward shift from the intravascular to the extravascular space as equilibrium is established between the plasma and interstitial fluid. The terminal portion of the decline is more gradual, corresponding to clearance of the drug (Figure 7.7). On this "semilog graph," the vertical axis is the *natural* logarithm of the plasma concentration; the horizontal axis is linear, showing time following administration. Note that the decline in plasma levels over time is a straight line on a semilog graph. (We will return to this concept later.)

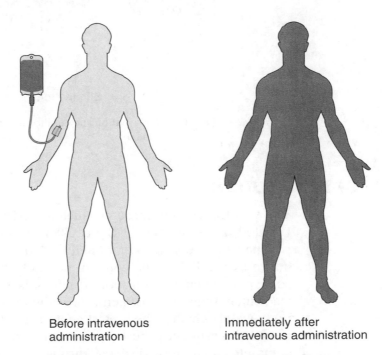

Before intravenous administration

Immediately after intravenous administration

Figure 7.4 One-compartment model of drug distribution.

⚠ [6]*Compartment,* as used here, does not correspond to the body's anatomic fluid compartments—the extracellular fluid (ECF) and intracellular fluid (ICF).

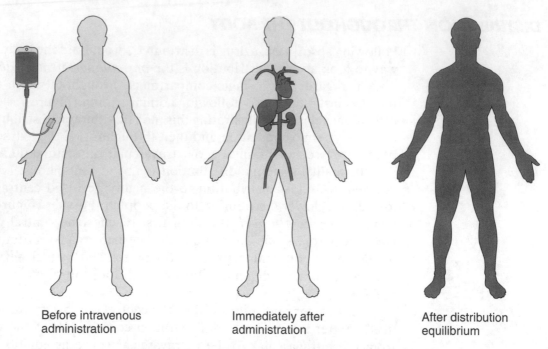

Before intravenous
administration

Immediately after
administration

After distribution
equilibrium

Figure 7.5 Two-compartment model of drug distribution.

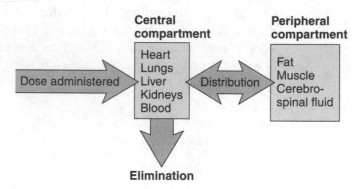

**Central
compartment**

Heart
Lungs
Liver
Kidneys
Blood

Dose administered

Distribution

**Peripheral
compartment**

Fat
Muscle
Cerebro-
spinal fluid

Elimination

Figure 7.6 Distribution and equilibrium in the two-compartment
model.

VOLUME OF DISTRIBUTION

The smallest volume through which a systemically acting drug may be distributed is the plasma. At the capillary level, some of the drug leaves the
plasma and enters the interstitial fluid. Now, the volume of distribution is the
entire extracellular fluid compartment. Of the drug in the interstitial fluid,
some portion may be transported across the cell membranes, entering the intracellular fluid compartment; the volume of distribution is now the total
body water, which is the largest anatomic volume. In the one-compartment
model, this sequence of events is assumed to take place instantaneously, with
complete and uniform dispersal throughout all body fluids. (A drug's volume
of distribution may be given in units of L/kg as a way to adjust for differences
in body size.)

We must distinguish between these *actual* volumes and a drug's *apparent* volume of distribution. The latter (symbolized V_d) does not correspond

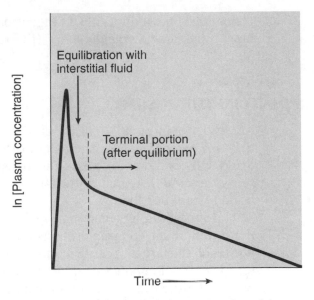

Figure 7.7 Decline in plasma concentration of drug.

to any defined anatomic fluid space or compartment; as we will see, V_d can be many times larger than total body water. To understand this seeming paradox, let's look at a concept discussed in Chapter 2—determining the volume of water in a tank by adding a known quantity of substance and then measuring the resulting concentration. In that model, volume is the quotient of the amount added divided by the concentration. If 5 grams yields a concentration of 200 milligrams per liter, there must be 25 liters of water (because $25 \text{ L} \cdot 0.2 \text{ g/L} = 5 \text{ g}$). The lower the measured concentration, the larger the volume must be, and vice versa.

We can use an analogous method to determine V_d for a drug:

$$V_d = \frac{\text{dose administered}}{\text{plasma concentration}}$$

However, this method gives only the *apparent* volume of distribution. The reason? Unlike distribution in a tank of water, drug distribution in the body fluids is *not* uniform. A drug that is tightly bound to plasma proteins (see the next section) remains largely inside the intravascular space, yielding a high plasma concentration and a small V_d. A loosely bound drug readily leaves the plasma to enter the interstitial fluids, yielding a lower concentration and a larger V_d (attachment to cell membrane receptors further reduces the fluid concentration and expands V_d). And a minimally bound drug that is highly soluble in lipid will readily enter storage in fat, yielding an extremely low concentration in the plasma and an enormous V_d; for such a drug, V_d can be literally hundreds of liters!

In general, a large V_d indicates that the drug quickly and easily reaches its site of action in the tissues; therefore, therapeutic effects are seen at relatively low plasma concentrations. Conversely, a small V_d indicates that the drug has limited distribution to its site of action; therefore, it takes higher plasma concentrations to achieve therapeutic effects. Maintaining the plasma concentration within its ideal range is particularly important with drugs that have a narrow *therapeutic index* (the range between therapeutic and toxic levels).[7]

PROTEIN BINDING IN THE PLASMA

Many drugs in circulation are carried bound to plasma proteins, forming *drug-protein complexes*. The extent of protein binding varies with different drugs. However, in most cases, the measured concentration in plasma represents the *total drug* concentration—the sum of the protein-bound and unbound (free) portions.

Bound drug molecules are inert for two reasons. First, drug molecules that are bound to plasma proteins remain in the intravascular space and do not reach the target organ. Second, even if drug-protein complexes could enter the extravascular space intact, a drug molecule must be free of its carrier protein before it can bind to a receptor site at the target organ or tissue to bring about therapeutic activity.[8]

> **REMEMBER:** It is the *free* (unbound) drug that is pharmacologically active—because only the free drug can enter the interstitial fluid and bind to target organ receptor sites.

The protein-bound fraction, often representing the greatest portion of the total amount in circulation, serves as a reservoir for the free fraction. Take-up at cell membrane binding sites leaves a transient deficit of free drug in the interstitial fluid; to maintain the equilibrium between the free drug levels in the plasma and interstitial fluid, some free molecules leave the intravascular space, and the resulting transient deficit in the plasma is corrected by the dissociation of some drug-protein complexes. Conversely, lack of available binding sites leaves excess free drug in the interstitial fluid; some of these free molecules reenter the intravascular space, and the resulting excess in the plasma is corrected by the formation of new drug-protein complexes.

So we see that the formation and breakdown of drug-protein complexes is a dynamic, reversible process. Variations in the amount of carrier protein in circulation can alter the measured plasma concentration of total drug without necessarily disturbing the concentration of pharmacologically active free drug. With an excess in plasma proteins, more of the drug absorbed from each dose becomes bound—leaving less unbound drug to be filtered out at the kidney (see Clearance, later in the chapter). As a result, the free concentration remains stable, and therapeutic activity is normal even though the plasma concentration of total (bound plus free) drug is high. Conversely, with a deficit in plasma proteins, less of the drug absorbed from each dose becomes bound—but the excess free drug is quickly filtered at the kidney. As a result,

---

[7]Therapeutic index is also given as the toxic/therapeutic dose *ratio*. A high index indicates a relatively safe drug, because the doses given to achieve therapeutic plasma levels are much lower than the doses necessary to produce toxicity; a low ratio indicates that therapeutic doses are dangerously close to toxic doses, which means little or no margin for error in the dosage.

[8]Similarly, various hormones are carried in the circulation partially bound to plasma proteins, but only the free hormone is physiologically active.

therapeutic activity is normal even though the plasma concentration of total drug is low.

However, abnormalities in protein binding can have serious clinical consequences if the resulting alteration in the free fraction is not quickly offset by a corresponding change in clearance at the kidney. This situation can occur as a result of impaired renal function or as a consequence of the drug's pK, which powerfully affects clearance, as we will see.

HALF-LIFE AND STEADY-STATE

As previously stated, the body works constantly to remove drugs. For different drugs, the rate of removal may be faster or slower, and that rate determines the drug's *half-life*. Half-life (*t* ½) is defined as the time required for the total amount of drug in the body to decrease by 50%, as determined by the time required for the plasma concentration to decrease by 50%. Once equilibrium is reached between drug levels in the plasma and interstitial fluid, and the plasma level has declined by half, the plasma concentration at any moment is half what it was one half-life earlier, and twice what it will be one half-life later (as always, with the stipulation that no further doses have been or will be administered).

For example, if it takes ten hours for the plasma concentration to fall from 12 mg/L to 6 mg/L, it takes another ten hours to fall from 6 mg/L to 3 mg/L, another ten hours to fall from 3 mg/L to 1.5 mg/L, and so on, until the amounts left in the body are vanishingly small.

A drug's half-life is determined by administering a single intravenous dose and then taking serial measurements of the plasma concentration. Plotting these data on a semilog graph, the terminal segment is a straight-line decline. Take two points on that straight-line segment such that the plasma level at one point is twice the level at the other; *t* ½ is the time interval between those points (Figure 7.8).

Drugs with a long half-life take a long time to build up in the body after initiation of dosage, and a correspondingly long time to be eliminated from the body upon discontinuation of therapy; such drugs are often called "long-acting." Drugs with a short half-life are "short-acting." Even if long-acting

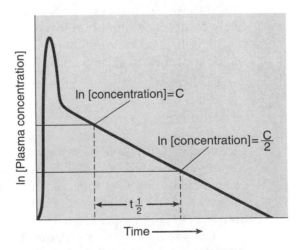

Figure 7.8 Determination of half-life.

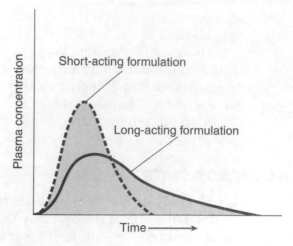

Figure 7.9 If bioavailability is the same, AUC is the same.

and short-acting formulations of the same drug are bioequivalent, their plasma concentration curves over time look different—the curve for the long-acting formulation rises and falls gradually, while the curve for the short-acting formulation rises faster and reaches a higher peak, but also disappears more quickly. Yet if bioavailability is the same, the *area* under the curve (AUC) is the same, even if the *shape* or contour of the curves are different (Figure 7.9).

With ongoing therapy, it is usually desirable to achieve a *steady-state* condition—the plasma concentration is in equilibrium with the tissue concentration, and the amounts of drug being absorbed match the amounts being eliminated. Half-life determines how long it takes, at a constant dosage, to build up the amount of drug in the body to achieve this steady-state level after therapy is initiated. With appropriate doses given at appropriate intervals (usually, the dosing interval is the half-life), it takes one half-life to reach 50% of the steady-state level, two half-lives to reach 75%, three half-lives to reach 87.5%, and so on. Steady-state concentrations are effectively attained after four to five half-lives.

Half-life also determines how long it takes for the drug to be eliminated from the body after therapy is discontinued. It takes one half-life for 50% washout, two half-lives for 75%, three half-lives for 87.5%, and so on. Effectively complete elimination usually takes four to five half-lives. This pattern is called *first-order elimination*. The feature we have noted—a terminal-segment decline in plasma levels forming a straight line on a semilog graph—is a key characteristic of drugs that exhibit first-order elimination.

CLEARANCE

Clearance is a method of *excretion*. It is the *rate* at which blood or plasma is completely cleared of a specific substance—that is, the volume of blood or plasma cleared of that substance per unit of time. In other words, clearance varies with different substances; it also varies with the patient's renal function, as reflected by the glomerular filtration rate (GFR).

Let's review some basic concepts from Chapter 5. For a substance that is neither reabsorbed nor actively secreted, the clearance is equal to GFR.

For a substance that undergoes any significant degree of reabsorption, the clearance is lower than GFR; with total reabsorption, clearance is zero. For a substance that undergoes any significant degree of active tubular secretion, the clearance is higher than GFR; with total secretion, clearance is maximal (that is, equal to total renal plasma flow).

As blood flows through the kidney, some of the free drug in the plasma is filtered at the glomerulus (protein-bound drug remains in circulation). Recall that ionized molecules are more water-soluble and nonionized molecules are more lipid-soluble. Of the drug that entered the glomerular filtrate, the ionized portion resists reabsorption and tends to remain in the tubular fluid moving through the nephron; this portion will therefore be excreted through the urine. The nonionized portion is more easily reabsorbed, moving out of the tubular fluid and back into the bloodstream; this portion will therefore be retained in the body.

Earlier, we saw that for a drug with pK = 7.1, about two-thirds of the amount present is ionized in a normal plasma environment, with pH = 7.4. However, within the renal system, the pH is lower (the range is wide, but let's take a typical value of 6.2). Applying the Henderson-Hasselbalch equation:

$$\log\left(\frac{[\text{nonionized}]}{[\text{ionized}]}\right) = pK - pH = 7.1 - 6.2 = 0.9$$

Antilog $0.9 \approx 7.9$, which means that there is almost eight times as much nonionized drug as ionized drug—that is, about 88% of the drug in the tubular fluid is nonionized and therefore subject to reabsorption. As a result, its clearance is low and its half-life is long.

Now suppose we administer a *urinary alkalinizing agent*—a compound that raises the pH in the kidney. With the same drug, notice the effect on ionization when the renal pH is raised to 8.0:

$$\log\left(\frac{[\text{nonionized}]}{[\text{ionized}]}\right) = 7.1 - 8.0 = -0.9$$

Toxicity Without Overdose

The anticonvulsant drug phenytoin (pK = 8.3) is about 90% protein-bound. What if a patient taking this drug develops hypoalbuminemia?

If the total phenytoin concentration is 15 mg/L, the free concentration (about 10% of the total) is about 1.5 mg/L. But a deficit in carrier protein creates a higher-than-normal free concentration—perhaps 3 mg/L. Unbound molecules are readily filtered at the glomerulus. However, in the environment of the kidney (pH = 6.2):

$$\log\left(\frac{[\text{nonionized}]}{[\text{ionized}]}\right) = 8.3 - 6.2 = 2.1$$

Antilog $2.1 \approx 126$, which means that almost all of the amount present is nonionized and subject to reabsorption. So the excess in free fraction created by the hypoalbuminemia is not offset by an increase in clearance, and the patient may experience toxic effects at a normal dosage.

Antilog $-0.9 \approx 0.13$, which means that now only one-eighth as much nonionized drug as ionized drug is present—that is, about 88% of the drug in the tubular fluid is ionized and will therefore be excreted. Its clearance is now much higher and its elimination half-life is much shorter.

This technique of alkalinizing the urine can be used to accelerate the excretion of certain drugs in cases of overdose and toxicity. It is appropriate only for drugs with all three of the following characteristics: minimal protein-binding, with distribution throughout the extracellular fluid compartment; elimination mainly through the kidney; and pK in the appropriate range. A classic example is phenobarbital (pK $= 7.4$). In the kidney:

$$\log\left(\frac{[\text{nonionized}]}{[\text{ionized}]}\right) = 7.4 - 6.2 = 1.2$$

Antilog $1.2 \approx 16$, which means that the concentration of nonionized phenobarbital is 16 times greater than that of the ionized drug; in other words, only about 6% of the drug is ionized. As a result, reabsorption is extensive, and elimination is slow. But if we raise the renal pH to 8.0:

$$\log\left(\frac{[\text{nonionized}]}{[\text{ionized}]}\right) = 7.1 - 8.0 = -0.6$$

Antilog $-0.6 \approx 0.25$, which means that the concentration of nonionized phenobarbital is only about one-fourth that of the ionized drug; about 80% of the drug is now ionized. Much less is reabsorbed, and elimination is therefore much faster.

Next, let's look at the mathematical computation of clearance. The standard formula for clearance is as follows:

$$Cl = \frac{[\text{drug}]_{\text{urine}} \cdot \text{urine formation rate}}{[\text{drug}]_{\text{plasma}}}$$

Again, the square brackets denote concentrations. Here we see why reabsorption diminishes clearance; the greater the amount reabsorbed, the lower the drug concentration in the urine (in the numerator) and the smaller the value of the quotient. This method requires determinations of drug concentrations in both urine and plasma, and a timed urine collection. (See Chapter 5 for a closer look at this equation, including clinical applications and actual calculations.)

Another mathematical approach depends only on serial measurements of the plasma concentration:

$$Cl = K_e \cdot V_d$$

Clearance is the product of the *elimination rate constant* (K_e) and the volume of distribution (V_d). That is, clearance is directly proportional to V_d—higher for drugs with a larger V_d, and lower for drugs with a smaller V_d (as we expect, because a large V_d implies that the drug is not extensively protein-bound, and a large free fraction implies high clearance; and vice versa). The exact proportionality is given by the constant, K_e, which is the percentage of drug in the body removed per unit of time.

Note the relationship between K_e and $t\,\frac{1}{2}$: K_e is the percentage of drug removed per unit of time (usually, per hour), while $t\,\frac{1}{2}$ is the time required for removal of a specific percentage of drug (50%):

$$K_e = \frac{0.693}{t_{1/2}}$$

The factor 0.693 is the three-digit approximation of the natural logarithm of 2 (see the box). If half-life is 8 hours:

$$K_e = \frac{0.693}{8\ hr} \approx 0.087\ hr^{-1}$$

At $t\,\frac{1}{2} = 8$ hours, about 8.7% of whatever amount is present is removed hourly.

REMEMBER: The *percentage rate* of drug removal is constant, but the *amount* of drug removed per hour declines as the total amount remaining in the body falls.

Half-Life, the Elimination Constant, and the Natural Logarithm of 2

The natural logarithm of 2 ($\ln 2 = 0.693147\ldots$, often rounded off to 0.693) is useful in computing the time required for a value to be doubled or halved (assuming a constant rate of change). For example, half-life ($t\,\frac{1}{2}$) is the time required for a drug concentration to be reduced by half. If the rate of decrease (that is, the elimination constant, K_e) is 8.7% hourly:

$$t_{1/2} = \frac{\ln 2}{K_e} = \frac{0.693}{0.087 \cdot hr^{-1}} \approx 7.97\ hr$$

The half-life is almost 8 hours. To check, let's start with 500 mg:

1st hour:	$500 - (0.087 \cdot 500) = 500 - 44 = 456$ mg
2nd hour:	$456 - (0.087 \cdot 456) = 456 - 40 = 416$ mg
3rd hour:	$416 - (0.087 \cdot 416) = 416 - 36 = 380$ mg
4th hour:	$380 - (0.087 \cdot 380) = 380 - 33 = 347$ mg
5th hour:	$347 - (0.087 \cdot 347) = 347 - 30 = 317$ mg
6th hour:	$317 - (0.087 \cdot 317) = 317 - 28 = 289$ mg
7th hour:	$289 - (0.087 \cdot 289) = 289 - 25 = 264$ mg
8th hour:	$264 - (0.087 \cdot 264) = 264 - 23 = 241$ mg

Allowing for approximations and rounding off, the amount left after one half-life of 8 hours is about half of the original 500 mg, as predicted.

Here is another way to think about the elimination constant: K_e is the *slope* of the straight-line segment (on a semilog graph) showing the terminal decline in plasma levels after equilibrium (Figure 7.10). Recall that the slope of a straight line, as defined by any two points on the line, equals the change on the vertical axis divided by the change on the horizontal axis. On this graph, the vertical axis is the natural logarithm of the plasma level, and the horizontal axis is time. Therefore, taking any two points on that line:

$$K_e = \text{slope} = \frac{\ln [\text{drug}]_2 - \ln [\text{drug}]_1}{\text{time}_2 - \text{time}_1}$$

Suppose the two points represent one half-life. As shown in Figure 7.10, for a drug with $t\,\frac{1}{2} = 8$ hr, the first point could be 12 mg/L at 2 hours after administration, and the second point could be 6 mg/L at 10 hours after administration:

$$K_e = \frac{\ln 6 - \ln 12}{10 \text{ hr} - 2 \text{ hr}} = \frac{1.79 - 2.48}{8 \text{ hr}} = \frac{-0.69}{8 \text{ hr}} \approx -0.086 \text{ hr}^{-1}$$

The negative sign merely indicates that the change in plasma concentration over time is a decrease. Notice that the numerator in the next-to-last step, 0.69, is approximately the natural logarithm of 2 (it would have been more exact if we had used more precise figures to derive it). It is not a coincidence—we deliberately selected two points that were separated by one half-life. However, we can get the same result (that is, the same K_e) by taking *any* two points on the straight-line segment. Referring again to Figure 7.10, the plasma concentration is approximately 10 mg/L at 4 hours after administration, and 6.5 mg/L at 9 hours:

$$K_e = \frac{\ln 6.5 - \ln 10}{9 \text{ hr} - 4 \text{ hr}} = \frac{1.87 - 2.30}{5 \text{ hr}} = \frac{-0.43}{5 \text{ hr}} \approx -0.086 \text{ hr}^{-1}$$

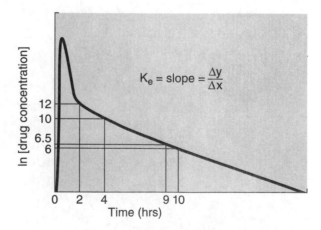

Figure 7.10 Slope of terminal decline is the elimination constant, K_e.

Nonlinear Pharmacokinetics

Drugs that exhibit linear pharmacokinetics exhibit a direct and consistent relationship between dose size and resulting plasma concentration.

However, some drugs do not exhibit linear pharmacokinetics. A common cause of nonlinearity is the limited capacity of metabolic enzyme systems involved in drug elimination. Saturation of these systems sets a limit to the rate of removal—the amount of drug removed per unit of time reaches a maximum at a certain plasma concentration, and will not increase no matter how much higher the plasma concentration may be.

Beyond that point, larger doses produce disproportionately large increases in plasma concentration and AUC, and the half-life (which is a constant for drugs with linear pharmacokinetics) increases as the plasma level rises ever higher.

Returning to the formula $Cl = K_e \cdot V_d$, note that both K_e and V_d are constants, which means that clearance is a constant.

REMEMBER: Clearance is a constant that defines the *volume of plasma* cleared of a substance per unit of time. The *amount of substance* removed from plasma per unit of time is *not* constant; it is highest when the plasma level is highest and declines as the plasma level falls.

The formulas presented in this section may appear in varied guises:

$$K_e \approx \frac{0.693}{t_{1/2}} \qquad t_{1/2} = \frac{0.693}{K_e}$$

$$Cl = K_e \cdot V_d \qquad K_e = \frac{Cl}{V_d}$$

$$t_{1/2} \approx \frac{0.693}{Cl \cdot V_d^{-1}} = \frac{0.693 \cdot V_d}{Cl}$$

The concepts are the same, just rearranged.

POLYPHARMACY AND THE RISK OF INTERACTIONS

Finally, let's look at a mathematical fact about drug therapy that has great clinical relevance, although it does not involve pharmacokinetics. Many patients, especially the elderly, have a variety of illnesses and therefore take several medications concomitantly ("polypharmacy"). Aside from the fact that each drug has its own side effects, and that the presence of one drug in the body can affect the pharmacokinetics of another, the simultaneous use of different drugs poses a risk of potentially serious interactions between drugs.

The number of potential interactions between any two drugs—not even counting the possibility of multidrug interactions—can be computed from the number of different drugs being taken. For any number of drugs, N > 2:

$$\text{Number of possible interactions} = \frac{N!}{2 \cdot (N-2)!}$$

The exclamation point denotes the *factorial* of a number:

$$N! = N \cdot (N-1) \cdot (N-2) \cdot (N-3)\ldots$$

If a patient is taking five different drugs at once, the number of possible interactions would be:

$$\frac{5!}{2 \cdot 3!} = \frac{5 \cdot 4 \cdot 3 \cdot 2}{2 \cdot (3 \cdot 2)} = \frac{120}{12} = 10$$

Adding just one more drug to the regimen adds five more possible interactions:

$$\frac{6!}{2 \cdot 4!} = \frac{6 \cdot 5 \cdot 4 \cdot 3 \cdot 2}{2 \cdot (4 \cdot 3 \cdot 2)} = \frac{720}{48} = 15$$

This computation does not tell us which if any interactions will occur, or how serious they may be. What it tells us is that the potential for interactions escalates rapidly as the number of simultaneously used medications is increased.

QUIZ

1. If the elimination constant for a drug with first-order elimination is $0.063 \cdot hr^{-1}$, what is its half-life? If the half-life for another drug is 15 hours, what is the elimination constant?

2. What is the half-life of a drug whose V_d is given as 0.48 L/kg and whose clearance is given as 0.04 L/hr/kg?

3. What is the elimination constant of a drug whose concentration in the plasma is 15 mg/L at 1 hour after intravenous administration, and 10 mg/L at 9 hours?

4. For a drug with pK = 6.7, what percentage is ionized in plasma at pH 7.4? What percentage is ionized in the renal tubular fluid at pH 6.2? If a different drug is 33.3% ionized in plasma, what is its pK?

5. Predict which drug would have a larger apparent volume of distribution (V_d): Drug A, with pK = 5.0 and 90% protein-binding; or Drug B, with pK = 8.2 and 50% protein-binding. (The two drugs are otherwise similar.)

6. After rapid intravenous administration of 0.5 g of a drug, the plasma concentration (immediately after equilibrium with interstitial fluid is established) is 12.5 mg/L. What is the drug's V_d? (Just the volume, in liters, without adjustment for body size.)

7. Suppose that the plasma concentration of a drug is measured to be 30% above the therapeutic range, yet the patient still shows an appropriate clinical response with no evidence of toxicity. If the explanation is that the patient has an abnormality in the amount of plasma binding protein, is the abnormality too much or too little of the binding protein?

8. Which drug formulation has a larger AUC? Slow-acting with 98% bioavailability, or fast-acting with 85% bioavailability?

9. If 1.0 g of a drug whose half-life is 9 hours is administered intravenously, how much drug is still left in the body after 3 hours?

10. In what two ways does the liver contribute to drug elimination?

Answers and explanations at end of book.

Energy and Metabolism

Chapter Outline

Mathematical Tools Used in This Chapter

In this chapter, we will briefly review the digestion of nutrients, and their role in the synthesis of ATP—the source of energy for all active cellular processes. Then we will look at the mathematical expression of oxygen utilization as a key indicator of metabolic function.

DIGESTION OF NUTRIENTS

Protein, carbohydrates, and fat are *nutrients*—substances that can be oxidized in the cells (specifically, in the mitochondria) to provide energy. Foods also contain *vitamins* and *minerals*—substances that do not provide any energy but are needed as enzymes (catalysts) in various biochemical processes. Finally, foods contain *fiber*—material that is not absorbed from the gastrointestinal tract but aids in normal motility and digestion.

Digestion refers to the mechanical and chemical breakdown of food into fragments and then molecules that can be absorbed via the gastrointestinal tract. Mechanical digestion (chewing before swallowing; churning during passage through the alimentary canal) is common to all nutrients;

chemical digestion (action of catabolic substances) is more specific for each of the nutrients.

Carbohydrates, consisting of groups of saccharides (sugars), are chemically broken down to simple units (chiefly glucose) by salivary amylase (also called ptyalin) and pancreatic amylase enzymes. Proteins, consisting of long chains of amino acids, are chemically broken down in the stomach, by hydrochloric acid and an enzyme called pepsin; and in the small bowel, by proteolytic enzymes from the pancreas. Certain fats, consisting of chains of fatty acids connected in groups of three to a glycerol framework (forming triglycerides), are chemically broken down by hydrochloric acid, lipolytic enzymes from the pancreas, and bile salts in fluid from the liver.

Passage of material from the stomach into the *duodenum* (the proximal portion of the small bowel, about 25 cm) is regulated by the *pyloric sphincter*. Pancreatic and hepatic fluids enter the duodenum at the same location—*Vater's ampulla*, where the main pancreatic duct and the common bile duct meet. Distal to the duodenum is the *jejunum* (about 2.5 m) and then the *ileum* (about 3.6 m).

As the digested food passes down the length of the small bowel, nutrient molecules are absorbed into the capillaries inside the villi; this event marks the actual entry of nutrients from food into the body's systems. Some of the salts from the bile fluid are reabsorbed and recycled. By the time the intestinal contents reach the *cecum* (the proximal portion of the large bowel), most of the nutrients have been absorbed. Distal to the cecum are the portions of the colon designated *ascending, transverse, descending,* and *sigmoid,* and finally the rectum.

As material passes down the length of the large bowel, much of the water that remains is reabsorbed, along with additional bile salts. Feces is a mixture of fiber, undigested food, brown pigments from the bile fluid, and bacteria, along with a variable amount of water.

Blood from the alimentary canal goes to the liver by way of a set of vessels that form the *portal system.* In the liver, the bile salts are taken up to be reused as more bile fluid is produced, and toxic substances such as ammonia are neutralized and removed. Then, cleared of bile salts and toxins, the blood (carrying nutrients) returns to the heart and systemic circulation.

USABLE ENERGY

Energy is needed to carry out all physical activity and all active physiologic functions; passive processes, such as osmosis, do not require energy. Usable energy is supplied by *adenosine triphosphate* (ATP). Energy is stored in the bonds between the three phosphate groups within the ATP molecule. Breaking a bond splits off a phosphate group, releasing usable energy and leaving the rest of the molecule as adenosine diphosphate (ADP):

$$ATP \rightarrow ADP + phosphate + energy$$

In different tissues, various factors trigger this process to release energy exactly when and where it is required for carrying out cellular activities.

Because the actual supply of ATP in the body is very limited, it must be constantly replenished. Two pathways lead to the formation of ATP. The *anaerobic* pathway quickly generates a small amount of ATP when the body's demands for oxygen exceed the available supply, such as during brief but intensive physical exertion; this pathway can operate for only a limited time. The *aerobic* pathway generates a much larger amount of ATP when oxygen supply exceeds demand; this pathway operates as long as the oxygen supply is maintained.

Both pathways require energy from some outside source to form the high-energy bond between phosphate and ADP, thereby resynthesizing ATP. Like any other active physiologic process, it takes ATP to initiate the production of more ATP; the net gain in ATP is the amount produced minus the amount used.

THE ANAEROBIC PATHWAY

In the anaerobic pathway, either of two substances can serve as the energy source. One is a compound called *creatine phosphate* (CP; also called *phosphocreatine* or *phosphorylcreatine*). CP consists of creatine (a molecule synthesized from certain amino acids) and a phosphate group, linked by a high-energy bond. In muscle cells at rest, energy is stored in these bonds. During physical exertion, the bonds break down to release energy, which is used to synthesize ATP (Figure 8.1). In essence, breaking the phosphate bond of

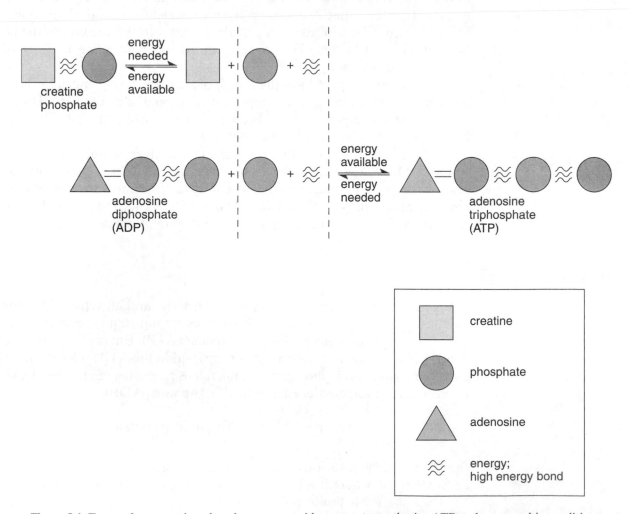

Figure 8.1 Energy from creatine phosphate can provide energy to synthesize ATP under anaerobic conditions.

CP supplies the energy needed to form the phosphate bond of ATP. Thus, the process constantly shifts back and forth between CP formation and breakdown, depending on the level of muscular activity.[1]

The other source of energy for the anaerobic pathway is glucose (or its tissue storage form, glycogen). Glucose is a sugar containing six carbon atoms ($C_6H_{12}O_6$). In the process of *glycolysis*, a molecule of glucose is degraded into two molecules of *pyruvic acid*, a three-carbon intermediate. From each of the two pyruvic acid molecules, 2 (molecules of) ATP are formed, for a total of 4 ATP produced. The energy cost is 2 ATP if the process starts with glucose, or 1 ATP if it starts with glycogen; therefore, the net gain is 2 or 3 ATP.

In the process, pyruvic acid is reduced to *lactic acid*. The accumulation of lactic acid in muscle tissue halts further glycolysis. The process is self-limiting, because it creates an "oxygen debt" that must be repaid—as soon as more oxygen is available, it will be used to oxidize accumulated lactic acid back to pyruvate.

THE AEROBIC PATHWAY

The aerobic pathway is far more efficient than the anaerobic pathway in terms of the amount of ATP generated. The complete aerobic conversion of a molecule of glucose or glycogen to six molecules each of carbon dioxide and water results in the formation of 40 ATP. Again, the energy cost is 2 ATP if the process starts with glucose, or 1 ATP if it starts with glycogen; so the net gain is 38 or 39 ATP. Thus, the aerobic pathway accounts for 95% of the total gain in ATP, while the anaerobic pathway accounts for just 5%.

The aerobic pathway also begins with glycolysis. Pyruvic acid is the turning point. Under anaerobic conditions, pyruvic acid is reduced to lactic acid; under aerobic conditions, it is transformed and incorporated into *acetyl-coenzyme A* (acetyl-CoA). Acetyl-CoA provides the point of entry into the main sequence of events in the aerobic pathway—the *Krebs cycle*.[2]

The Krebs cycle takes place in the mitochondria. The final product in the cycle is the very compound that reacts with acetyl-CoA to start the cycle; thus, the cycle keeps turning as long as acetyl-CoA is available. The Krebs cycle generates hydrogen atoms as well as carbon dioxide as a waste product. The hydrogen, each linked to an extra electron to form *hydride* ion, is utilized with oxygen in a series of electron transfers that provide the energy needed for the resynthesis of ATP from ADP:

$$ADP + phosphate + energy \rightarrow ATP$$

This process, called *oxidative phosphorylation*, represents the major use of oxygen in the body.

---

[1]Some CP undergoes conversion to *creatinine*, which is excreted by the kidney; see Chapter 5.

[2]In 1953, German-born English biochemist and physician Sir Hans Adolf Krebs (1900–1981) was one of two scientists awarded the Nobel Prize in Medicine or Physiology. The Krebs cycle is also called the *citric acid cycle* (because citric acid is the first compound generated in the sequence) or the *tricarboxylic acid cycle* (because citric acid contains three carboxyl groups, –COOH).

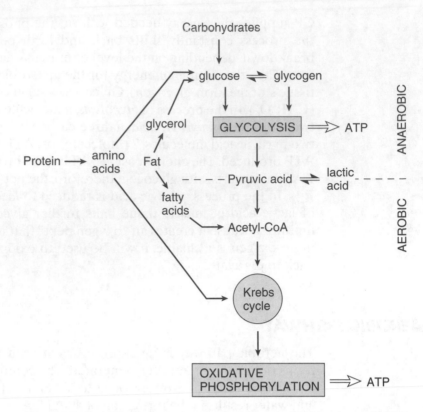

Figure 8.2 Simplified schematic of anaerobic and aerobic pathways in ATP synthesis.

REMEMBER: Energy for all active cellular processes comes from ATP, which must constantly be replenished. The anaerobic pathway (via glycolysis or breakdown of creatine phosphate) operates very quickly to provide a limited amount of ATP. The aerobic pathway (via the Krebs cycle and oxidative phosphorylation) operates as long as adequate oxygen is present, producing 19 times as much ATP as is produced anaerobically.

Under anaerobic conditions, the only nutritional energy source is glucose or glycogen. Under aerobic conditions, any of the nutrients can be utilized. From carbohydrate digestion, glucose enters the aerobic pathway via pyruvate. From protein digestion, excess amino acids may be converted to carbohydrate for entry into the pathway via pyruvate, or they may enter directly into certain sequences of the Krebs cycle. From fat digestion, glycerol is converted to carbohydrate, while fatty acids are converted to acetyl-CoA (Figure 8.2).

(The foregoing description of ATP synthesis offers only the briefest possible highlights of several extremely complex processes. A more detailed review is deferred to a comprehensive course in physiology or biochemistry.)

CALORIC INTAKE

Energy balance refers to the relationship between the amount of energy taken in as nutrients versus the amount expended for physical activity and cellular function. Surplus intake of *any* type of nutrient ultimately results in storage as fat. When tissue storage of glycogen is at maximum, additional glucose is converted to fat. Excess amino acids are deaminated (the amine group, $-NH_2$, is split off to form ammonia, from which urea is derived), and the remainder of the molecule can be converted to glucose; again, excess glucose is eventually stored as fat.

With inadequate nutrient intake (negative energy balance), energy needs are met by mobilizing stored fat. Free fatty acids are converted to acetyl-CoA, which enters the Krebs cycle or, in some tissue, is converted to another energy source called ketone bodies.[3]

The conventional unit of energy derived from nutrients is the *kilocalorie;*[4] the SI unit is the *kilojoule.* Relating these units, 1 kcal $\approx$ 4.2 kJ; 1 kJ $\approx$ 0.24 kcal. Carbohydrates and proteins each provide about 4.0 kilocalories per gram; fat, about 9.4 kilocalories per gram. Do not confuse the weight of the food portion with the weight of the nutrient; food contains variable amounts of water and fiber as well as nutrients; the caloric value per gram refers to the pure nutrient.

Good nutrition requires not only an appropriate number of kilocalories but also an appropriate division of that total into percentages derived from each type of nutrient. By current recommendations, no more than 30% of total caloric intake should come from fat; protein intake is based on a standard formula of 1 gram per kilogram of lean body weight; the remainder of caloric intake should come mainly from complex carbohydrates (starches rather than simple sugars).

Caloric Calculator

To determine the percentage of kilocalories derived from each nutrient in a meal, multiply the number of grams by the number of kilocalories per gram. Then divide the product by the total number of calories in the entire meal.

Suppose a 598-kcal meal contains 58 grams of carbohydrate, 21 grams of protein, and 30 grams of fat. The percentage of kilocalories from each nutrient is calculated as follows:

$$\textit{Carbohydrate:}\quad 58 \text{ g} \cdot \frac{4.0 \text{ kcal}}{\text{g}} = 232 \text{ kcal}; \quad \frac{232 \text{ kcal}}{598 \text{ kcal}} \approx 0.39 = 39\%$$

$$\textit{Protein:}\quad 21 \text{ g} \cdot \frac{4.0 \text{ kcal}}{\text{g}} = 84 \text{ kcal}; \quad \frac{84 \text{ kcal}}{598 \text{ kcal}} \approx 0.14 = 14\%$$

$$\textit{Fat:}\quad 30 \text{ g} \cdot \frac{9.4 \text{ kcal}}{\text{g}} = 282 \text{ kcal}; \quad \frac{282 \text{ kcal}}{598 \text{ kcal}} \approx 0.47 = 47\%$$

This meal is too high in fat content. The best way to keep daily caloric intake from fat under 30% is to maintain that same limit within each meal. (Some authorities recommend that dietary fat should be restricted to 20%, 15% or, for certain patients, just 10% of total caloric intake.)

---

[3]Normally, ketones are oxidized as soon as they are formed. However, diabetes mellitus, starvation, or a diet high in fat but deficient in carbohydrate can lead to an accumulation of ketones, which is a severe metabolic disturbance.

[4]1 Calorie = 1000 calories = 1 kcal.

The Metabolic Rate and Body Weight

Some of the factors that influence the metabolic rate are intrinsic to the individual:

- *Body build:* A taller, thinner habitus represents a larger body surface area per unit-weight; body heat is lost more rapidly, and metabolism must therefore be faster to maintain temperature.

- *Age:* The metabolic rate is highest in young children, and it gradually decreases throughout adulthood.

- *Gender:* The metabolic rate is slightly higher in men than in women.

Other factors are variable:

- *Climate:* Living in a cold environment results in faster loss of body heat; metabolism must be faster than would be needed in a warm environment.

- *Nutrition:* Starvation reduces the metabolic rate to a minimum, so as to conserve energy (which is why starvation diets are self-defeating).

- *Exercise:* Physical activity not only uses up energy, but it also raises the metabolic rate (which is why the combination of sensible diet and exercise is so efficient for achieving weight reduction).

- *Stress:* Stress tends to raise the metabolic rate; however, some people respond to stress by compulsive eating, leading to weight gain.

METABOLIC RATE

Metabolism is the general term for all the energy-related processes that take place in a living organism. Metabolic processes may be classified as *anabolic* (using energy to build larger molecules from smaller ones) or *catabolic* (releasing energy by breaking down larger molecules into smaller ones).

Most metabolic processes depend on *catalysts* to drive them at a rate sufficient to serve physiologic needs. In living systems, *enzymes* are proteins that act as catalysts.

The *basal metabolic rate* (BMR) is the minimum rate of energy expenditure necessary to sustain life during complete rest. Even at rest, it takes energy to drive cell membrane ion pumps, to create complex molecules, and to maintain normal function in all the internal organs.

BMR is determined under controlled conditions (*not* during sleep; at rest, at a comfortable temperature, 12 hours after the last meal). One method involves measuring the rate of oxygen utilization and then applying the formula 4.825 kcal per liter of oxygen consumed, per unit-time. If the basal rate of oxygen consumption is 220 mL/min = 0.22 L/min:

$$\text{BMR} = 0.22 \, \frac{\text{L}}{\text{min}} \cdot \frac{1440 \, \text{min}}{\text{day}} \cdot \frac{4.825 \, \text{kcal}}{\text{L}} \approx 1529 \, \text{kcal} \cdot \text{day}^{-1}$$

OXYGEN UTILIZATION

Because most of ATP synthesis occurs via the aerobic pathway culminating in oxidative phosphorylation, the rate at which oxygen is used ($\dot{V}o_2$—the dot indicates a rate) is one of the key indicators of metabolic function.

In theory, we can predict the oxygen consumption rate via the *Fick principle* (see Chapter 3):

$$\dot{V}_{O_2} = \text{arteriovenous difference} \cdot \text{blood flow rate}$$

The rate of consumption of oxygen is obtained as the product of the difference in arterial versus venous oxygen content and the blood flow rate (cardiac output). If arterial oxygen content is 20 mL/dL and venous oxygen content is 15 mL/dL, the A-V difference is 5 mL/dL. Then if cardiac output is 4.9 L/min = 49 dL/min:

$$\dot{V}_{O_2} = 5\,\frac{\text{mL}}{\text{dL}} \cdot 49\,\frac{\text{dL}}{\text{min}} = 245\,\text{mL} \cdot \text{min}^{-1}$$

To allow for differences in body size, the basal oxygen consumption rate is divided by body weight in kilograms. For a 70-kg man:

$$\frac{245\,\text{mL} \cdot \text{min}^{-1}}{70\,\text{kg}} = 3.5\,\text{mL} \cdot \text{kg}^{-1} \cdot \text{min}^{-1}$$

This specific basal rate of oxygen consumption ($3.5\,\text{mL} \cdot \text{kg}^{-1} \cdot \text{min}^{-1}$) is defined as 1 MET ("metabolic equivalent"). In most people, basal oxygen consumption is about 1 MET. An individual whose oxygen consumption rate during exercise is six times higher ($6 \cdot 3.5 = 21\,\text{mL} \cdot \text{kg}^{-1} \cdot \text{min}^{-1}$) is working at 6 METS.

The increase in oxygen consumption during exercise results from increases in both factors—the arteriovenous oxygen difference and cardiac output. With increased muscular activity, more oxygen is extracted from the blood; thus, venous oxygen content is lower and the A-V difference is larger. Cardiac output rises due to increases in both stroke volume and heart rate.[5]

In actual practice, the rate of oxygen consumption (at rest or during exercise) is determined by measuring the ventilatory volume per minute and the oxygen content of expired air. The difference in oxygen content between inspired air and expired air reveals the volume of oxygen extracted per liter of air. Multiplying that difference by the number of liters of air moved in 1 minute reveals the total volume of oxygen consumed, per minute:

$$\dot{V}_{O_2} = \frac{\text{ventilatory volume}}{\text{minute}} \cdot (\text{inspired O}_2 \text{ content} - \text{expired O}_2 \text{ content})$$

Under normal conditions, the oxygen content in the *inspired* air is almost 21%. If the ventilatory volume per minute (at complete rest) is 6 liters and

---

[5]Cardiac output = stroke volume · heart rate (see Chapter 3). Increased muscular activity also generates more carbon dioxide, which reduces pH (see Chapter 6); and excess heat, which is dissipated through perspiration. Higher temperature and lower pH both cause a rightward shift in the oxygen-hemoglobin dissociation curve, thereby enhancing oxygen delivery to the cells (see Chapter 4).

the oxygen content in the *expired* air is 17%, then basal $\dot{V}_{O_2}$ is calculated as follows:

$$6\,\frac{L}{min}\cdot(0.21-0.17)=\frac{6\,L}{min}\cdot(0.04)=\frac{0.24\,L}{min}=240\,mL\cdot min^{-1}$$

Suppose the test is repeated during exercise; the ventilatory volume per minute increases to 64 liters, and oxygen content in the expired air is, again, 17%:[6]

$$64\,\frac{L}{min}\cdot(0.21-0.17)=\frac{64\,L}{min}\cdot(0.04)=\frac{2.56\,L}{min}=2560\,mL\cdot min^{-1}$$

If these results are obtained in a 70-kg man, the basal rate of oxygen consumption per kilogram is $240/70 \approx 3.43$ mL·kg^{-1}·min^{-1} (just under 1 MET); the rate during exercise is $2560/70 \approx 36.6$ mL·kg^{-1}·min^{-1} (over 10 METS).

Various activities and exercises are rated in METS and energy utilization (kcal/min). For example, walking at a rate of 2.5 miles per hour might be rated as a 3-MET activity that burns about 4 kilocalories per minute; jogging at 5 miles per hour, as a 7.5-MET activity that burns about 9 kilocalories per minute. However, such ratings are just averages. Two people performing the same activity may be consuming oxygen and utilizing energy at different rates, depending on such variables as individual basal oxygen consumption rate and fitness level.

The rate of oxygen consumption at maximum physical exertion ($\dot{V}_{O_2}$max) is a standard measure of cardiorespiratory fitness. A healthy individual should be able to perform at 10–12 METS (35–42 mL·kg^{-1}·min^{-1}). However, fatigue often limits performance. In place of direct measurement, $\dot{V}_{O_2}$max may be extrapolated from the oxygen utilization rate during submaximal exertion. Submaximal exertion is defined by heart rate; the usual goal is 75% to 85% of the predicted maximum heart rate, which can be estimated as $220 - $ age. For example, in a 54-year-old man, $220 - 54 = 166$ beats per minute; 75% of this rate is 125 beats per minute. Because heart rate and the rate of oxygen consumption increase in linear fashion up to a maximum, readings of oxygen consumption at three different submaximal workloads (for example, at 50%, 65%, and 80% of maximum heart rate) will fall along a straight line. Extending that line to 100% of maximum heart rate reveals the maximum oxygen consumption rate—that is, $\dot{V}_{O_2}$ max.

[6]It may seem surprising that the oxygen content in the expired air could be the same during vigorous exercise when the body needs more oxygen. With an increase in pulmonary ventilation (volume of air moved per minute), the total amount of oxygen entering the blood each minute does increase, yet the percentage of oxygen left in each liter of expired air is not necessarily lower. Do not confuse the percentage of oxygen left in expired air with the oxygen content in venous blood; venous oxygen content definitely decreases during exercise as more oxygen is extracted from arterial blood (creating a larger A-V difference).

In contrast to *actual* (or extrapolated) $\dot{V}o_2max$, *expected* $\dot{V}o_2max$ can be predicted by formulas based on gender, activity level, and age in years:

$$\text{Males, active:} \quad 69.7 - (0.612 \cdot \text{age})$$

$$\text{Males, sedentary:} \quad 57.8 - (0.445 \cdot \text{age})$$

$$\text{Females, active:} \quad 42.9 - (0.312 \cdot \text{age})$$

$$\text{Females, sedentary:} \quad 42.3 - (0.356 \cdot \text{age})$$

$\dot{V}o_2max$ is somewhat lower in women than in men of the same age and activity level; and $\dot{V}o_2max$ decreases slightly with each year of age in both genders.

These formulas[7] are derived statistically from population-wide data; however, it is always wise to be cautious in applying population-based data too rigidly to an individual patient. For example, no allowance is made for variability in body weight, and the simple classification as active versus sedentary does not consider gradations of activity. Nevertheless, if the observed $\dot{V}o_2max$ in an individual is less than expected, it suggests a degree of functional impairment, which can be quantified as follows:

$$\text{Impairment} = \frac{\text{expected } \dot{V}o_2max - \text{actual } \dot{V}o_2max}{\text{expected } \dot{V}o_2max}$$

For example, expected $\dot{V}o_2max$ in a 52-year-old sedentary woman is:

$$42.3 - (0.356 \cdot 52) \approx 42.3 - 18.5 = 23.8 \text{ mL} \cdot \text{kg}^{-1} \cdot \text{min}^{-1}$$

Note that this value is equivalent to 6.8 METS. If her actual $\dot{V}o_2max$ is measured at $20.3 \text{ mL} \cdot \text{kg}^{-1} \cdot \text{min}^{-1}$ (or 5.8 METS), the degree of functional impairment is:

$$\frac{23.8 - 20.3}{23.8} = \frac{3.5}{23.8} \approx 0.15 = 15\%$$

Although genetics accounts for much of the difference in aerobic capacity between people, a program of cardiorespiratory fitness training can significantly boost a person's $\dot{V}o_2max$—and the lower the baseline $\dot{V}o_2max$, the greater the percentage of improvement that can be achieved at a given level of exercise. Younger, habitually sedentary individuals derive the most benefit.

Exercise is prescribed in terms of intensity, duration, and frequency. The usual recommended intensity is 75% of peak performance; the duration at

---

[7]Source: R. A. Bruce, "Principles of exercise testing." In J. P. Naughton, H. K. Hellerstein, and I. C. Mohler, eds., *Exercise Testing and Exercise Training in Coronary Heart Disease* (New York: Academic Press, 1973).

this workload might start low and build toward a goal of 30 minutes; and a typical frequency is every other day. (Note that if intensity is measured in METS, the product of METS and the duration in minutes yields total oxygen consumed per kilogram, which reflects the total work performed.)

REMEMBER: Fitness is assessed by the rate of oxygen consumption at maximum physical exertion. $\dot{V}o_2$max (in milliliters of oxygen consumed per kilogram per minute) is extrapolated from $\dot{V}o_2$ at submaximal workload.

RESPIRATORY QUOTIENT

To conclude this topic, let's look at the utilization of oxygen with each of the nutrient types. In the aerobic pathway, each nutrient requires a certain amount of oxygen and produces a certain amount of carbon dioxide. For glucose, the overall reaction can be summarized as:

$$C_6H_{12}O_6 + 6O_2 \rightarrow 6CO_2 + 6H_2O$$

The *respiratory quotient* is the ratio of carbon dioxide produced to oxygen utilized:

$$\text{Respiratory quotient} = \frac{CO_2 \text{ produced}}{O_2 \text{ used}}$$

For the complete oxidation of 1 mole of glucose, it takes 6 moles of oxygen to produce 6 moles of carbon dioxide. On a mole-for-mole basis, the ratio for glucose is 6/6, so the respiratory quotient is 1.0.

With palmitin (as a representative fat), the reaction is:

$$2C_{51}H_{98}O_6 + 145O_2 \rightarrow 102CO_2 + 98H_2O$$

Note that this reaction starts with 2 moles of palmitin. For the complete oxidation of 1 mole of palmitin, it takes 72.5 moles of oxygen to produce 51 moles of carbon dioxide. The ratio is $51/72.5 \approx 0.7$. (Protein is not typically employed as an energy source, but its respiratory quotient would be approximately 0.8.)

Although fat provides more than twice the energy per gram that carbohydrate provides, it takes substantially more oxygen to oxidize a gram of fat than to oxidize a gram of carbohydrate. In fact, the energy derived per unit-volume of oxygen consumed is slightly higher for carbohydrate than for fat (see box).

Energy from Carbohydrate vs. Fat

On average, carbohydrates deliver about 4.0 kcal/g; the specific yield for glucose is about 3.8 kcal/g. Fat delivers about 9.4 kcal/g. In addition to looking at energy per gram, we can theoretically predict the energy derived from each nutrient *per liter of oxygen consumed.*

Carbohydrate: It takes 6 moles of oxygen to oxidize 1 mole of glucose (formula weight = 180):

$$180 \text{ g} \cdot \frac{3.8 \text{ kcal}}{\text{g}} = 684 \text{ kcal}$$

$$\frac{684 \text{ kcal}}{6 \text{ mol}_{\text{oxygen}}} = 114 \text{ kcal} \cdot \text{mol}_{\text{oxygen}}^{-1}$$

The standard molar volume for an ideal gas under ideal conditions is about 22.4 liters. Oxygen in the body does not represent an ideal gas or ideal conditions, but we can still use this value as an estimate:

$$\frac{114 \text{ kcal} \cdot \text{mol}_{\text{oxygen}}^{-1}}{22.4 \text{ L} \cdot \text{mol}^{-1}} \approx 5.09 \text{ kcal} \cdot \text{L}_{\text{oxygen}}^{-1}$$

Fat: It takes 72.5 moles of oxygen to oxidize 1 mole of palmitin (formula weight = 806):

$$806 \text{ g} \cdot \frac{9.4 \text{ kcal}}{\text{g}} \approx 7576 \text{ kcal}$$

$$\frac{7576 \text{ kcal}}{72.5 \text{ mol}_{\text{oxygen}}} \approx 104.5 \text{ kcal} \cdot \text{mol}_{\text{oxygen}}^{-1}$$

$$\frac{104.5 \text{ kcal} \cdot \text{mol}_{\text{oxygen}}^{-1}}{22.4 \text{ L} \cdot \text{mol}^{-1}} \approx 4.67 \text{ kcal} \cdot \text{L}_{\text{oxygen}}^{-1}$$

In terms of oxygen utilization, carbohydrate is more efficient than fat as an energy source. Note that the formula used in estimating BMR—4.825 kcal expended per liter of oxygen consumed—reflects a mix of carbohydrate and fat as energy sources in the individual at rest.

The respiratory quotient for each nutrient is a theoretical concept—a model in which all energy is obtained from perfectly efficient oxidation of that nutrient alone. The actual respiratory quotient (volume of carbon dioxide produced divided by volume of oxygen consumed, per minute) in an individual reflects the mix of different nutrients being oxidized and the fact that metabolic processes do not necessarily match an idealized model.

In addition, the respiratory quotient will be different at rest and at peak performance. The reason is that prolonged peak performance leads to an *anaerobic threshold* at which the aerobic pathway cannot produce ATP as fast as it is being used to supply energy to muscle tissue. At that point, the anaerobic pathway begins working simultaneously, for the relatively brief time until lactic acid accumulation causes such fatigue and discomfort that further effort is impossible.

Under these conditions, with a metabolic acidosis caused by the buildup in lactic acid, there is a sudden and disproportionate increase in carbon dioxide exhalation (creating a compensatory respiratory alkalosis to minimize the pH change; see Chapter 6). Yet oxygen utilization is already at maximum. The result is a rise in the ratio of carbon dioxide produced to oxygen consumed—that is, a higher respiratory quotient.

QUIZ

1. A patron at a diner orders a burger and a soda. The burger weighs 116 grams (just over 4 ounces); half of this weight is water, and the other half is divided equally between protein and fat. The burger bun contains 27 grams of carbohydrate, 1 gram of protein, and 1 gram of fat. The soda (about 15 fluid ounces) contains 45 grams of sugar. What is the total caloric content of this meal?

2. From the same meal as in Question 1, what are the percentages of kilocalories derived from each nutrient?

3. At 75% of maximum workload, a patient's ventilatory volume in 1 minute is 45 liters, and the oxygen content in the expired air is 16%. What is the oxygen consumption rate?

4. If $\dot{V}o_2$max in a 35-year-old sedentary male is 37.5 mL $\cdot$ kg^{-1} $\cdot$ min^{-1}, rate his cardiorespiratory fitness (percent of impairment).

5. What is the basal metabolic rate (BMR) in an individual whose rate of oxygen consumption at rest is measured at 200 mL/min?

6. If basal $\dot{V}o_2$ is 240 mL/min and ventilatory volume is 6 L/min, what is the percentage of oxygen in expired air?

7. What is the expected maximum heart rate in a 38-year-old man? At 75% of maximum heart rate, oxygen consumption in an active 38-year-old man who weighs 154 lb is 2.0 L/min. What is his performance in METS?

8. What is the $\dot{V}o_2$max in a sedentary 35-year-old woman who is functioning at 106% of expected performance? What is her performance in METS?

9. If basal oxygen consumption is measured at 206 mL/min in a woman who weighs 52 kg, what is her BMR in METS?

10. Consider a short, heavy-set, sedentary man living in a cold environment, and a tall, thin, active woman living in a warmer climate. Other things being equal, is it possible to predict which of these two people has the higher basal metabolic rate?

Answers and explanations at end of book.

chapter 9

Diagnostic Tests

Chapter Outline

Mathematical Tools Used in This Chapter

This chapter deals with some basic mathematical concepts relating to diagnostic testing in general, but *not* with the clinical interpretation of various findings on specific tests; test data interpretation is a standard part of the educational curriculum in most professional training programs.

DEFINITIONS

Certain terms are often used in a vague, generic, or interchangeable way to convey a sense of how far we can trust the results obtained from diagnostic tests. In fact, these terms have distinct meanings. Here are the terms as defined in the language of quality control within the laboratory.

Specificity refers to the ability of a test to measure only what it is supposed to measure—that is, to recognize the substance to be measured and to distinguish it from everything else. If a test cannot distinguish between the substance to be measured and some other substance, it will yield a result that represents the combined values of the two substances.

Accuracy refers to the correctness of a test result—how closely that result matches the true value of the measurement as determined by a virtually infallible "gold standard" test (more about gold standard tests later in this chapter). Because no test is perfectly accurate at all times, accuracy is usually defined as a result that falls within an acceptably narrow range surrounding the true value.

Precision refers to reproducibility or consistency of results if the procedure is performed repeatedly—by the same or different personnel, at the same or different laboratories. A test that yields consistent results on repeated trials is considered precise. A test may yield a result that is accurate; but if that accurate result is not reproducible, the test is not considered precise.

Validity refers to the overall technical quality of test results, in terms of accuracy, precision, and specificity. Test results are valid if they are specific (measuring only the substance that was intended to be measured), accurate (close to the true value), and precise (reproducible).

SENSITIVITY AND SPECIFICITY

No laboratory test is perfect. Whether we are trying to determine the presence or absence of a marker of disease, or the amount of some substance in the body fluids, there is always a degree of uncertainty concerning the result.

Let's look first at tests for disease markers. Normal test results are "negative"—if the marker is absent, we *assume* the disease is absent. Abnormal results are "positive"—if the marker is present, we *assume* the disease is present.

Among all people showing positive (abnormal) results on a test, the greatest number have the disease in question; these results are called *true positives*. The remaining few are people who are free of the disease and yet show positive results on the test; a positive test result in the absence of the disease for which the test is performed is called a *false positive*.

Conversely, among all people showing negative (normal) results on a test, the greatest number are free of the disease in question; these results are called *true negatives*. The remaining few are people who have the disease and yet show negative results on the test; a negative test result in the presence of the disease for which the test is performed is called a *false negative*.

These definitions can be summarized as follows:

	Disease Present	Disease Absent
Positive Test	True positive	False positive
Negative Test	False negative	True negative

Almost all diagnostic tests yield some false negatives and false positives, which is why diagnosis is seldom based solely on laboratory data. A good test has relatively low rates of false results. In many cases, tests with unusually low rates of false negatives have somewhat higher rates of false positives, and vice versa.

The *sensitivity* and *specificity* of a diagnostic test reflect its rates of true and false positive and negative results. *Sensitivity* is the ability of a test to de-

Concepts and Terminology

The trustworthiness of data is of concern in biostatistics as well as in laboratory quality control (QC). However, certain concepts are unique to one discipline or the other, and the terminology differs.

Concept	QC Term	Statistics Term
Correctness of result; conformity to an accepted standard	Accuracy	Accuracy
Reproducibility; consistency in data from repeated tests	Precision	Reliability
Relative size of measurement unit	—	Precision
Ability to isolate the object of measurement	Specificity	—
Clinical meaningfulness of test result	—	Validity
Overall technical quality of test results	Validity	—

Note that consistency or reproducibility is called *precision* in the language of QC, and *reliability* in the language of biostatistics. By these definitions, QC precision reflects the technical integrity of the test procedure or apparatus, whereas statistical reliability reflects the intrinsic stability versus variability of the object of measurement. For example, serum glucose levels can vary widely, depending on the type and amount of food last consumed, and the time between the last meal and the blood test. Thus, "random glucose testing" shows poor statistical reliability even if the procedure for measuring glucose shows high QC precision.

Statistical precision refers to the relative size of the unit of measurement. For example, measuring the weight of an object in grams is more precise than measuring its weight in kilograms. A test is considered statistically precise if its unit of measurement is the smallest attainable with available technology.

In QC terms, *validity* means that the result meets the technical standards for specificity, accuracy, and precision. In statistical terms, *validity* means that the result is clinically meaningful in the light of other information about a patient.

tect a given condition when that condition is present in a patient; therefore, a test that is highly sensitive has a low rate of false negatives. The rated sensitivity is the percentage of cases in which the test yields positive results in people who have the condition in question. If a test is administered to 1000 people in whom the condition has already been confirmed by a "gold standard" test (see the box), and the results are positive in 840 cases:

$$\text{Sensitivity} = \frac{840 \text{ positive results}}{1000 \text{ people with the condition}} = 0.84 = 84\%$$

Notice that an 84% sensitivity rate implies a 16% rate of false negatives *in this selected population of people known to have the condition* (1000 tests − 840 true positives = 160 false negatives; 160/1000 = 16%). Because every

Gold Standard Tests

To determine the rates of true and false results from a diagnostic test, we need a surefire method of establishing the presence or absence of the condition in question. But if the tests themselves can yield false results, how do we know who really does or does not have the disease?

For certain conditions, there are diagnostic tests that are so rarely wrong that they are accepted as the *gold standard.* In clinical usage, this term refers to a test that is, for all practical purposes, infallible. The classic example is *contrast venography*—a special type of chest x-ray in which an intravenously administered radio-opaque dye outlines blood vessels in the lungs, to confirm or rule out pulmonary embolism.

Often, the gold standard test is more difficult, invasive, and costly than other diagnostic tests, which are less often correct but more often used. Other tests are evaluated by comparing their results to those obtained with the gold standard.

member of this selected population has the condition, we see no false positives or true negatives in this study. By itself, sensitivity tells us nothing about the results we would see in the *general* population.

Conversely, *specificity* is the ability of a test to differentiate between the condition in question and other conditions; a test that is highly specific will have a low rate of false positives. The rated specificity is the percentage of cases in which the test yields negative results in people who do not have the condition. If a test is administered to 1000 people in whom the condition has already been excluded or "ruled out" by the gold standard, and the results are negative in 930 cases:

$$\text{Specificity} = \frac{930 \text{ negative results}}{1000 \text{ people without the condition}} = 0.93 = 93\%$$

Again, 93% specificity implies 7% false positives *in this selected population of people known to be free of the condition* (1000 tests − 930 true negatives = 70 false positives; 70/1000 = 7%). Because no member of this selected population has the condition, we see no true positives or false negatives in this study. By itself, specificity tells us nothing about results in the *general* population.

Optimally, a diagnostic test would have *both* the highest possible sensitivity (that is, the highest rate of true positives and the lowest rate of false negatives) *and* the highest possible specificity (the highest rate of true negatives and the lowest rate of false positives). In reality, it is often a trade-off, as extremely sensitive tests tend to be less specific, while extremely specific tests tend to be less sensitive.

PREDICTIVE VALUES

The sum of true positives and false negatives represents 100% of test results in people who have the condition (by definition, if the condition is present, a positive result is true and a negative result is false). Likewise, the sum of true negatives and false positives represents 100% of results in people who do not have the condition (if the condition is absent, a negative result is true and a positive result is false).

Then we can define a diagnostic test's sensitivity and specificity rates (percentages) as follows:

$$\text{Sensitivity} = \frac{\text{true positives}}{\text{true positives} + \text{false negatives}} \cdot 100$$

$$\text{Specificity} = \frac{\text{true negatives}}{\text{true negatives} + \text{false positives}} \cdot 100$$

The *predictive values* of a test are rearrangements of these ratios, but their derivation and implications are quite different from those of sensitivity and specificity. The positive predictive value is the likelihood that a positive test result is true—that is, the percentage of *all* positives that are *true* positives:

$$\text{Positive predictive value} = \frac{\text{true positives}}{\text{true positives} + \text{false positives}} \cdot 100$$

Similarly, the negative predictive value is the likelihood that a negative test result is true—the percentage of *all* negatives that are *true* negatives:

$$\text{Negative predictive value} = \frac{\text{true negatives}}{\text{true negatives} + \text{false negatives}} \cdot 100$$

THE IMPACT OF PREVALENCE

Sensitivity and specificity reveal the rates of true results obtained when a diagnostic test is used in a selected population in which we already know that the disease in question is either universally present or universally absent. Sensitivity reveals the test's intrinsic ability to detect a disease that is definitely present; specificity reveals the test's ability to exclude a disease that is definitely absent. However, these measures do *not* reveal how often the results are correct when the test is used for actual diagnostic purposes in the clinical setting.

In contrast, positive and negative predictive values measure the frequency with which a given result is true when the test is used in a nonselected population in which we have no prior knowledge about the presence or absence of the disease in any individual patient. The positive predictive value reveals the likelihood that a positive test result is a true positive; the negative predictive value reveals the likelihood that a negative test result is a true negative. Unlike sensitivity and specificity, predictive values *do* reveal how often positive and negative results are correct when the test is used for diagnostic purposes in the clinical setting.

Because sensitivity and specificity are rates seen in *preselected* populations, they are unaffected by the prevalence of the disease in the *general* population (prevalence is the portion of the at-risk population in whom that disease is present at a given time; see Chapter 10). In contrast, predictive values *are* affected by prevalence. The positive predictive value varies *directly* with prevalence (a higher prevalence yields a higher positive predictive value, and vice versa), while the negative predictive value varies *inversely* with prevalence (a higher prevalence yields a lower negative predictive value, and vice versa).

Here is an illustration of how the prevalence of a condition affects the predictive values of a diagnostic test. A new test has been designed to detect a condition for which a gold standard test already exists. The new test is faster, easier, safer, and cheaper than the gold standard test—but we must determine how often it will be right or wrong. First, we identify a population of patients suspected (on clinical grounds—that is, signs and symptoms) of having the condition in question. Next, we employ the gold standard test to find out who really has the condition; suppose that among 600 individuals tested, 450 have the condition while the other 150 do not. Finally, the new test is performed in the same 600 individuals, with results as follows:

	Disease Present*	**Disease Absent***
New Test: Positive	True pos. $= 360$	False pos. $=\ 25$
New Test: Negative	False neg. $=\ 90$	True neg. $= 125$
Totals:	450	150

*As determined by the gold standard test.

Since the gold standard test has already shown who does and does not have the condition, the sensitivity and specificity of the new test can be computed:

$$\text{Sensitivity} = \frac{\text{true pos.}}{\text{true pos.} + \text{false neg.}} = \frac{360}{360 + 90} = 0.8 = 80.0\%$$

$$\text{Specificity} = \frac{\text{true neg.}}{\text{true neg.} + \text{false pos.}} = \frac{125}{125 + 25} \approx 0.833 = 83.3\%$$

We can also compute the predictive values *in this selected population:*

$$\text{Pos. pred. value} = \frac{\text{true pos.}}{\text{true pos.} + \text{false pos.}} = \frac{360}{360 + 25} \approx 0.935 = 93.5\%$$

$$\text{Neg. pred. value} = \frac{\text{true neg.}}{\text{true neg.} + \text{false neg.}} = \frac{125}{125 + 90} \approx 0.581 = 58.1\%$$

Notice the impact of prevalence. Among 600 patients already suspected of having the condition, 450 really do have it; so the prevalence rate *in this selected population* is 450/600 = 75%, which is high. Recall that a high prevalence is associated with a relatively high positive predictive value but a low negative predictive value—exactly as we see here.

Now suppose we employ the new test for screening purposes; and further suppose that epidemiologic research has established a 5% prevalence rate for the condition *in the general population*. Among 6000 individuals randomly screened, we expect the condition to be present in $6000 \cdot 0.05 = 300$ people. We already know the sensitivity of the test, 80.0%, and the specificity, 83.3%. With this information, we can estimate the number of positive and negative results the test will yield. Sensitivity is the rate of positive results in people who have the condition:

$$\text{Sensitivity} = 0.8 = \frac{\text{number of positive results}}{300 \text{ people with the condition}}$$

$$\text{Number of true positives} = 0.8 \cdot 300 = 240$$

$$300 \text{ with condition} - 240 \text{ true positives} = 60 \text{ false negatives}$$

If the condition is present in 300 of 6000 people, then it is absent in the other 5700 people. Specificity is the rate of negative results in people who do not have the condition:

$$\text{Specificity} = 0.833 = \frac{\text{number of negative results}}{5700 \text{ people without the condition}}$$

$$\text{Number of true negatives} = 0.833 \cdot 5700 \approx 4750$$

$$5700 \text{ without condition} - 4750 \text{ true negatives} = 950 \text{ false positives}$$

Now we can summarize these expected results as follows:

	Disease Present*	Disease Absent*
New Test: Positive	True pos. $= 240$	False pos. $= 950$
New Test: Negative	False neg. $= 60$	True neg. $= 4750$
Totals:	300	5700

*As predicted by epidemiologic data on prevalence.

And when we compute the predictive values *in the general population:*

$$\text{Pos. pred. value} = \frac{\text{true pos.}}{\text{true pos.} + \text{false pos.}} = \frac{240}{240 + 950} \approx 0.202 = 20.2\%$$

$$\text{Neg. pred. value} = \frac{\text{true neg.}}{\text{true neg.} + \text{false neg.}} = \frac{4750}{4750 + 60} \approx 0.988 = 98.8\%$$

Compare these predictive values to those we saw earlier. In a selected population of patients considered to be at high risk based on clinical evidence, the prevalence was 75%; at that high prevalence, the positive predictive value was high and the negative predictive value was low. In the general population, the prevalence is 5%; at this low prevalence, the positive predictive value is much lower and the negative predictive value is much higher.

REMEMBER: The sensitivity and specificity of a diagnostic test do not reveal the likelihood that the result will be true when the test is used for diagnostic purposes in the clinical setting. The predictive values *do* reveal the likelihood of a true result, but these values vary with the prevalence of the condition; a positive test result is more likely to be true when the prevalence is higher rather than lower, while a negative result is more likely to be true when the prevalence is lower rather than higher.

SCREENING VS. DIAGNOSTIC TESTING

We must make a clear distinction between *screening* in the general population and *diagnostic testing* in individuals who are suspected of having a condition.

For purposes of screening, the goal is detection. Therefore, highly sensitive tests are used; since a sensitive test has a low rate of false negatives, it is unlikely to miss a condition that is present. However, highly sensitive tests tend to be less specific. Moreover, the positive predictive value of the test is relatively low in these circumstances (because the prevalence in the general population is relatively low). Therefore, positive results from high-sensitivity screening tests are generally subject to confirmation by more specific tests.

For purposes of diagnostic testing, the goal is not detection but confirmation—confirmation of positive results from screening, or of clinical findings that are suggestive of the condition. Therefore, highly specific tests are used; since a specific test has a low rate of false positives, a positive result here is very likely to be true.

The Appropriate Use of Diagnostic Tests

Diagnostic tests should never be used indiscriminately, even if cost were not a concern. Indiscriminate use drastically lowers a test's positive predictive value, because the prevalence of any condition is significantly lower in the general population than in a high-risk population.

For example, on a test for evidence of the presence of the human immunedeficiency virus (HIV), a positive result is much more likely to be true in an intravenous drug abuser (who is representative of a high-risk population) than in someone with no known risk of exposure to HIV (representative of a low-risk population). Conversely, a negative result on the same test is more likely to be true in the person with no known risk of exposure than in the drug abuser.

Diagnostic testing is most appropriate when the likelihood of disease is higher than average—that is, in patients with a known history of exposure to the cause of a disease, in patients with a strong family history of a disease, and in those who are currently exhibiting clinical signs and symptoms suggestive of a disease. The proper purpose of a diagnostic test (as distinct from screening in the general population) is to confirm a clinical suspicion—not to fish blindly for abnormalities.

TITERS

Certain diseases are detected by the presence of specific types of immune antibodies in serum or other physiologic fluids. These antibodies are disease markers, but some such marker antibodies are almost universally present; then a positive result is not determined by the presence or absence of the antibody, but on the *extent* of its presence.

We measure the presence of antibodies by *titer* (or *titre,* for those who prefer the British spelling). As the word suggests, the procedure is akin to titration—a volumetric analysis involving the addition of measured amounts of fluid until a certain reaction is detected or is no longer detected.

Titer is analogous to concentration—the amount present per volume of body fluid. However, whereas concentration is measured as the weight or particle-count of substance per unit volume of fluid (see Chapter 2), titer is measured as the maximum dilution of fluid at which the antibody is still detectable. A titer of 1:16 means that the marker is still detectable when the fluid is diluted to one sixteenth of its original concentration.[1] A titer of 1:16 is higher than a titer of 1.8, which is higher than a titer of 1:4, and so on. *Obviously, the greater the amount of antibody present in the fluid, the higher the titer will be.*

For various conditions, different titers are accepted as criteria for defining negative (normal) versus positive (abnormal) results. For certain conditions, titer is also used to monitor disease progression or the response to treatment. In those conditions in which antibody levels rise and fall sensitively with changes in the severity or extent of disease, a rising titer suggests a worsening condition, while a falling titer suggests spontaneous resolution or a favorable response to therapy.

MEASUREMENTS OF CONCENTRATION

Many laboratory tests measure the concentration of various substances in body fluids—endogenous substances such as glucose or electrolytes, or exogenous substances such as drugs or toxic chemicals. The fluids most often tested are serum or plasma[2] and urine; less often, measurements are made from cerebrospinal fluid or joint space fluid.

Many substances in the body are measured in standard units of weight per volume (such as milligrams per deciliter; mg/dL) or particle-count concentration (such as milliequivalents per liter; mEq/L). As discussed in Chapters 1 and 2, measurements may also be made in the special form of particle-count concentration known as SI units (such as millimoles or

---

[1]In most cases, dilutions are 1-to-1. By adding an equal amount of diluent to a measured serum sample, the volume is doubled and the amount of antibody per unit volume is halved. A second 1-to-1 dilution yields fluid at one-fourth of the original concentration; a third such dilution yields fluid at one-eighth of the original concentration; and so on.

[2]For technical reasons, the concentrations of different substances may be given specifically as the amount per unit-volume of *serum* or per unit-volume of *plasma;* for this review of the entry-level mathematical basis of laboratory testing, the distinction may be ignored.

micromoles per liter: mmol/L, μmol/L). In some listings of reference laboratory values (that is, normal ranges), a multiplication factor is given for immediate conversion of standard measures to SI units.

For example, if serum glucose is reported as 96 mg/dL and we want the value in SI units, we first multiply the weight-per-milliliter by 10 to obtain the weight-per-liter, and then divide by the molecular weight (for glucose, 180) to get the number of millimoles represented by that weight:

$$96 \, \frac{mg}{dL} \cdot \frac{10 \, dL}{L} = 960 \, \frac{mg}{L}$$

$$960 \, \frac{mg}{L} \cdot \frac{mmol}{180 \, mg} \approx 5.33 \, mmol \cdot L^{-1}$$

In a single step, we could divide the original reading by 18 to get the equivalent value in SI units; or we could multiply the original reading by the reciprocal of 18, which is approximately 0.0556:

$$96 \, \frac{mg}{dL} \cdot 0.0556 \approx 5.34 \, mmol \cdot L^{-1}$$

That is, 0.0556 is the multiplication factor for direct conversion of a glucose concentration in standard units of mg/dL to the equivalent value in SI units of mmol/L.

Likewise, to convert a serum cholesterol level of 210 mg/dL to the equivalent value in SI units of mmol/L, we again multiply by 10 and then divide by the molecular weight of this compound, which is 386 (the formula is $C_{27}H_{45}OH$; multiply it out, using 12 for the weight of the carbon, 1 for hydrogen, and 16 for oxygen):

$$210 \, \frac{mg}{dL} \cdot \frac{10 \, dL}{L} = 2100 \, \frac{mg}{L}$$

$$2100 \, \frac{mg}{L} \cdot \frac{mmol}{386 \, mg} \approx 5.44 \, mmol \cdot L^{-1}$$

In a single step, divide the original reading by 38.6; or multiply it by the reciprocal of 38.6, which is approximately 0.0259:

$$210 \, \frac{mg}{dL} \cdot 0.0259 \approx 5.44 \, mmol \cdot L^{-1}$$

Multiplication factors for direct conversion of standard measures to SI units have been computed for many laboratory measurements. Just be sure that the laboratory data to be converted are reported in the same standard units as those used in the reference listing.

The Units Called "Units"

Certain measurements are given in special units called *Units* (abbreviated U), rather than as weight or particle count per volume. This type of determination may be made when it is difficult to isolate the object or substance for direct measurement. Instead, its presence is quantified by a laboratory test that causes some measurable effect that varies with the amount or activity of the substance. The extent or strength of that effect is reported on a scale calibrated in Units. (Thus, the actual meaning of the term depends on the type of effect being measured.)

For a test that has been accepted as a worldwide standard, measurements are given in International Units (IU). In some cases, Units are divided into metric fractions, such as milli-International Units (mIU), representing 0.001 IU.

Do not confuse International Units with SI units (Système International d'Units)—the scientific standard by which the concentration of a solution is measured in terms of particle-count concentration per liter.

ARE DIAGNOSTIC TESTS THE SAME AS PROOF?

With most diagnostic tests, the criteria that define certain results as normal and other results as abnormal are statistical in nature—that is, they indicate probabilities, not certainties. It is crucial to understand that an abnormal laboratory result does not by itself prove that the patient is sick; nor does a normal result prove that the patient is healthy.

A gold standard test gives a definitive answer—yes, the disease is present; or no, it is not present. On other tests for disease markers, we must deal with a certain likelihood that a positive or negative result may be false. With tests that measure concentrations or titers, we get a value that must be clinically interpreted.

In general, a test result is considered abnormal if it is outside the range of values found in 95% of the population (the minimum criterion most often used to define a difference or correlation as being meaningful or "statistically significant"; see Chapter 10). By that same criterion, up to 5% of the population may have abnormal test results in the absence of the condition in question. We are dealing with statistical likelihoods, not hard certainties.

Short of having a positive finding on a gold standard test, diagnosis is really a highly educated guess based on *all* the evidence at hand—known risks, clinical signs and symptoms, and laboratory data. And a working diagnosis that seems obvious at the outset may have to be reconsidered, revised, or replaced completely if contrary evidence pointing away from that diagnosis or toward a different diagnosis is later found.

Lesson: Laboratory testing is part of the diagnostic process, but it does not obviate the need for sound clinical judgment.

QUIZ

1. A patient's weight is confirmed at 68,236 grams. Then on Scale A, three consecutive readings (in pounds) are 149.8, 150.1, and 150.3; on Scale B, the three readings (in kilograms) are 68.841, 68.841, and 68.842. Compare Scales A and B in terms of quality control accuracy and precision.

2. In the language of quality control, is it possible for a test result to be nonspecific yet accurate? Accurate yet imprecise? Imprecise yet valid?

3. A new diagnostic test for a certain condition is evaluated in a population of 500 people known to have the condition, and the results are positive in 398 cases. The test is then evaluated in 500 people known to be free of the condition, and the results are negative in 428 cases. Compute the test's sensitivity and specificity.

4. On the same test as in Question 3, what do the results tell us with respect to the test's positive and negative predictive values?

5. The prevalence of a certain condition is $6.6 \cdot 10^{-3}$ in the general population. The diagnostic test for this disease is rated at 85% sensitivity and 72% specificity. What are the positive and negative predictive values if this test is used to screen for the disease in the general population?

6. For the same disease as in Question 5, the prevalence is $4.0 \cdot 10^{-1}$ in patients with signs and symptoms of the disease. What are the predictive values if the same test is used to confirm the diagnosis in these patients?

7. An antibody test is considered positive as a disease marker if the titer exceeds 1:16. Does a titer of 1:32 prove that the disease is present?

8. A patient's serum iron level is 94 μg/dL. If the atomic weight of iron is 55.85, what would this reading be in SI units (μmol/L)? Compute the multiplication factor for direct conversion from the standard measurement to SI units.

9. Which serum creatinine concentration is higher? 1.7 mg/dL or 95 μmol/L? (The molecular weight of creatinine is approximately 113.)

10. If highly specific tests are used to confirm positive screening results anyway, why not use these tests in the first place for screening?

Answers and explanations at end of book.

chapter 10

Biostatistics

Chapter Outline

Mathematical Tools Used in This Chapter

Biostatistics is the science of analyzing and interpreting data from clinical research.[1] The mathematical tools used for this purpose range from simple to sophisticated. Selecting and applying these tools is the responsibility of the

---

[1]*Data* is a plural word; the singular is *datum*. Properly speaking, we say, "By itself, *this datum is* meaningless," and "All together, *these data are* meaningful."

researcher, but students and practicing clinicians are expected to understand the fundamental principles of biostatistics.

This chapter will focus on relatively simple concepts and methods; the rest is left to a full-length textbook and comprehensive course of study. The scope is deliberately limited because it is better for entry-level students in the clinical sciences to gain a firm grasp of basic principles than to have a vague awareness of a much wider range of more sophisticated methods.

TWO MAJOR CATEGORIES OF RESEARCH

Biomedical researchers usually study a *sample*—a group of subjects who are collectively representative of a much larger defined population. Typical data consist of numbers of subjects displaying some characteristic, or measurements taken from those subjects. Two major categories of research are *epidemiology* (profiling the defined population) and *therapeutics* (responses to treatment).

Epidemiologic research often involves the *incidence* and *prevalence* of various clinical conditions. *Incidence* refers to the rate of occurrence of new cases of a given condition within a defined population, usually on an annual basis:

$$\text{Incidence} = \frac{\text{number of new cases reported}}{\text{population}} \cdot \text{year}^{-1}$$

Three points to observe: First, "occurrence" is the rate at which new cases are diagnosed, not the rate at which they actually arise; some conditions exist for years before they are recognized. Second, the population basis must be identified as national, worldwide, by state or other local jurisdiction, or by demographic group. Third, if no other time unit is specified, assume the rate is per year.

The population basis in the denominator is given as an integer power of 10, as in, "The incidence of (name of disease) in the United States is 5.6 per 100,000." That is, for every 100,000 people in this country, 5.6 new cases of that condition are diagnosed each year.[2] If the total population is known, the actual number of new cases per year can be computed. In a population approaching 280 million:

$$\frac{5.6 \text{ new cases} \cdot \text{year}^{-1}}{10^5 \text{ people}} \cdot (2.8 \cdot 10^8 \text{ people}) \approx 1.6 \cdot 10^4 \frac{\text{new cases}}{\text{year}}$$

Prevalence refers to the number of existing cases among all individuals who potentially could have the condition. In contrast to incidence, all that matters for prevalence is that the case exists—not when it first developed or when it was first recognized:

$$\text{Prevalence} = \frac{\text{number of people with condition}}{\text{population at any degree of risk}}$$

---

[2]If the denominator in an annual incidence rate is selected so that the numerator is less than 10, the rate will be in scientific notation, as in the hypothetical example given in the text: 5.6 per $100,000 = 5.6/10^5 = 5.6 \cdot 10^{-5}$.

Prevalence can be reported in various ways. Prevalence itself means, "Right now, at this very moment, what portion of the defined population has the condition?" Lifetime prevalence means, "What portion of the population will *ever* have the condition at *any* time during the course of their lives?" Prevalence based on a specified time period (for example, a certain month or year) means, "What portion of the population had the condition during that time period?"

Prevalence varies *directly* with incidence (rising as new cases are added to the total) and *inversely* with rates of recovery or death (falling as existing cases are subtracted from the total). Therefore, the right-now prevalence is typically higher for chronic conditions than for transient or rapidly fatal conditions (except perhaps during an epidemic of influenza or other acute illness). Lifetime prevalence is unaffected by the duration or acute-versus-chronic nature of a condition.[3]

An example of therapeutic data is "Response to an antihypertensive drug." We would collect data from individuals with high blood pressure, showing what their systolic and diastolic pressures were before, during, and after treatment with the drug being tested, to document the magnitude of the change in pressure (presumably, a decrease) achieved in each case.

CLASSIFICATION OF DATA

There are four types or levels of data. From least to most mathematically specific, they are: *nominal, ordinal, interval,* and *absolute.*

Nominal data are lists or groups of related objects, such as "Plant sources of protein." Numbers used for identification (1. Soy; 2. Wheat; 3. Peas; etc.) are arbitrary; that is, they do not imply any natural order or defined magnitude.

Ordinal data reveal the natural order or ranking of items in a group—first, second, third, etc.—but they do not reveal or even imply any defined magnitude. Ordinal data may be rankings of actual numeric values; for example, in a set of test scores arranged in ascending order, the lowest is first and the highest is last. Ordinal data may also be rankings of nonnumeric objects that have some natural order; for example, in a list of the taxonomy categories (for classification of living things) arranged from broadest to narrowest, kingdom is first, phylum is second, and so on through class, order, family, genus, and species.

Interval data are measurements made on a scale calibrated in dimensional units of defined magnitude. However, zero on an interval scale does *not* indicate a total absence of the measured characteristic. Therefore, we can see the difference but not the ratio between two values on an interval scale. The Celsius temperature scale is an interval scale, because 0°C is *not* absolute zero (the temperature that represents, in theory, a total absence of heat). Thus, 30°C is 10 degrees less than 40°C; but 30°C is *not* three-fourths of 40°C. The Fahrenheit scale is also an interval scale, because 0°F is not absolute zero.

Absolute data, like interval data, can be in the form of measurements on a calibrated scale. However, zero on an absolute scale (also called a "ratio scale") *does* indicate a total absence of the measured characteristic. We

---

[3]Incidence and prevalence are distinct concepts, unlike references to the "rate" of a condition, which could mean almost anything.

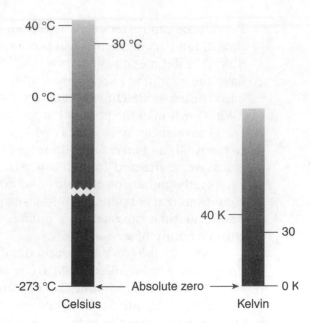

Figure 10.1 Obviously, the ratio 30/40 is *not* 75% on the Celsius scale, but it *is* on the Kelvin scale.

can see not only the difference but also the ratio between two values on an absolute scale. The Kelvin temperature scale is a ratio scale, because 0 K *is* absolute zero. Thus, 30 K is 10 kelvin less than 40 K; and also, 30 K *is* three-fourths of 40 K (Figure 10.1).

Another form of absolute data is a count or tally of separate objects or events. The scale is the number line itself (calibrated in integers), and zero indicates a total absence of the things being counted.

Many quantifiable characteristics (such as weight or temperature) can be measured to any desired degree of precision; they are therefore called *continuous variables.* In contrast, separate objects or events (such as the number of people with a certain characteristic) can be counted only in whole numbers; they are therefore called *discontinuous variables.*

The different types of data require different statistical methods. Interval and absolute data in a *normal distribution* (see Patterns of Data Distribution, later in the chapter) call for *parametric* tests. Ordinal data and data that are not in a normal distribution call for *nonparametric* tests. (We will return to these concepts later.)

GRAPHS

Suppose we are studying "Weight of infants born in the United States." We collect data showing how many babies are born each year in these weight categories (in grams): below 1000, 1000–1499, 1500–1999, . . . 4000–4499, 4500–4999, and 5000 or greater. Now we create a graph of the *distribution* of these data—birth-weight categories of uniform interval as the horizontal axis, and the numbers of babies in each weight group as the vertical axis (Figure 10.2). This type of graph is a *histogram,* a bar graph in which the rectangular area of each bar is proportional to the number of data within that range. The horizontal scale is calibrated in intervals of 500 grams in the range from 1000 to 5000, which includes most births. If we used a much larger interval,

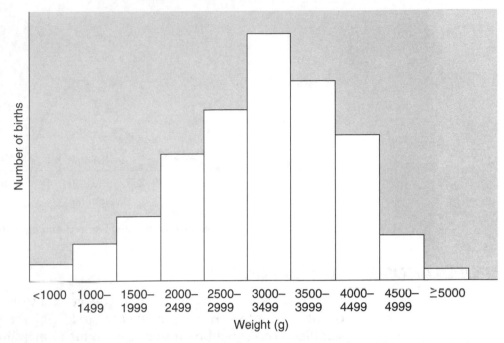

Figure 10.2 Histogram showing numbers of babies born in different weight ranges; (interval = 500 g).

such as 1 kilogram, we would have a poorer sense of the distribution of values (Figure 10.3).

Conversely, if we used a much smaller interval, such as 5 grams, we would have an excessive number of bars. The nature of the data will often suggest an appropriate scale. Alternatively, we can make a line graph by letting the horizontal scale vary continuously rather than being divided into interval groups (Figure 10.4).

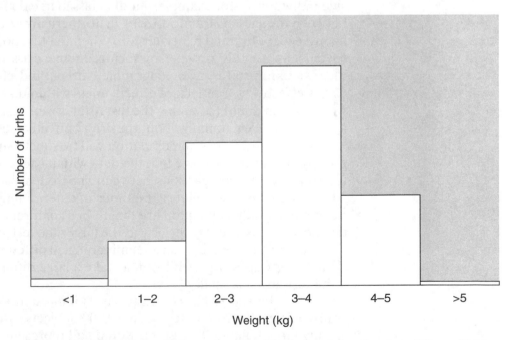

Figure 10.3 Same data, with interval = 1 kg. (Vertical scale is twice that of Figure 10.2.)

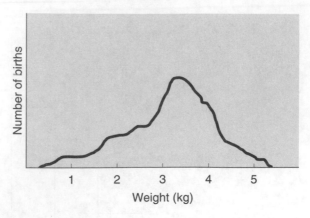

Figure 10.4 Birth weight as a continuous variable.

POPULATIONS AND SAMPLING

Because it is not practical to try to collect data from the whole population of interest, we study a *representative sample* of the population. But what sample? Who should be tested if we want to measure the effectiveness of our antihypertensive agent?

One way would be to include anyone with high blood pressure—men and women, young and old; people of every race, living in communities of every size, from every part of the nation; people at every level of education and income; people whose hypertension is mild, moderate, or severe. But even if the drug seems effective in this all-inclusive population, it might not work in every small subset of the population. Conversely, even if it doesn't work in the general population, it might still be effective within a limited subset. In short, studying an all-inclusive general population cannot reveal differences in small subgroups.

Altenatively, we might focus our study on a narrow subset of hypertensive individuals—for example, middle-class married male black nonsmokers between the ages of 40 and 50, living in northern urban environments, in good health except for mild hypertension. When the test population is defined so specifically—in this example, by socioeconomic class, marital status, gender, race, smoking status, age, geographic setting, and clinical condition—the drug's effectiveness or lack of effectiveness within that population subgroup will not be in doubt (assuming the test itself is well designed). However, such a study will reveal nothing about the drug's effectiveness in other subgroups.

Frequently, a general population will be subclassified into defined subgroups. However, to ensure that the data within each subgroup are meaningful and to allow comparisons between subgroups, each subgroup must be adequate in size. The technical methods of determining the minimum size of a sample or study population are taught in a comprehensive course in statistics; however, as a general principle, the larger the test group, the more meaningful the results. That is, even a seemingly small difference may be significant if the test group is large, while it may take a large difference to be significant if the test group is small.

How the test group is selected is as important as its size. Even if our antihypertensive drug is tested in 10,000 subjects, which would be an unusually large number, that group would still represent less than 0.1% of the tens of millions of Americans with hypertension. How can we possibly extrapolate out to the whole population of hypertensive individuals from such a small fraction?

The answer is *random sampling*. As long as a sample is representative of the population of interest, the data are applicable even if the sample seems small. Random sampling avoids *bias*, which is a systematic selectivity toward or away from a certain type of subject, resulting in a sample that is not representative of the population. Bias in the sample distorts the data and invalidates the study.[4]

STUDY DESIGN

Suppose we want to assess the effects of an antihypertensive drug in middle-aged males with moderate to severe arterial hypertension. From this defined population, a representative sample is recruited. Use of the drug is the *independent variable* (what we are testing); blood pressure is the *dependent variable* (what we measure to assess the effect of the independent variable). We try to prevent extraneous variables ("confounding factors") from contaminating the results. To that end, most therapeutic studies are *placebo-controlled* and *double-blinded*.

Placebo-controlled means that the subjects are randomly assigned to receive either the drug being tested or a placebo (an inert substance prepared in a pill or other formulation that looks exactly like the real drug). If the sample size is adequate and the participants in the study have been randomly assigned to receive the drug or the placebo, the "treatment wing" and "placebo wing" of the study will be well matched. Then, any significant difference in outcome cannot be attributed to the composition of groups (which would be a confounding factor); we would therefore attribute the outcome difference to the use of the drug.

Double-blinded means that neither the subjects nor the researchers know who is receiving active drug and who is receiving placebo. Each subject receives pills from a coded set, and it is only after the study is finished that the codes are "broken" to reveal who has been taking the drug and who has been taking placebo. Because no one knew who was taking active drug and who was taking placebo, differences in outcome between the groups cannot be attributed to any expectation of success or failure among the subjects, or to any difference in the way the researchers interacted with the subjects in the two groups.

With placebo-control, double-blinding, and confounding factors excluded, the intrinsic activity of a drug is fairly tested. If blood pressures come down by a significantly greater amount in the treatment group than in the control group, we conclude that our antihypertensive drug has genuine therapeutic value.[5]

---

[4]A random sample gives every individual and every combination of individuals an equal chance of being selected. Random sampling is achieved through use of a table or computer program that provides a set of random numbers. The simple device of taking every tenth or hundredth name from a listing does not give an equal chance to every individual or combination of individuals.

[5]In many clinical situations, placebo effects can be surprisingly large; that is, patients feel better at least in part because they expect to feel better.

THE CENTRAL TENDENCY OF A DATA SET

Once a set of data is collected, we can apply a number of descriptive measures to characterize the set. Suppose we have recorded heights from a sample comprising five men and six women. The men's heights are (in increasing order) 173, 176, 178, 181, and 186 cm. The women's heights are 155, 156, 160, 168, 169, and 175 cm. As a group, the men are obviously taller—but by how much?

We must look at the *central tendency* of the data—a single value to which the collected data seem to point, or around which they seem to cluster. The three measures of central tendency are the *mean, median,* and *mode.*

The *mean* (or average) of a set of data is computed as the sum of all the individual values divided by the number of values recorded. A mean is indicated by a horizontal bar over the symbol for that measurement.[6] For any set of values, X, obtained from a sample containing any number of subjects, N, the mean, $\overline{X}$ (pronounced "mean of X" or "X-bar"), is computed as follows:

$$\overline{X} = \frac{X_1 + X_2 + X_3 + \dots + X_N}{N} = \frac{\Sigma X_i}{N}$$

In the shorthand form, Σ (uppercase Greek sigma) symbolizes "sum of" and X_i indicates the individual values of X.

---

[6]The symbol for the actual but unknown mean for a certain characteristic within the whole defined population of interest is μ (Greek mu).

Let's compute the mean for each of the two sets of heights:

$$\overline{H}_{\text{men}} = \frac{173 + 176 + 178 + 181 + 186}{5} = \frac{894}{5} \approx 179 \text{ cm}$$

$$\overline{H}_{\text{women}} = \frac{155 + 156 + 160 + 168 + 169 + 175}{6} = \frac{983}{6} \approx 164 \text{ cm}$$

We now have one number for each group. And because height is recorded on an absolute scale, we can compute the difference and the ratio between the means:

$$\text{Difference} = \overline{H}_{\text{men}} - \overline{H}_{\text{women}} = 179 - 164 = 15 \text{ cm}$$

$$\text{Ratio} = \frac{\overline{H}_{\text{men}}}{\overline{H}_{\text{women}}} = \frac{179 \text{ cm}}{164 \text{ cm}} \approx 1.09$$

As a group, the men are 15 cm or about 9% taller than the women. Alternatively:

$$\text{Ratio} = \frac{\overline{H}_{\text{women}}}{\overline{H}_{\text{men}}} = \frac{164 \text{ cm}}{179 \text{ cm}} \approx 0.916$$

The women are about 8% shorter than the men. (The arithmetic difference, 15 cm, is 9% of the women's mean height and 8% of the men's mean height.)

A different method is used when data are grouped within intervals. To find the mean of a set of grouped data, multiply the midpoint value of each interval by the number of data within that interval, and then divide the sum of all those products by the total number of data in the entire set. We'll use birth weight as an example, with 0.5 kg as the interval. Borderline data are included in the lower interval (2.5 kg would be included in the interval 2.0–2.5 rather than in 2.5–3.0). Here are the data on 100 babies at birth weights of 1 to 5 kilograms:

Weight (kg)	Interval Midpoint	·	Number of Data	=	Product
1.0–1.5	1.25		2		2.50
1.5–2.0	1.75		4		7.00
2.0–2.5	2.25		10		22.50
2.5–3.0	2.75		31		85.25
3.0–3.5	3.25		26		84.50
3.5–4.0	3.75		17		63.75
4.0–4.5	4.25		7		29.75
4.5–5.0	4.75		3		14.25

The sum of the products is 309.5, and the number of data in the entire set is 100. Therefore, the mean for this set is $309.5/100 \approx 3.1$ kg.

Our second measure of central tendency is the *median,* which is defined as the midpoint value in a set of data arranged in ascending numeric sequence; that is, the median is that datum above and below which are found equal numbers of higher and lower data. There is no computational formula; but in a set containing N items arranged in ascending sequence, the median is the value found at:

$$\frac{N + 1}{2}$$

If N is an odd number, the set has a single midpoint datum. In the set of men's heights (173, 176, 178, 181, 186 cm), $N = 5$ and $(N + 1)/2 = 6/2 = 3$. So the median is the third datum: 178 cm. If N is an even number, the set has no single midpoint. In the set of women's heights (155, 156, 160, 168, 169, 175 cm), $N = 6$ and $(N + 1)/2 = 7/2 = 3.5$. The median lies midway between the third and fourth data, and its value is the mean of those two centermost data: $(160 + 168)/2 = 164$ cm.

A concept closely related to the median is *percentile.* Percentile identifies a score or value that is higher than that same percentage of all the individual scores in the entire set. The 30th percentile identifies the score that is higher than 30% (or lower than 70%) of all the scores recorded. Similarly, only 5% of all scores are higher than that representing the 95th percentile, and the highest score in a set defines the 100th percentile. Therefore, a score representing the 50th percentile is in the middle of the set, with half of all scores above it and the other half below it. *That is, the 50th percentile is the same thing as the median of a set.*

Our third measure of central tendency is the *mode,* which is the most commonly recorded value in a set. Suppose we are studying the size of families; we survey 100 families and record the following data, showing the number of families with no children, the number with one child, with two children, and so on.

Number of Children	Number of Families with This Number of Children
0	11
1	14
2	22
3	20
4	17
5	10
6	6

The most frequent value is "2 children" (reported by 22 families), so 2 is the mode of this set. Note that the mean is 2.72 children per family (272 children among 100 families) and the median is 3 (the number of children in the 50th and 51st families when the set of 100 families is arranged in order by number of children).

Not all sets of data have a meaningful mode. With data that show no tendency to cluster around certain values, such as "Number of births per month," the data are distributed evenly (ignoring small, random variability). The presence of a mode in such a set would be an artifact, suggesting that the sample is too small. With data that tend to cluster, it works the opposite way—the absence of a mode would suggest that the sample is too small. In the tiny set of five men and six women, all the heights are different, so we see no mode; but a larger sample would reveal a mode, because heights do cluster around typical values for men and women.

Some sets of data contain more than one mode. If the same peak number of individual data is associated with two different values, the set has two modes.

Measures of central tendency are classified as "descriptive statistics." By themselves, they merely characterize a given set of data. However, these and other descriptive measures may be used as input data for "inferential statistics"—analytic methods of drawing conclusions and making comparisons.

ATYPICAL DATA VS. CENTRAL TENDENCY

In any set of data, the *range* is the interval from the lowest score or value in the set to the highest score. As a descriptive statistic, the range defines the extremes but tells us little about the tendency of the data—exactly because the extreme high and low values are often atypical of the set as a whole.

Here are three sets of test scores from a sample group of 11 students selected randomly from an arbitrarily large class:

Set A: 12, 29, 64, 66, 66, 67, 68, 70, 70, 74, 99

The mean score is 62.3 (the sum of all 11 scores, divided by 11). The median is 67 (the sixth of the 11 scores in ascending order). There are two modes, at 66 and 70 (each with two scores).

Set B: 12, 14, 20, 20, 21, 22, 22, 26, 26, 26, 99

The mean is 28.0, the median is 22, and there is one mode, at 26.

Set C: 12, 19, 30, 46, 55, 60, 66, 74, 83, 88, 99

The mean is 57.5, the median is 60, and there is no mode.

Sets A, B, and C all have the same range—87 (the interval between the lowest score, 12, and the highest, 99). Yet these data sets are very different. Set A depicts an ordinary situation, with most of the scores in the 60s and 70s, plus two atypically low scores and one atypically high score. In Set B, ten students have abysmal scores (suggesting a tough exam), and one student has a wildly atypical high score (suggesting prior knowledge of the questions). Set C shows a strangely even distribution of scores across the full range; and although we can compute the mean and find the median, we see no obvious clustering in this set.

"Outliers" are isolated, atypical data, far from the median. They may be misleading, especially in small sets of data.[7] Let's form Set A' (A-prime) by replacing the two low outliers from Set A with higher scores—still the lowest in the set, but not atypically low:

Set A': 59, 63, 64, 66, 66, 67, 68, 68, 70, 74, 99

---

[7]An outlier may be a valid (if atypical) datum, or it could represent an error in measurement or the accidental inclusion of a measurement from a different set.

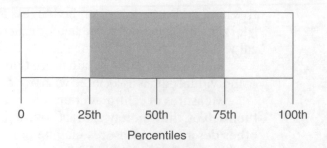

Figure 10.5 The interquartile range is the central 50% of the data in a set.

Now the mean is 69.6 (increased from 62.3, the mean in Set A). But note that the median remains at 67, because it depends only on its position-in-order within the sequence, not on the actual values of the scores above and below it.

A large set of data may be divided into four *quartiles,* partitioned at the 25th, 50th, and 75th percentiles. The inner quartiles, comprising all scores between the 25th and 75th percentiles, is the *interquartile range* (Figure 10.5). This range encompasses the central half of all scores, straddling the median and excluding low and high outliers. It is sometimes more useful to compare two sets by their interquartile ranges than by their full ranges.

PATTERNS OF DATA DISTRIBUTION

Statistical analysis depends on the type or level of data and the way the data are distributed—the number of cases associated with each value or score. The basic pattern of distribution is the normal or Gaussian[8] curve, also called a *bell-shaped curve.* In this distribution, the greatest number of data are found at a mid-range value where the curve reaches its peak.

An example is IQ score, where the greatest number of people have a score of 100 (in other words, 100 is the mode).[9] The number of people with higher and lower scores keeps decreasing the further we move in either direction from the peak (Figure 10.6).

A key feature of a normal curve is that it is symmetric around the peak; that is, equal numbers of scores fall within any given interval above and below the peak (for example, in the general population, the number of people with an IQ of 108 would be about the same as the number with an IQ of 92, because those numbers represent 100 ± 8). *Because of this symmetry, the peak simultaneously represents the mean, median, and mode.* That is, all three indicators of central tendency are equal when the distribution is perfectly normal.

Note the shape of the curve—a large group of scores clustered around the median, and then a relatively sharp decline on either side to low numbers of data, followed by a flattening out at both extreme ends of the curve, where the few remaining data are outliers. *This bell-shaped configuration is charac-*

---

[8]Karl Friedrich Gauss (1777–1855), one of the greatest mathematicians in history, was a pioneer in exploring the mathematical properties of the normal curve.

[9]An IQ score of 100 indicates an intellectual level that is 100% of the mean level of the age-peer population. A score of 120 indicates a level 20% above the age-peer mean. A score of 95 indicates a level 5% below the age-peer mean, and so on.

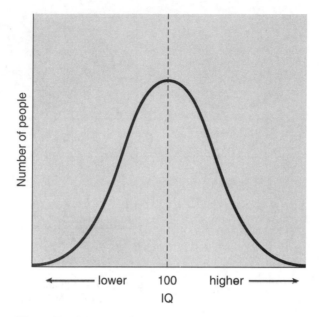

Figure 10.6 IQ scores as an example of normal distribution.

teristic of all normal distributions, regardless of the type of data, and regardless of the actual numeric values obtained.

The same data can be used to produce a different type of graph. Instead of showing the numbers of individuals at each score within the range, set the vertical axis to show the *cumulative percentage* of individuals with scores *up to and including* each value represented on the horizontal axis (Figure 10.7). This characteristic "S-shaped" graph starts rising slowly from the low end of the range, where there are only a few outliers. It rises more quickly around the middle of the range, where the greatest number of scores are found, and tapers off to an almost flat segment at the high end, where again there are only a few outliers. The curve rises from 0% at the left to 100% at the right (no one has a score below the lowest, and everyone has a score up to and including the highest). The rate of rise may vary, and the slope can even be zero

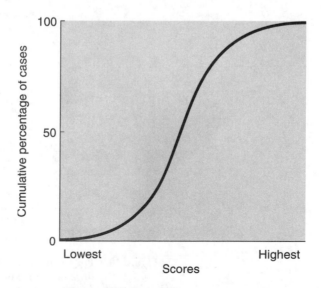

Figure 10.7 Cumulative percentages.

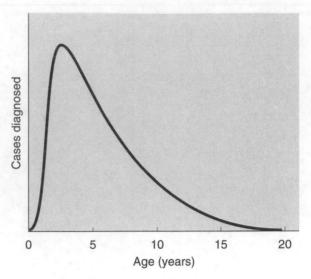

Figure 10.8 Incidence of acute otitis media in children is positively skewed.

in certain segments (if no data fall within that interval), but a cumulative graph can never show a decline.

However, not all data sets are normally distributed. In a *positively skewed distribution,* the peak is closer to the low end. This pattern might be seen in data on "Incidence of acute otitis media in children," with the horizontal scale of age ranging from birth to the oldest preadult age, 20 years (Figure 10.8). Acute otitis media occurs more often in infants and younger children than in older children, so the peak is skewed to the left.

In a *negatively skewed distribution,* the peak lies closer to the high end of the range, such as in "Frequency of use at different dosages" for a drug that is most often given at or near the maximum recommended dosage (Figure 10.9). The peak incidence is to the right of center, toward the high end of the dosage range, reflecting the higher frequency of use at maximum dosage.

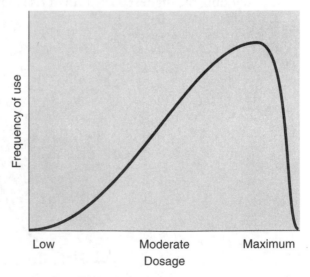

Figure 10.9 For a drug that is usually given at maximum dosage, the frequency-of-use curve is negatively skewed.

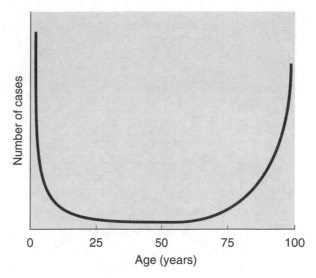

Figure 10.10 Urinary incontinence by age shows a bimodal distribution.

The mean is most useful with normally distributed data. With severely skewed data, the median and mode are better measures of central tendency.

A *bimodal distribution* features two clusters of data separated by a valley, such as in "Prevalence of urinary incontinence, by age" (Figure 10.10). Urinary incontinence is most common in younger children, but it also occurs in some older adults as a result of certain chronic diseases or surgical procedures; hence, there are peaks at each end of the age range.

THE SHAPE OF THE NORMAL CURVE

In the following discussion, we assume that the measured values from the sample show a normal distribution. As stated earlier, the bell-shaped curve is characteristic of any type of normally distributed data. However, the curve can vary in terms of its width-to-height ratio (Figure 10.11). Both of

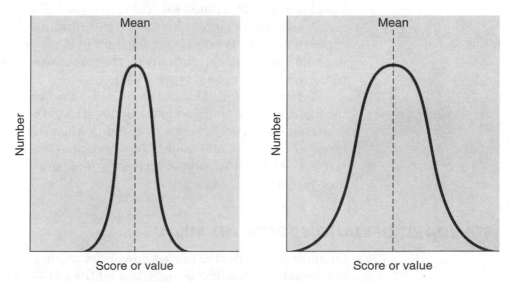

Figure 10.11 Two normal curves, differing in steepness.

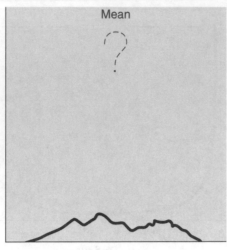
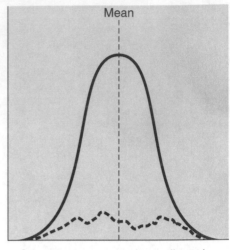

| Stage 1: few data | Stage 2: many more data collected |

Figure 10.12 The accumulation of data reveals a normal curve.

these graphs show normal distributions of data; but the curve on the left is narrow and steeply peaked, while that on the right is spread out and shallow. *The narrower and steeper the curve, the tighter the clustering of data about the mean.*

Assuming that the measured characteristic is, in fact, normally distributed, it may take more data to produce a clear normal curve when the data are widely scattered. Think about it: If the first 50 scores all fall within a narrow range, we anticipate that most subsequent scores will fall within or near that same range. In contrast, if the first 50 scores are scattered all over a wide range, who knows where the next score will fall? However, after 100 or 500 or 1000 scores, we may finally see a clustering tendency around a single value (Figure 10.12). The first graph shows a wide scattering of data; the shape is vague. The second shows the effect of including additional data; now we can see a normal curve about a central peak. As a general principle, the more widely the data are scattered, the greater the number of data needed to reveal a normal curve, and vice versa.

Certain data show inherently tight clustering. For example, the normal range of arterial blood pH (see Chapter 6) is very narrow. If we measure pH in people on the street, virtually all the values would lie within or quite close to the 7.35–7.45 range (Figure 10.13).

Arterial blood pressure does not show such tight clustering. If we record pressures from people on the street, the first ten data might be scattered over a wide range; or if the first ten data *do* fall within a narrow range, the *next* 10 data might fall within an overlapping or totally separate range. Only when we obtain a larger sample does a clear clustering pattern emerge (Figure 10.14).

STANDARD DEVIATION FROM THE MEAN

Quantifying the degree of clustering or scattering of data requires the use of mathematical tools more sophisticated than those we have seen so far. The first step is the calculation of *variance* (s^2). The symbol is squared because, as

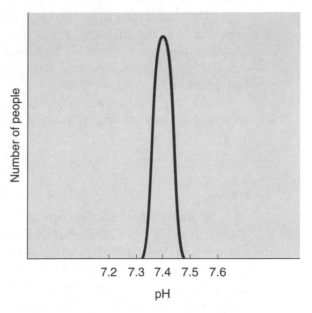

Figure 10.13 Arterial blood pH, showing very tight clustering within a narrow range.

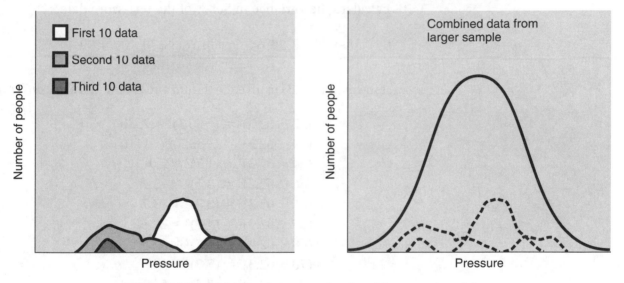

Figure 10.14 Arterial blood pressure, showing wider scattering of data.

we will see, what we really care about is *s*, which we obtain by taking the square root of the variance. For any set of values, *X*, obtained from a sample containing any number of data, *N*, we calculate variance as follows:

$$s^2 = \frac{(X_1 - \overline{X})^2 + (X_2 - \overline{X})^2 + (X_3 - \overline{X})^2 + \ldots + (X_N - \overline{X})^2}{N - 1}$$

In the numerator, $\overline{X}$ denotes the mean or average of all the individual values of *X*. Then $X - \overline{X}$ is the difference between an individual datum and the mean of the set. Each difference is squared, and the numerator is the sum of these squares.

The reason for squaring each difference is that $X - \overline{X}$ will be positive when $X > \overline{X}$ but negative when $X < \overline{X}$, and the algebraic sum of all these positive and negative differences would then be zero, which would tell us precisely nothing. However, with the differences squared, all the results are positive.[10]

We can also show the formula in a shorthand version; again, Σ is the summation symbol and X_i indicates the individual values of X in the set:

$$s^2 - \frac{\sum (X_i - \overline{X})^2}{N - 1}$$

Note that the denominator of the variance equation is $N - 1$ rather than N. We are dealing with probabilities based on samples, not certainties based on whole populations. To offset any bias associated with sample variability, we diminish the denominator, thereby augmenting the quotient, which is the calculated variance. The smaller the set, the greater the impact of subtracting 1 from the denominator. Think about it: With any given numerator, dividing by denominator $N - 1 = 99$ instead of $N = 100$ increases the quotient by only about 1%; however, dividing by $N - 1 = 9$ instead of $N = 10$ increases the quotient by 11%. Again we see why it is precarious to rely on small samples.

Let's calculate the variance in Set A of the test-score data:

Set A: 12, 29, 64, 66, 66, 67, 68, 70, 70, 74, 99

We subtract the mean (62.3) from each datum and then square the difference:

$$(12 - 62.3)^2 = (-50.3)^2 \approx 2530$$
$$(29 - 62.3)^2 = (-33.3)^2 \approx 1110$$
$$(64 - 62.3)^2 = (1.7)^2 \approx 2.9$$
$$(66 - 62.3)^2 = (3.7)^2 \approx 13.7$$
$$(66 - 62.3)^2 = (3.7)^2 \approx 13.7$$
$$(67 - 62.3)^2 = (4.7)^2 \approx 22.1$$
$$(68 - 62.3)^2 = (5.7)^2 \approx 32.5$$
$$(70 - 62.3)^2 = (7.7)^2 \approx 59.3$$
$$(70 - 62.3)^2 = (7.7)^2 \approx 59.3$$
$$(74 - 62.3)^2 = (11.7)^2 \approx 137$$
$$(99 - 62.3)^2 = (36.7)^2 \approx 1350$$

The sum of these squared differences is approximately 5330, and this number forms the numerator of the variance equation. Since the set contains 11

---

[10]We could also take the absolute magnitude of each difference (subtracting the smaller from the larger) to ensure that the answer is always positive. The absolute magnitude of A minus B is written $|A - B|$. Regardless of which is greater, A or B, $|A - B| = |B - A|$. For example, $3 - 5 = -2$, but $|3 - 5| = |5 - 3| = 5 - 3 = 2$. The result obtained by squaring the differences has greater applicability in statistics.

scores, the denominator is $11 - 1 = 10$. Now we can just plug these numbers into the variance equation:

$$s^2 - \frac{\sum (X_i - \overline{X})^2}{N - 1} \approx \frac{5330}{10} = 533$$

When we square the difference between the mean and each datum, the dimensional units are squared along with the numeric factors. For example, if we are studying changes or differences in the serum concentration of sodium, which is reported in milliequivalents per liter, variance would be reported in the bizarre units "squared milliequivalents per squared liter." To restore meaningful numbers and units, we must take the square root of the variance. The positive square root gives us the descriptive measure we have been aiming for all along—the *standard deviation from the mean* (abbreviated *s*).[11]

$$\text{Standard deviation} = \sqrt{\text{Variance}} = (s^2)^{1/2} = s$$

In the example just given:

$$s = \sqrt{533} \approx 23.1$$

This result characterizes the dispersion of data in Set A: its mean is 62.3 ± 23.1.

Set B had 10 of 11 scores clustered within the range 12 to 26, and one atypical score of 99. The mean is 28.0, and the sum of the squared differences is approximately 5720. To compute the variance (s^2) and standard deviation (*s*):

$$s^2 = \frac{5720}{10} = 572$$

$$s = \sqrt{572} \approx 23.9$$

To see the effect of that single high outlier, 99, we'll replace it with 66—a score that is still atypically high but less extreme—the mean drops to 25.0, the sum of the squared differences is approximately 2050, and *s* becomes notably smaller:

$$s^2 = \frac{2050}{10} = 205$$

$$s = \sqrt{205} \approx 14.3$$

---

[11]The symbol for the actual but unknown standard deviation for the whole population is σ (Greek lowercase sigma).

And if we replace 66 with 33—still the highest score, but one that is at least consistent with the rest of the data—the mean becomes 22.0 and the sum of the squared differences is 342:

$$s^2 - \frac{342}{10} = 34.2$$

$$s = \sqrt{34.2} \approx 5.85$$

By replacing the outlier with values closer to the other data, we diminished the degree of scattering. Obviously, we would never manipulate real data in this fashion; we did it here simply to show that the less scattered or more tightly clustered the data, the smaller s will be, and vice versa.

A routine application of s is in calculating the *standard score* (also called the *z-score*) of a datum. The z-score measures the distance between any individual datum and the mean of the set in units of s—that is, the mean is defined as zero, and intervals above and below the mean are measured in units of $s = 1$. Whereas variance, standard deviation, and the three measures of central tendency describe the clustering or scattering of data in a set as a whole, the z-score gives the location of an individual datum within the set.

To compute the standard score, subtract the mean, $\overline{X}$, from the individual datum, X_i, and divide the difference by the standard deviation of the set, s:

$$z = \frac{X_i - \overline{X}}{s}$$

In Set A of the test scores, the mean was 62.3 and s was 23.1. Let's compute the z-scores for the low outlier, 12, and for a more typical datum, 70, which was one of the modes of that set:

For the score of 12: $\quad z = \dfrac{12 - 62.3}{23.1} = \dfrac{-50.3}{23.1} \approx -2.18$

For the score of 70: $\quad z = \dfrac{70 - 62.3}{23.1} = \dfrac{7.7}{23.1} \approx 0.333$

That low outlier is more than $2s$ below the mean (note that the result is negative), whereas 70 is just one-third of s above the mean.

Another routine application of s is the *coefficient of variation,* which lets us compare the scattering in separate sets of data that measure the same feature. For each set, compute the ratio of s to the mean:

$$\text{Coefficient of variation} = \frac{s}{\overline{X}}$$

(Multiply by 100 to see the answer as a percent.) The lower the coefficient, the tighter the clustering of data. The coefficient is lower when the standard deviation is smaller and/or the mean is larger. Suppose we compare two sets of lab-

The Importance of the Standard Deviation

The standard deviation from the mean (s), perhaps the most frequently encountered of all statistical measures, quantifies the degree of scattering of data within a set. By itself, s is merely descriptive of that set. However, s is used as input for other statistical formulas by which we can make inferences about the population and comparisons with other data sets.

Other things being equal, when s is a smaller value (that is, when the sample data are more tightly clustered), we can often make such inferences and comparisons with more certainty.

oratory data on serum electrolytes; the mean concentration of sodium is 139 ± 12 meq/L in one sample, and 142 ± 8 meq/L in another. The coefficient of variation from the first sample is $12/139 \approx 8.6\%$; from the second sample, $8/142 \approx 5.6\%$. The data from the second sample are more tightly clustered.

STANDARD DEVIATION AND THE NORMAL CURVE

Now we come to a crucial point: *in a normal distribution, standard deviations above and below the peak are associated with specific percentiles.*

The mean $\pm 1s$ defines the range between the 84.1 and 15.9 percentiles, which denote the median $\pm 34.1\%$. Therefore, this range encompasses approximately the central 68% of all the data in the set. Likewise, the mean $\pm 2s$ (the 97.7 and 2.3 percentiles) encompasses more than 95% of all the data. The mean $\pm 3s$ encompasses over 99.7% of the data, and the mean $\pm 4s$ encompasses over 99.9% (the entire set except for extreme high and low outliers).

These demarcations can be shown on the normal curve (Figure 10.15). *These same percentile demarcations are characteristic of all normal distributions,*

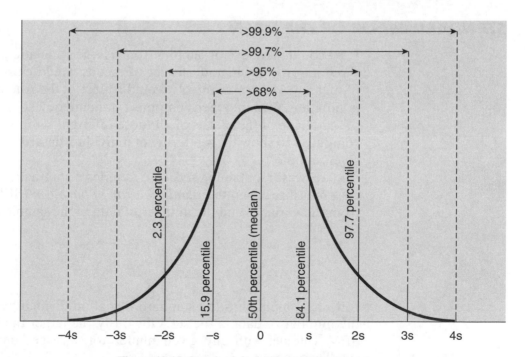

Figure 10.15 Standard deviations on the normal curve.

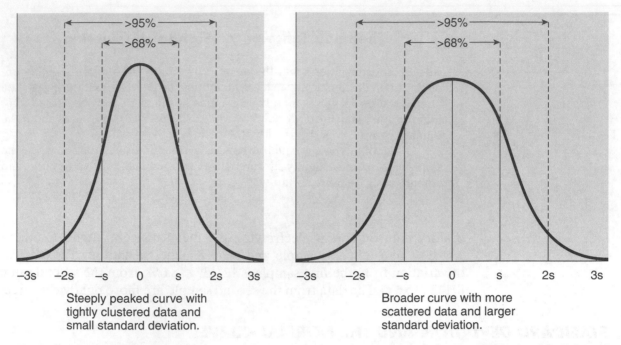

Figure 10.16 Regardless of the steepness of the normal curve, the percentage of scores within ±s, ±2s, ±3s, and ±4s of the mean is the same.

regardless of the nature of the data. The numeric value of *s* depends on the clustering or scattering of the data—but in a true normal distribution, the percentage of scores found within the ranges of 1*s*, 2*s*, 3*s*, and 4*s* above and below the mean will be essentially the same for any type of data (Figure 10.16). In fact, 1*s*, 2*s*, 3*s*, and 4*s* are simply whole-number *z*-scores. Established tables give the predicted percentages of data lying above or below any given *z*-score, whether it is an integer multiple or a fractional multiple of *s*.

STANDARD ERROR OF THE MEAN

If we could take several samples from a large population, each sample would show its own mean, and this set of means would cluster around a "mean of means," which presumably would reflect μ, the true mean for the whole population. We could then compute a standard deviation showing the degree of scattering of the means of all those different samples about μ, just as we compute *s* to show the scattering of individual data about the mean within a given sample.

However, obtaining separate samples is impractical; therefore, we rely on a measure called the *standard error of the mean* (SEM). SEM is derived from the variance and the number of data in the sample at hand:

$$\text{SEM} = \sqrt{\frac{s^2}{N}} = \frac{\sqrt{s^2}}{\sqrt{N}} = \frac{s}{\sqrt{N}}$$

SEM is computed as the standard deviation divided by the square root of the number of data in the set. Obviously, the larger the sample, the smaller SEM will be; with any given numerator, a larger denominator yields a smaller quotient.

We'll again use Set A of the test-score data, recalling the relevant features: $s \approx 23.1$, $N = 11$. Computing SEM for that set:

$$\text{SEM} = \frac{s}{\sqrt{N}} = \frac{23.1}{\sqrt{11}} \approx \frac{23.1}{3.32} \approx 6.96$$

Now suppose our sample had 22 students instead of 11. Even if s were still 23.1, the SEM becomes substantially smaller:

$$\text{SEM} = \frac{23.1}{\sqrt{22}} \approx \frac{23.1}{4.69} \approx 4.93$$

In reality, a larger sample would probably be associated with a smaller s because the impact of outliers would be reduced. With a smaller value for s in the numerator, SEM would be even smaller.

SEM quantifies our uncertainty about the correlation between the mean of our sample and μ, the true but unknown mean of the whole population of interest. The smaller SEM is, the more likely it is that the sample mean is close to μ; and the larger the sample, the smaller SEM will be.[12]

CONFIDENCE INTERVAL

We must distinguish between the mean of a set of data obtained from a sample, and μ, the true mean of the population represented by that sample. We do not know the value of μ, but we assume that it lies within the data range of the sample.

If we exclude the high and low outliers from the sample set, it is still likely that μ lies within that narrower range, but this likelihood is now less than 100%. Obviously, as we consider ever-narrower ranges, the likelihood that μ is still included keeps decreasing. In a data set with a normal distribution, the *confidence interval* (CI) identifies a certain range of values such that the likelihood of including μ is exactly 95%.

Recall that slightly *more than* the central 95% of normally distributed data lie within the range defined by the mean $\pm 2s$—that is, between those values whose z-scores are ± 2.00. Computation of CI is based on the sample mean, the standard error of the mean, and the z-score for the values encompassing *exactly* the central 95% of all the data in the set.[13] The CI is calculated as follows:

$$\text{CI}_{95} = \text{mean} \pm (1.96 \cdot \text{SEM})$$

---

[12]SEM is sometimes used inappropriately. Reporting data tendency as mean $\pm$ SEM instead of mean $\pm s$ falsely suggests a tighter clustering than the data really display. (Since SEM $= s/N^{1/2}$, SEM is always smaller than s, and a smaller plus-or-minus number suggests tighter clustering of data.) When data tendency is reported as mean $\pm$ SEM, multiply SEM by $N^{1/2}$ to get s.

[13]The factor 1.96 is the z-score that yields the 95% CI. Less often, CI is based on a 90% or 99% probability of including μ; the 95% CI may be assumed unless specified otherwise. The z-score for the 90% CI is 1.64; for the 99% CI, 2.57.

Again using Set A of the test-score data (mean $\approx$ 62.3, SEM $\approx$ 6.96):

$$CI = 62.3 \pm (1.96 \cdot 6.96) = 62.3 \pm 13.6$$

The likelihood that μ lies somewhere within this range (48.7–75.9) is 95%.

We do not know the value of μ, but it is more likely to be at the middle of the CI than anywhere else. The further we go from the midpoint, the less likely it is that a given value reflects μ. By definition, the midpoint of the CI is the sample mean itself, which merely indicates that the mean is more likely than any other given value to reflect μ. However, it is *only* a likelihood; the fact that we measure SEM at all demonstrates the uncertainty about the value of μ.

Again, notice the effect of the sample size. When we doubled the sample to 22 students, SEM dropped from 6.96 down to 4.93. And because SEM is smaller:

$$CI = 62.3 \pm (1.96 \cdot 4.93) = 62.3 \pm 9.66$$

With the larger sample, the 95% CI is correspondingly narrower (52.6–72.0).

THE NULL HYPOTHESIS AND STATISTICAL SIGNIFICANCE

Suppose we are performing a series of coin tosses, and the coin comes up heads on each of the first three tosses. We cannot conclude that it is a trick coin—not yet. But with a larger number of tosses, we expect to get approximately equal numbers of heads and tails. Now suppose that after 100 tosses, the tally is 51 heads and 49 tails. This result is so close to the expected 50/50 split that the slight difference can easily be attributed to chance (random variability). In contrast, a heads-tails ratio of 99/1 cannot reasonably be attributed to chance.[14]

A *statistically significant* difference is one that cannot reasonably be attributed to chance. However, it is not always possible to determine whether an observed difference is significant simply by glancing at the data; we must employ specific statistical tools, for a seemingly large difference might not meet the formal criteria for significance while a seemingly small difference might.

We start by assuming that there is no meaningful or systematic difference between the outcome data from the test and control groups—that is, we assume that any difference we may observe is due to random variability. This assumption is called the "Null Hypothesis." Then we see whether there is sufficient statistical evidence to reject that assumption. The "Alternative Hypothesis" is that the observed difference between the groups is meaningful rather than random. (In some cases, we do expect some difference, and the Null Hypothesis is that the value representing the control group is somewhat greater or somewhat less than the value representing the test group.)

For any hypothesis, each possible outcome is represented by a probability, $p \leq 1.00$. Some hypotheses have only one possible outcome. For example,

---

[14]With a fair coin, the odds are equal on every toss. Therefore, if a fair coin *has already* come up heads 99 times in a row (somehow), there is still an equal chance of heads or tails on the 100th toss. However, before the first toss, we look at the likelihood that it *will* come up heads 99 times in a row—and the likelihood of that happening purely by chance is vanishingly small.

"A solid one-kilogram cube of pure lead will sink when placed into a tank of water." The probability of that outcome is 100% ($p = 1.00$); conversely, the probability that the cube will float is 0% ($p = 0.00$). Most situations have more than one possible outcome, and each outcome then has a probability greater than 0 but less than 1; however, the sum of all the probabilities is always 1.00.

In our coin-toss model, the probability for heads or tails is 50% on each toss (for each, $p = 0.5$). The sum of these probabilities is $0.5 + 0.5 = 1.0$. The Null Hypothesis states, "In any sufficiently large sequence of tosses of a fair coin, there is no difference between the numbers of heads and of tails." If we do 100 tosses and the actual numbers of heads and tails closely match the expected numbers (50 and 50), we conclude that the observed difference is random (the Null Hypothesis). If the actual numbers of heads and tails are sufficiently different from the expected numbers, we reject the Null Hypothesis and conclude that the observed difference is meaningful (the Alternative Hypothesis; in this case, a trick coin). How large that difference must be is determined by specific statistical tests based on the type of data.

Statistical significance is most often defined as a Null Hypothesis probability below 5% ($p < 0.05$); that is, less than 1 chance in 20 that an observed difference is random. The same criterion also applies to normally distributed data: a difference is significant if the value representing the test group is outside the range encompassing 95% of the data from the reference or control group.

For the Null Hypothesis "There is no meaningful difference in the data from the test and reference groups," we look at the *central* 95% of data from the reference group and see whether the value representing the test group is outside that range. In this situation, the remaining 5% of data is split between both ends of the range, and an observed difference must be at least 1.96s above or below the reference mean to be significant (recall that the range between ±1.96s encompasses the central 95% of normally distributed data). This situation is the most commonly applicable, and the criterion we use is called a *two-tailed test* (Figure 10.17).

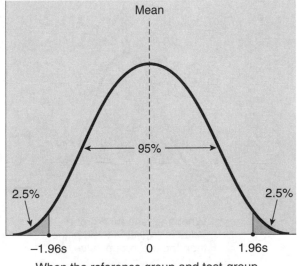

Figure 10.17 Two-tailed test criteria.

For the Null Hypothesis "There *is* a meaningful difference," we look at the uppermost *or* lowermost 95% of data from the reference group and see if the value representing the test group is outside that range. The remaining 5% of data is all at one or the other end of the range, and an observed difference must be at least 1.64s in a specified direction from the reference mean to be significant (the range between ±1.64s encompasses the central 90% of normally distributed data, leaving 5% at either end). This criterion is called a *one-tailed test* (Figure 10.18). On a one-tailed test, any difference in the direction opposite from the expected direction is insignificant. If the reference value is expected to be *higher,* the test value must be at least 1.64s below the reference mean (z-score $\leq -1.64s$) to be significant. If the reference value is expected to be *lower,* the test value must be at least 1.64s above the reference mean (z-score $\geq +1.64s$) to be significant.

Failing to meet the criteria for significance does not prove that the observed difference is random; it merely indicates a lack of compelling evidence that the difference is meaningful, and so the Null Hypothesis stands (formally, we "fail to reject" the Null Hypothesis). By analogy, a verdict of Not Guilty in a criminal trial does not prove that the defendant is innocent; it merely indicates a lack of compelling evidence that the defendant is guilty, and so the presumption of innocence stands.

The smaller the value of p, the less likely it is that an observed difference is random. That is, $p < 0.05$ indicates that the difference is meaningful (probability that it is due to chance is below 5%), but $p < 0.01$ provides stronger evidence (probability below 1%), and $p < 0.001$ provides even stronger evidence (probability below 0.1%). However, the p value does *not* reflect the actual impact of an independent variable. Suppose we test two antihypertensive drugs, A and B, against placebo. For each drug, the Null Hypothesis is, "There is no difference in outcome between this drug and placebo." Now suppose both drugs turn out to be significantly better than

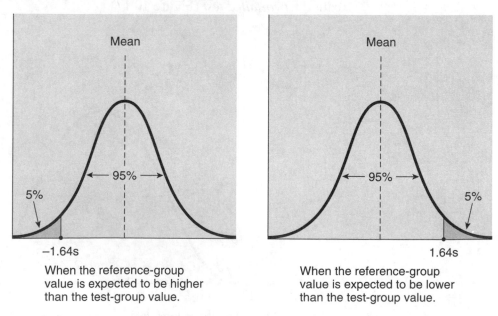

When the reference-group value is expected to be higher than the test-group value.

When the reference-group value is expected to be lower than the test-group value.

Figure 10.18 One-tailed test criteria.

placebo: for Drug A, $p < 0.01$; for Drug B, $p < 0.05$. *The fact that p is smaller for A than for B does not prove that A is better than B.* That determination requires a study specifically designed as a direct comparison of A versus B, for which the Null Hypothesis is, "There is no difference in outcome between these two drugs."

Statistical significance is not the same thing as clinical significance. Statistical significance merely indicates that an observed difference is unlikely to be due to chance. But it is possible for an observed difference to be meaningful rather than random, and yet trivial or irrelevant to health.

SIGNIFICANT DIFFERENCE: NONPARAMETRIC DATA

A count of events is a form of absolute data; but if the distribution is not normal (such as in coin-toss trials), we must use a nonparametric test to measure statistical significance. One frequently used nonparametric test is called *chi-squared* (x^2; the Greek chi is pronounced "kie" to rhyme with "tie"). This test compares the observed results with the expected results:

$$x^2 = \sum \frac{(\text{expected} - \text{observed})^2}{\text{expected}}$$

For each possible outcome, subtract the observed number of occurrences from the expected number, square the difference, and divide by the expected number; x^2 is the sum of these computations.

Suppose we toss a coin 100 times and get 51 heads and 49 tails instead of the expected 50 heads and 50 tails:

$$x^2 = \frac{(50 - 51)^2}{50} + \frac{(50 - 49)^2}{50} = \frac{(-1)^2}{50} + \frac{1^2}{50}$$

$$= \frac{1}{50} + \frac{1}{50} = 0.02 + 0.02 = 0.04$$

When we have only two numbers to compare, we can employ a simplified equation that yields the same result:

$$x^2 = \frac{(\text{greater} - \text{lesser})^2}{\text{number of trials}}$$

$$\frac{(51 - 49)^2}{100} = \frac{2^2}{100} = \frac{4}{100} = 0.04$$

Now we consult a table of probabilities (p) that this outcome could be random. The x^2 values on the table are listed according to the "degrees of freedom" (df), which is 1 in this case. (On many tests, df is 1 less than the number of outcome groups; a coin-toss study, has only two outcomes, so if one is known, the other is automatically determined.) Here is the relevant portion of the table:

Value of p:

df	0.90	0.80	0.70	...	0.20	0.10	0.05	0.01	0.001
1	0.0158	0.0642	0.148	...	1.642	2.706	3.841	6.635	10.827

For df $= 1$, the χ^2 value 0.04 lies between the values for $p = 0.90$ and $p = 0.80$; therefore, we say that $p > 0.80$. In other words, the probability that this outcome could be due to chance exceeds 80%, and the Null Hypothesis stands.

In contrast, suppose the result of 100 tosses is 35 heads and 65 tails:

$$\chi^2 = \frac{(65 - 35)^2}{100} = \frac{30^2}{100} = \frac{900}{100} = 9.0$$

$\chi^2 = 9.0$ lies between $p = 0.01$ and 0.001; then since $p < 0.01$ (less than 1%), the Null Hypothesis is rejected.

Notice how the sample size (in this case, the number of coin tosses) affects the statistical interpretation. Suppose we toss the coin 40 times and get 14 heads and 26 tails instead of the expected 20 and 20; the head-tails ratio is the same as that seen in the previous example ($14/26 = 35/65$). But when we compute χ^2:

$$\chi^2 = \frac{(26 - 14)^2}{40} = \frac{12^2}{40} = \frac{144}{40} = 3.6$$

For $\chi^2 = 3.6$, p is slightly over 0.05, so the Null Hypothesis stands—but we came close to statistical significance, and further investigation may be warranted.

Finally, with the same ratio obtained from only 20 trials (7 heads, 13 tails):

$$\chi^2 = \frac{(13 - 7)^2}{20} = \frac{6^2}{20} = \frac{36}{20} = 1.8$$

For $\chi^2 = 1.8$, $p > 0.10$, so the Null Hypothesis stands. In general, the smaller the sample or number of trials, the larger the difference must be to achieve statistical significance on the χ^2 test.

Now that we have seen how the χ^2 test works in the simplest of models, let's consider a clinical example of its use. Suppose we are testing three analgesic drugs—Painender, Acheaway, and Sufferstop—in patients with headaches. We want to know whether there is any significant difference in the percentages of patients who experience complete, partial, or no relief with each drug. The Null Hypothesis states, "There is no difference in efficacy between these three drugs." That is, any observed difference in outcome with each drug is random.

The study involves 450 patients, and the following data are collected:

	Painender	Acheaway	Sufferstop	Totals
Relief				
Complete	39	48	31	118
Partial	70	56	75	201
None	55	44	32	131
Totals	164	148	138	450

Since the Null Hypothesis states that the drugs have equal efficacy, the combined results of all three drugs are the expected results for each drug. Of the 450 patients in the study, 118 (26.2%) experienced complete relief, 201 (44.7%) experienced partial relief, and 131 (29.1%) experienced no relief. We use these percentages to predict the number of patients in each relief category with each drug (for example, the expected number of patients experiencing complete relief from Painender is 26.2% of 164, or 43.0). Now we compare the observed results to these expected results (shown in parentheses):

	Painender	Acheaway	Sufferstop
Complete relief	39 (*43.0*)	48 (*38.8*)	31 (*36.2*)
Partial relief	70 (*73.3*)	56 (*66.2*)	75 (*61.7*)
No relief	55 (*47.7*)	44 (*43.1*)	32 (*40.2*)

This presentation of data is a 3-by-3 *contingency table*—three rows and three columns, forming nine "cells." Each cell represents an entry in the computation:

$$\chi^2 = \sum \frac{(\text{expected} - \text{observed})^2}{\text{expected}} = \frac{(43.0 - 39)^2}{43.0} + \frac{(73.3 - 70)^2}{73.3}$$

$$+ \frac{(47.7 - 55)^2}{47.7} + \frac{(38.8 - 48)^2}{38.8} + \frac{(66.2 - 56)^2}{66.2} + \frac{(43.1 - 44)^2}{43.1}$$

$$+ \frac{(36.2 - 31)^2}{36.2} + \frac{(61.7 - 75)^2}{61.7} + \frac{(40.2 - 32)^2}{40.2}$$

And after much arithmetic (for each fraction, squaring the difference in the numerator and dividing by the denominator, and then adding up all nine quotients), we obtain $\chi^2 \approx 10.697$. For data displayed in a contingency table, df is the product of 1 less than the number of rows times 1 less than the number of columns. In this 3-by-3 table, df $= (3-1) \cdot (3-1) = 2 \cdot 2 = 4$. Now we consult the probability table and find the *p* value associated with $\chi^2 \approx 10.697$ at df $= 4$. Here is the relevant line from the table:

Value of *p*:

df	*0.90*	*0.80*	*0.70*	...	*0.20*	*0.10*	*0.05*	*0.02*	*0.01*
4	1.064	1.649	2.195	...	5.989	7.779	9.488	11.668	13.277

$\chi^2 \approx 10.697$ lies below the value for $p = 0.05$, and the result is statistically significant. So there *is* a meaningful difference between the three drugs; but because it was not a placebo-controlled trial, we do not know whether any of the drugs are significantly better than placebo or which drug is best.[15]

SIGNIFICANT DIFFERENCE: PARAMETRIC DATA

To assess the difference between two sets of parametric data, we may use *Student's t-test*.[16] To employ the *t*-test, we must first obtain a combined standard deviation (s_c) from the two groups. This quantity is calculated as follows:

$$s_c = \sqrt{\frac{\sum_1 (X_i - \overline{X}_1)^2 + \sum_2 (X_i - \overline{X}_2)^2}{N_1 + N_2 - 2}}$$

Again, X_i represents any individual datum; $\overline{X}_1$ and $\overline{X}_2$, the two group means; N_1 and N_2, the number of data in each group. Note that the numerator of the radicand is the sum of the numerators used in calculating variance in each group; likewise, the denominator is the sum of the denominators. The quotient is therefore a combined variance, whose square root is the desired combined standard deviation. Now plug this value into the denominator of the formula for the *t*-test:

$$t = \frac{\overline{X}_1 - \overline{X}_2}{s_c \sqrt{\frac{1}{N_1} + \frac{1}{N_2}}}$$

The result obtained is then interpreted according to a *t*-table to see whether the difference between the two means is statistically significant.

As an example, let's look again at the data on men's and women's heights. (Ignore the small sample size; height is ratio data, with a normal distribution for each gender.) The five men's heights (in centimeters) are 173, 176, 178, 181, and 186 (mean 179); the six women's heights are 155, 156, 160, 168, 169, and 175 (mean 164). The sum of the squared differences from the mean is 99 for the men, 323 for the women. Now we compute the combined standard deviation:

$$s_c = \sqrt{\frac{99 + 323}{5 + 6 - 2}} = \sqrt{\frac{422}{9}} \approx \sqrt{46.9} \approx 6.85$$

---

[15]If the sample is small and any cell of a contigency table has an expected value below 5, a different test is used, called Fisher's exact test; discussion of this method is deferred to a comprehensive course in statistics.

[16]Comparison of three or more sets requires a method called ANOVA (analysis of variance), discussion of which is deferred to a comprehensive course. By the way, Student's *t*-test is not a test for students; Student was the pseudonym of British scientist and statistician William S. Gossett (1876–1937).

And we use this result in computing t:

$$t = \frac{179 - 164}{6.85\sqrt{\dfrac{1}{5} + \dfrac{1}{6}}} = \frac{15}{6.85\sqrt{\dfrac{11}{30}}} \approx 3.62$$

For the t-test, $df = N_1 + N_2 - 2$ (the same as the denominator in computing s_c), which in this case is 9. Now look at the relevant portion of the t-table to see whether the observed difference in height is statistically significant:

df	Value of p:			
	0.10	*0.05*	*0.01*	*0.001*
9	1.833	2.262	3.250	4.781

With $df = 9$ and $t = 3.62$, $p < 0.01$; so we conclude that the observed gender-related difference in heights is meaningful.

We can see the effect of sample size just by glancing at the formulas for s_c and t. If N_1 or N_2 is larger, s_c is smaller. Also, larger samples yield a smaller radicand in the denominator of the t-test formula. Because both factors in the t-test denominator are smaller, t is larger; and at any given df, when t is larger, p is smaller. In short, the larger the groups, the more likely it is that a given difference between their means is significant, and vice versa—as we would expect.

The example just shown is the "unpaired" t-test, in which the means of two distinct groups are compared. In the "paired" t-test, each test datum is paired with a matched control datum. Pairing occurs in *repeated measures* studies (each subject is his or her own control, as baseline data are compared with data recorded later—typically, after treatment) and in *case-control* studies (each test subject is matched with a like individual in a control group). The test data constitute one group and the matched controls constitute the other, and we compare the means of the two groups. Here is the formula for the paired t-test:

$$t = \frac{\overline{X} - \mu}{\text{SEM}}$$

$\overline{X}$ is the mean of the observed differences in the data pairs, and μ is the expected or theoretical mean difference. In the usual Null Hypothesis ("There is no difference between the members of the pairs"), $\mu = 0$, by definition; therefore, we can often drop μ from the numerator. SEM is the standard error of the mean difference. Since $\text{SEM} = s/N^{1/2}$, we can rewrite the formula as follows:

$$t = \frac{\overline{X}}{\text{SEM}} = \frac{\overline{X}}{s \cdot N^{-1/2}} = \frac{\overline{X}\sqrt{N}}{s}$$

As an example, suppose we record the following data on mean arterial pressure (in mm Hg) before and after treatment, in ten hypertensive patients (identified by their initials, as is customary in clinical research reports):

Subject	Baseline	Final	Change
AB	120	117	−3
CD	123	129	+6
EF	124	119	−5
GH	124	120	−4
IJ	126	120	−6
KL	130	132	+2
MN	132	132	0
OP	134	133	−1
QR	137	140	+3
ST	141	136	−5

Each patient's pressure at baseline is compared to his or her final pressure. In this repeated-measures study, the patients serve as their own controls, and the data pairs are pretreatment versus post-treatment in each patient.

First, we compute the mean change in pressure. The algebraic sum of the ten signed-number differences is −13, so the mean change is −1.3 mm Hg. Next, we compute variance and standard deviation for the change data. The sum of the squared differences between each change and the mean change is approximately 144, and this total divided by $N-1=9$ gives the variance, 16. Standard deviation is the square root of the variance, 4. Now plug the mean change, the square root of the number of data, and the standard deviation into the t-test formula:

$$t = \frac{\overline{X}\sqrt{N}}{s} = \frac{1.3 \cdot 3.16}{4} \approx 1.03$$

(Although the observed mean difference in blood pressure was a decrease, we show $\overline{X} = 1.3$ without any sign, because we are just looking at the absolute magnitude of the change.) With a paired t-test, df $= N-1$; in this case, 9. Now, using the same t-table as in the unpaired test, if $t = 1.03$ and df $= 9$, then $p > 0.1$. The mean difference between the pairs is not statistically significant, so the Null Hypothesis stands. At least in this trial, the treatment under study showed no significant impact on blood pressure, and the observed decrease could well be attributed to chance.[17]

MEASURES OF CORRELATION

So far, we have looked at research studying the effects of an independent variable on a dependent variable. Another type of clinical research assesses the possibility of a *correlation* between two variables. For example, the rela-

---

[17]A single study seldom if ever proves anything conclusively. The results of one study may be contradicted by the results of a different study on the same topic. Scientific proof demands *reproducible* results—that is, consistently similar data obtained independently by several separate investigators.

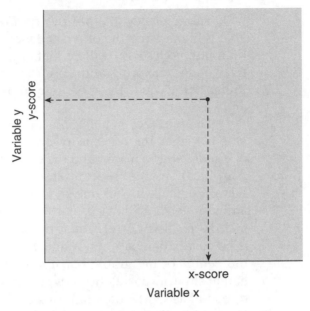

Figure 10.19 Two variables from the same subject.

tionship between smoking and lung cancer was established when a significant degree of correlation was noted between smoking history and risk of lung cancer.[18]

We use specific tools to assess the correlation between two variables, X and Y, in the same subjects. The information on both variables in each subject is used to plot a point on a coordinate graph (Figure 10.19). For each point, reading straight down to the horizontal axis (that is, the x-axis) gives the X score; reading across to the vertical axis (the y-axis) gives the Y score for that same individual. The collected set of data points representing each subject in the sample is called a *scattergram* (Figure 10.20).

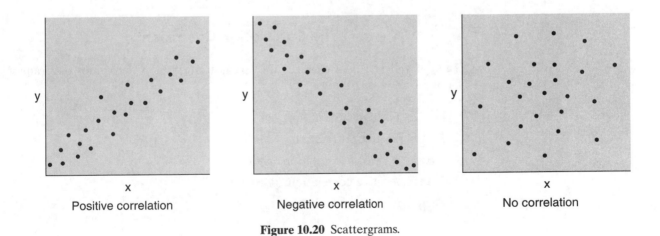

Figure 10.20 Scattergrams.

---

[18]A meaningful (consistent rather than coincidental) correlation between two variables, X and Y, does not prove that one causes the other. A correlation may be present in any of the following circumstances: X causes Y; Y causes X; or both X and Y are caused by some unsuspected third variable, Z. However, since no other factor causes smoking, the answer to the question "Does smoking cause lung cancer?" should be self-evident.

The tendency in a *positive* or *direct* correlation is that when one variable is higher, so is the other, and vice versa. In the same subject, more often than not, both scores will be higher or both will be lower than their respective means. Smoking and lung cancer show a positive correlation—other things being equal, the more one smokes, the higher the risk of lung cancer, and vice versa.

The tendency in a *negative* or *inverse* correlation is that when one variable is higher, the other is lower. In the same subject, more often than not, one score will be higher and the other will be lower than its respective mean. Aerobic fitness and heart disease show a negative correlation—other things being equal, the higher one's fitness level, the lower the risk of heart disease, and vice versa.

A negative correlation is not the same as a lack of correlation. If there is no correlation between X and Y, the data points may be clustered around the center of the graph, but not along any diagonal line. Intelligence and height show no correlation—a person with a high (or low) IQ is just as likely to be tall as short, and a tall (or short) person is just as likely to have a high IQ as a low IQ.

With a real positive or negative correlation, it is possible to draw a single diagonal line that best fits the set of data points.[19] A positive correlation is characterized by a line that rises as we move rightward on the graph (Y increases as X increases). A negative correlation is characterized by a line that falls as we move rightward (Y decreases as X increases).

To determine the correlation coefficient r, two common statistical methods are the *Pearson product-moment* (used when both variables are parametric) and the *Spearman rank test* (when either variable is nonparametric).

Here is the formula for the Pearson product-moment:

$$r = \frac{\sum [(X_i - \overline{X}) \cdot (Y_i - \overline{Y})]}{s_x \cdot s_y \cdot (N - 1)}$$

The Correlation Coefficient

The coefficient r characterizes the strength of correlation. Values of r range from +1 to −1, as follows:

Perfect positive correlation: $r = 1$

Partial positive correlation: $1 > r > 0$

Absolute lack of correlation: $r = 0$

Partial negative correlation: $0 > r > -1$

Perfect negative correlation: $r = -1$

---

[19]The correlation patterns described in this section are linear. A correlation in which the value of one variable is proportional to the square or higher power of the other is an example of nonlinearity. Also, some correlations are positive at low values of one of the variables, but negative at higher values, such as in the relationship between ventricular performance and preload. As discussed in Chapter 3, performance rises with increasing preload up to a certain point, but then begins falling with further increases in preload.

We'll deal first with the numerator. Obtain $\overline{X}$ and $\overline{Y}$ (the respective means of X and Y). Then, for each data point, subtract $\overline{X}$ from X and $\overline{Y}$ from Y, and multiply these two differences. Note that the order of subtraction counts, because we are not squaring the differences; that is, some of the differences will be positive and some will be negative. Also note that if X and Y are both greater or both less than their respective means, the product of the two differences for each data point will be positive; if one value is greater than its mean and the other is less than its mean, the product will be negative. The algebraic sum of all the positive and negative products—one product for each data point—is the numerator of the Pearson formula. The denominator is obtained by multiplying the product of the standard deviations for X and Y by 1 less than the number of data points.

As an example, let's see if there is any correlation between the variables in two of the data sets we have been using. Variable X is the test scores in Set A (mean, $\overline{X} = 62.3$). Variable Y is the set of men's and women's heights; but to avoid the confounding factor of gender-related height differences (which we saw in the unpaired t-test), we will now assume that these people are either all men or all women (mean, $\overline{Y} = 171$ cm). The Null Hypothesis is that there is no meaningful correlation between these two variables.

Conveniently, both sets contain the same number of data; so we will assume that the 11 people whose heights are recorded just happen to be the 11 students whose test scores form Set A. In no particular order, we can match up their test scores (X) and heights (Y) as follows:

Subject	X	$X - \overline{X}$	Y	$Y - \overline{Y}$	$(X - \overline{X}) \cdot (Y - \overline{Y})$
AB	70	+7.7	168	−3	−23.1
CD	12	−50.3	169	−2	+100.6
EF	64	+1.7	156	−15	−25.5
GH	66	+3.7	176	+5	+18.5
IJ	99	+36.7	181	+10	+367
KL	68	+5.7	160	−11	−62.7
MN	66	+3.7	186	+15	+55.5
OP	29	−33.3	175	+4	−133.2
QR	70	+7.7	178	+7	+53.9
ST	67	+4.7	173	+2	+9.4
UV	74	+11.7	155	−16	−187.2

For the numerator, the sum of the difference-products is approximately +173. For the denominator, the standard deviation (s) for Set A is already known—23.1. Now we must find s for the set of 11 heights; the variance is approximately 103 (work it out from the combined set of men's and women's heights), and s, the square root of the variance, is about 10.1. And $N = 11$. Therefore:

$$r = \frac{\sum [(X_i - \overline{X}) \cdot (Y_i - \overline{Y})]}{s_x \cdot s_y \cdot (N - 1)} \approx \frac{173}{23.1 \cdot 10.1 \cdot 10} \approx 0.074$$

The result is close to 0, indicating the lack of correlation. As with other measures, r must be interpreted on a table giving the probability that an observed correlation could be due to chance. In this case, suffice it to say that we are not even close to significance, and the Null Hypothesis stands. This result was predictable, because the notion of a correlation between height and test score is self-evidently preposterous.

The Spearman rank test can be used to assess correlation with ordinal data. The lowest score in a set containing N items is ranked 1; the highest, N. In case of tied ranks resulting from duplicate data, the rank for each of the duplicate scores is the average of their ordered position numbers. For example, if the third and fourth data in ascending order are the same, each would be ranked 3.5; if the seventh, eighth, and ninth data are the same, each would be ranked 8; and so on.

We can use the height and test-score data again; but here, the numeric values for each variable are given as ranks indicating their ordinal position within the numeric sequence. We take the difference, d, between the ranks of the two variables for each subject, and then square each difference, giving d^2. In this case, the positive and negative signs for the value of d are irrelevant, because each value is squared. Indeed, we could just as well say, "Square the absolute magnitude of the difference between the ranks for each subject." Here are the ordinal ranks of the same data we used in the Pearson product-moment example:

Subject	Score Rank	−	Height Rank	=	d	d^2
AB	8.5		4		+3.5	12.3
CD	1		5		−4	16
EF	3		2		+1	1
GH	4.5		8		−3.5	12.3
IJ	11		10		+1	1
KL	7		3		+4	16
MN	4.5		11		−6.5	42.3
OP	2		7		−5	25
QR	8.5		9		−0.5	0.25
ST	6		6		0	0
UV	10		1		+9	81

The Spearman rank coefficient is then calculated by the following formula:

$$r = 1 - \frac{6 \cdot \sum d^2}{N(N^2 - 1)}$$

The sum of the squared rank differences is multiplied by 6 and the product is then divided by $N(N^2 - 1)$, where N is the number of data pairs. The resulting quotient is subtracted from 1 to give the Spearman coefficient. In this example, the sum of the squared rank differences is about 207; $N = 11$. Therefore:

$$r = 1 - \frac{6 \cdot \sum d^2}{N(N^2 - 1)} = 1 - \frac{6 \cdot 207}{11(121 - 1)} = 1 - \frac{1242}{1320} \approx 1 - 0.941 = 0.059$$

This result is close to 0, the value associated with a total lack of correlation. As in the Pearson test, the probability table confirms that this value of r is not even close to being statistically significant.

Now let's try an experiment—once again working with the sets of test scores and heights, we deliberately arrange both sets in ascending order. As a result, subject AB now has the lowest score and is shortest, and subject UV has the highest score and is tallest. In other words, we are manipulating the data so as to fabricate a *positive* correlation (the shorter the height, the lower the test score, and vice versa). Again, we'll use the Spearman rank test to assess the strength of this artificially concocted correlation:

Subject	Test Score	Score Rank	Height	Height Rank
AB	12	1	155	1
CD	29	2	156	2
EF	64	3	160	3
GH	66	4.5	168	4
IJ	66	4.5	169	5
KL	67	6	173	6
MN	68	7	175	7
OP	70	8.5	176	8
QR	70	8.5	178	9
ST	74	10	181	10
UV	99	11	186	11

As before, we square the difference between the ranks for each subject:

Subject	Score Rank	−	Height Rank	=	d	d²
AB	1		1		0	0
CD	2		2		0	0
EF	3		3		0	0
GH	4.5		4		+0.5	0.3
IJ	4.5		5		−0.5	0.3
KL	6		6		0	0
MN	7		7		0	0
OP	8.5		8		+0.5	0.3
QR	8.5		9		−0.5	0.3
ST	10		10		0	0
UV	11		11		0	0

Because most of the differences are 0, the sum of the squared differences is small—only about 1.2. Therefore:

$$r = 1 - \frac{6 \cdot \sum d^2}{N(N^2 - 1)} = 1 - \frac{6 \cdot 1.2}{11(121 - 1)} = 1 - \frac{7.2}{1320} \approx 1 - 0.005 = 0.995$$

This result is very close to +1.00, the value associated with perfect positive correlation; not surprisingly, the probability table shows that this correlation is significant.

Finally, we manipulate these two sets yet again, this time arranging the data in *opposite* numeric sequence (test scores in ascending order, heights in descending order). By doing so, we fabricate a *negative* correlation (the shorter the height, the higher the test score, and vice versa):

Subject	Score Rank	−	Height Rank	=	d	d²
AB	1		11		−10	100
CD	2		10		−8	64
EF	3		9		−6	36
GH	4.5		8		−3.5	12.3
IJ	4.5		7		−2.5	6.3
KL	6		6		0	0
MN	7		5		+2	4
OP	8.5		4		+4.5	20.3
QR	8.5		3		+5.5	30.3
ST	10		2		+8	64
UV	11		1		+10	100

Now, because most of the differences are large, the sum of the squared differences is large—about 437. Therefore:

$$r = 1 - \frac{6 \cdot 437}{11(121 - 1)} = 1 - \frac{2622}{1320} \approx 1 - 1.99 = -0.99$$

This result is very close to −1.00, the value associated with perfect negative correlation; and again, the probability table shows that this correlation is significant.

The Quotient Subtracted from 1

In the formula for the Spearman rank test, the input data yield a fraction whose quotient is subtracted from 1 to give *r*, the correlation coefficient. Notice how that quotient determines the result.

Quotient	1 − Quotient	Interpretation
Close to 0	Close to +1	Positive correlation
Close to 1	Close to 0	Lack of correlation
Close to 2	Close to −1	Negative correlation

ERROR TYPES

Statistical analysis is based on probabilities, not certainties. Even if *p* is large, it is possible that an observed difference between two groups or a correlation between two variables is meaningful. Conversely, even if *p* is small, it is possible that an observed difference or correlation is random. We can never be absolutely sure. All we can do is apply the appropriate tests and then reject or fail to reject the Null Hypothesis according to the statistical "rules of evidence."

A *Type I error* is the false conclusion that an observed difference or correlation is meaningful when in fact it is random; that is, the Null Hypothesis is wrongly rejected. A *Type II error* is the false conclusion that an observed difference or correlation is random when in fact it is meaningful; that is, the Null Hypothesis should have been rejected. We can summarize these error types as follows:

| Observed | Null Hypothesis | |
Difference	*Rejected*	*Not rejected*
Meaningful	Correct	Type II error
Random	Type I error	Correct

The statistical likelihood of a Type I error is symbolized α (Greek alpha); the likelihood of a Type II error is symbolized β (beta). In general, α and β tend to vary inversely; that is, the greater the likelihood of a Type I error, the lower the likelihood of a Type II error, and vice versa.

Type I and II errors do not refer to computational blunders; α and β are inherent to the design of a study. Aside from the fact that research deals with probabilities based on samples rather than certainties based on whole populations, it is generally impossible to guarantee that all confounding factors are excluded.

QUIZ

Questions 1–3 are based on the following two sets of data on age at menarche from different populations of girls:

Set #1: 12.8, 14.0, 11.7, 11.9, 12.3, 13.5, 12.9, 12.8, 13.3

Set #2: 11.1, 10.8, 14.6, 12.3, 12.2, 13.0, 14.4, 11.9, 11.7, 13.1

Disregard the small size of these sets.

1. For each set, find the mean, median, and any mode(s) that may be present.

2. For each set, compute the standard deviation from the mean, the standard error, the coefficient of variation, and the z-score for the youngest age reported.

3. Apply an appropriate test to see if there is any significant difference between the two sets. Use this probability table excerpt to evaluate the test result:

| Value of p: | | | |
0.50	*0.10*	*0.05*	*0.01*
0.689	1.74	2.11	2.90

4. A study of serum cholesterol levels before and after treatment with a lipid-lowering drug shows the following data:

Number of subjects = 25
Pretreatment mean ± standard error = 243 ± 6 mg/dL
Posttreatment mean ± standard error = 190 ± 9 mg/dL

What is wrong with this report, and what should be done to correct it?

5. Within a defined population numbering 1,000,000 people, a total of 3200 new cases of a certain condition have been diagnosed over a period covering the past five years. Compute the annual incidence based on this information and give the result in scientific notation. What about the prevalence at a time halfway through that same five-year period?

6. Refer to the χ^2 table excerpt on page 226. What is the smallest statistically significant difference ($p < 0.05$) between the numbers of heads and tails in a trial consisting of 64 coin tosses?

7. What is the 95% confidence interval in a set of normally distributed data obtained from 53 subjects, if the mean is 76 ± 9.4?

8. Data are obtained from a population sample numbering 625 subjects; these data show a true normal distribution in which the combined mean, median, and mode is 100 ± 16. From this information, what value or score would have been obtained from a subject with a z-score of −3.00? Approximately how many subjects would have scores within the range of 84 to 132?

9. A researcher contacts ten psychiatrists in private practice in the community, requesting that they fill out a profile for each of the next ten patients they see. About half of the psychiatrists comply, and the researcher notes a higher divorce rate in patients diagnosed with depression than in patients with other diagnoses. The difference is statistically significant, and the researcher reports a correlation between divorce and depression. Is such a conclusion justified?

10. Recall that we illustrated the Pearson product-moment and the Spearman rank test with the 11 test scores and 11 heights randomly matched; we then used the Spearman test with the data artificially arranged so as to fabricate correlations. What happens when we use the artificially arranged data in the Pearson formula?

Answers and explanations at end of book.

Units Named for Scientists

Several concepts presented in this book are linked by name to the scientists and mathematicians who discovered or developed them—Einthoven's triangle, the Henderson-Hasselbalch equation, the Fick principle, Briggsian and Napierian logarithms, the Gaussian distribution, and others.

In addition, ten of the mathematical-physical units that appear in this book are named in honor of scientists who did pioneering work in the areas to which the units apply. It is worth acknowledging these individuals and reviewing the meaning of the units named for them:

Sir Isaac Newton (1642–1727) was an English mathematician and physicist, arguably the greatest scientist in history. Among his most famous contributions are the calculus, his theory of gravity, and the laws of force and motion. The *newton* (N) is a unit of *force*—the force that produces an acceleration of 1 meter per second, per second, on a 1-kilogram mass ($N = kg \cdot m \cdot sec^{-2}$).

Blaise Pascal (1623–1662) was a French religious philosopher, scientist, and mathematician. His far-ranging accomplishments include advances in the scientific understanding of the properties of fluids (Pascal's law is mentioned in Chapter 3) and in the measurement of pressure. The *pascal* (Pa) is a unit of *pressure* or *force per unit-area*—the pressure produced when a force of 1 newton is applied against an area of 1 square meter ($Pa = N \cdot m^{-2}$).

Evangelista Torricelli (1608–1647) was an Italian mathematician and physicist, secretary to Galileo, and then his successor as professor of mathematics and natural philosophy at the University of Florence. His main work was in optics and the measurement of pressure, and he is credited with inventing the earliest barometer. The *torr* is a unit of *pressure*—the pressure equal to one 760th part of atmospheric pressure (torr = atm/760). Since 1 atm ≈ 760 mm Hg, 1 torr = 1 Hg. Relating these units of pressure, 760 torr = 101,325 Pa; 1 Pa = 0.0075 mm Hg; and 1 mm Hg ≈ 133.3 Pa.

James Prescott Joule (1818–1889) was an English physicist. His main experimental work was in the areas of electricity and heat, and his discoveries were among the first in the emerging science of thermodynamics. The *joule* (J) is a unit of *work* or *energy*—the work done or energy expended when a force of 1 newton acts through a distance of 1 meter ($J = N \cdot m$).

James Watt (1736–1819) was a Scottish inventor. He is most famous for his role in the development of the steam engine. The *watt* is a unit of *power* or *work per unit-time*—the power equivalent to work done at a rate of 1 joule per second (watt = $J \cdot sec^{-1}$; J = watt · sec).

André Marie Ampère (1775–1836) was a French mathematician and physicist whose research focused on electricity and magnetism. The *ampere* (amp) is a unit of electrical *current* (flow of *charge;* see the following paragraph)—the current in parallel wires 1 meter apart such that the force per meter between them is $2 \cdot 10^{-7}$ N. More often, current is expressed as the rate at which electric charge is transferred along a wire.

Charles Augustin de Coulomb (1736–1806) was a French physicist who retired from his military career as an engineer to do research in electricity and magnetism. The *coulomb* (coul) is a unit of electric charge—the charge transferred in 1 second by a current of 1 amp (coul = amp $\cdot$ sec; amp = coul $\cdot$ sec^{-1}).

Michael Faraday (1791–1867) was a largely self-taught British scientist and inventor whose practical research in electricity, magnetism, and chemistry established some of the most fundamental laws of physical science. The *faraday* (F) is an electric-charge constant—the charge contained in 1 mole of electrons ($6.02 \cdot 10^{23}$ electrons). The faraday is almost five orders of magnitude greater than the coulomb: 1 F = 96,500 coul; coul = 1/96,500 F.

Count Alessandro Volta (1745–1827) was an Italian physicist and statesman, famous for his work in electricity. The *volt* (V) is a unit of electrical potential (also called *electromotive force* or *voltage*)—the gradient or difference in electrical potential between the ends of a conductor when the current in a circuit is 1 amp and the power dissipated is 1 watt (V = watt $\cdot$ amp^{-1}; watt = V $\cdot$ amp). Transmembrane electrical potential is measured in millivolts (mV = 10^{-3} V).

William Thomson, Lord Kelvin (1824–1907) was a British physicist and mathematician. As part of his research in thermodynamics, he introduced the absolute centigrade scale of temperature. The *kelvin* (K) is a unit of *temperature*—equal in magnitude to the Celsius degree, but numerically 273.15 above the corresponding Celsius temperature. Absolute zero is 0 K (equal to −273.15°C); the freezing point of water is 273.15 K (0°C); and the boiling point of water is 373.15 K (100°C). Since the kelvin *is* the unit of temperature, the word "degree" and its symbol are not used; a temperature of 300 on the Kelvin scale is written "300 K" and spoken as "three hundred kelvin."

appendix:

Basic Mathematical Concepts

This book presents a variety of mathematical tools applicable to clinical science. While a general knowledge of simple algebra and arithmetic operations involving integers and fractions is assumed, some students may benefit from a review of basic mathematical concepts and methods, as provided in the following sections:

- A. Significant Digits
- B. Scientific Notation
- C. Negative and Fractional Exponents
- D. Dimensional Units with Negative Exponents
- E. Unit Conversions
- F. Logarithms and Logarithmic Scales
- G. Order of Magnitude
- H. Quadratic Equations
- I. Vectors

This sequence forms a progressive, self-contained teaching unit; it does not follow the precise order in which these concepts appear in the text.

A. Significant Digits

Suppose we are measuring a box. Using a ruler calibrated in millimeters, we record its dimensions: length, 30.6 cm (that is, 306 mm); width, 24.5 cm (245 mm); height, 18.2 cm (182 mm). Since the volume of a rectilinear solid is the product of length times width times height:

$$\text{Volume} = 306 \text{ mm} \cdot 245 \text{ mm} \cdot 182 \text{ mm} = 13{,}644{,}540 \text{ mm}^3$$

That would be the answer *if* the dimensions were exact. In fact, they are precise only to the nearest millimeter. With a ruler calibrated in tenths of a millimeter, that 306-mm length could be anywhere from 305.5 to 306.4 mm; the width, anywhere from 244.5 to 245.4 mm; and the height, anywhere from 181.5 to 182.4 mm. Taking the lowest and highest of these values for each dimension:

$$\text{Least volume} = 305.5 \text{ mm} \cdot 244.5 \text{ mm} \cdot 181.5 \text{ mm} \approx 13{,}557{,}097 \text{ mm}^3$$

$$\text{Greatest volume} = 306.4 \text{ mm} \cdot 245.4 \text{ mm} \cdot 182.4 \text{ mm} \approx 13{,}714{,}758 \text{ mm}^3$$

The maximum result is more than 157,000 mm^3 greater than the minimum. In turn, still more precise results could have been obtained using a ruler calibrated in hundredths of a millimeter, in micrometers, and so on.

The limit of precision in taking measurements sets a limit to the number of digits we can use in computations. *We are allowed only as many digits as the number of digits in the given information.* This number of *significant digits* tells us where to round off the answer. The decimal point position does not matter—45, 4.5, and 0.45 are all regarded as having two digits. (On the other hand, 450 is an exact three-digit number.)

We approximated the dimensions of the box as a set of three-digit values in millimeters, so our volume computation can contain only three digits:

$$13{,}644{,}540 \text{ mm}^3 \approx 13{,}600{,}000 \text{ mm}^3$$

The first answer looks precise but it is misleading, for we have certainty only to the nearest hundred thousand. Similarly, in cubic centimeters, the three-digit answer is:

$$13{,}644.54 \text{ cm}^3 \approx 13{,}600 \text{ cm}^3$$

(Clinical examples appear throughout the remainder of the Appendix.)

B. Scientific Notation

In all areas of science, very large and very small numbers are often expressed in *scientific notation.* In this format, the quantity is rounded off to the allowed number of significant digits (Appendix A), and then multiplied by an integer (whole number) power of 10 to avoid a long string of zeros. (We'll look at large numbers here; the scientific notation of very small numbers is shown in Appendix C.)

Any number that is written as 1 followed by a string of zeros can also be expressed as the base 10 raised to a power equal to the number of zeros following the 1. Since 100 has two zeros after the 1, it can be expressed as 10^2; and indeed, $10^2 = 10 \cdot 10 = 100$. Likewise:

$$1{,}000 = 10 \cdot 10 \cdot 10 = 10^3$$

$$10{,}000 = 10^4$$

$$100{,}000 = 10^5, \text{ etc.}$$

To express a computed quantity in scientific notation, the first step is to round it off to the allowed number of significant digits. In Appendix A, the dimensions of a box were given in three-digit measurements, so the calculated volume was limited to three digits (in that example, 136). The volume in cubic millimeters was rounded off from 13,644,540 mm³ to 13,600,000 mm³, showing that we had certainty only to the nearest 100,000 cubic millimeters:

$$13{,}600{,}000 \text{ mm}^3 = 136 \cdot 100{,}000 \text{ mm}^3$$

Now place a decimal point after the first of the three digits, giving 1.36, which means we have divided this factor by 100. To preserve the value of the product, we must multiply the other factor by 100:

$$136 \cdot 100{,}000 \text{ mm}^3 = 1.36 \cdot 10{,}000{,}000 \text{ mm}^3$$

Finally, instead of writing out all those zeros:

$$1.36 \cdot 10{,}000{,}000 \text{ mm}^3 = 1.36 \cdot 10^7 \text{ mm}^3$$

In this example, 1.36 is the numeric factor and 10^7 is the exponential factor (and mm³ is the dimensional unit). Similarly, to the nearest 100 cubic centimeters:

$$13{,}600 \text{ cm}^3 = 136 \cdot 100 \text{ cm}^3$$

$$136 \cdot 100 \text{ cm}^3 = 1.36 \cdot 10{,}000 \text{ cm}^3 = 1.36 \cdot 10^4 \text{ cm}^3$$

This format is scientific notation—a quantity is rounded off to the allowed number of significant digits, the first of which goes to the left of the decimal point (so that the numeric factor is at least 1 but less than 10), and this factor is then multiplied by the appropriate power of 10. For any numeric factor X, the format is $X \cdot 10^n$, where $1 \le X < 10$ and the exponent n is an integer.

In computations involving quantities expressed in scientific notation, *the number of digits in the numeric factors gives the number of significant digits allowed in the answer.*

C. Negative and Fractional Exponents

A negative exponent indicates a reciprocal, whereby the same quantity (with a positive exponent) forms the denominator of a fraction whose numerator is 1:

$$10^2 = 100$$

$$10^{-2} = \frac{1}{10^2} = \frac{1}{100}$$

It is perfectly logical. Consider how we use exponents in multiplication when both factors are powers of the same base number—all we do is add the exponents:

$$X^a \cdot X^b = X^{a+b}$$

For example:

$$3^2 \cdot 3^3 = 3^{2+3} = 3^5$$

$$(3 \cdot 3) \cdot (3 \cdot 3 \cdot 3) = (3 \cdot 3 \cdot 3 \cdot 3 \cdot 3)$$

$$9 \cdot 27 = 243$$

Or:

$$10^2 \cdot 10^4 = 10^{2+4} = 10^6$$

$$(10 \cdot 10) \cdot (10 \cdot 10 \cdot 10 \cdot 10) = (10 \cdot 10 \cdot 10 \cdot 10 \cdot 10 \cdot 10)$$

$$100 \cdot 10,000 = 1,000,000$$

Note that *any* quantity to the first power is just itself: $X^1 = X$. Therefore, when we say that the square of 5 is 25, what it really means is:

$$5^1 \cdot 5^1 = 5^{1+1} = 5^2 = 25$$

And we can see why fractional exponents indicate radicals. The exponent 1/2 indicates a square root, 1/3 indicates a cube root, and so on:

$$\sqrt{5} = 5^{1/2} \; because \; 5^{1/2} \cdot 5^{1/2} = 5^{1/2+1/2} = 5^1 = 5$$

$$\sqrt[3]{5} = 5^{1/3} \; because \; 5^{1/3} \cdot 5^{1/3} \cdot 5^{1/3} = 5^{1/3+1/3+1/3} = 5^1 = 5, \text{etc.}$$

In division involving powers of the same base number, we subtract the exponent of the denominator from the exponent of the numerator:

$$\frac{X^a}{X^b} = X^{a-b}$$

For example:

$$\frac{2^6}{2^4} = 2^{6-4} = 2^2$$

$$\frac{64}{16} = 4$$

Or:

$$\frac{10^5}{10^2} = 10^{5-2} = 10^3$$

$$\frac{100,000}{100} = 1000$$

It works the same way when the exponent of the denominator is larger than that of the numerator:

$$\frac{10^1}{10^2} = 10^{1-2} = 10^{-1}$$

Now also note that:

$$\frac{10^1}{10^2} = \frac{10}{100} = \frac{1}{10} = 0.1$$

So $10^1/10^2 = 10^{-1}$; and also, $10^1/10^2 = 1/10$. Then because things equal to the same thing are equal to each other, $10^{-1} = 1/10$. Likewise:

$$\frac{10^1}{10^3} = 10^{1-3} = 10^{-2} = \frac{1}{10^2} = \frac{1}{100} = 0.01$$

$$\frac{10^1}{10^4} = 10^{1-4} = 10^{-3} = \frac{1}{10^3} = \frac{1}{1000} = 0.001$$

$$\frac{10^1}{10^5} = 10^{1-5} = 10^{-4} = \frac{1}{10^4} = \frac{1}{10,000} = 0.0001, \text{ etc.}$$

Here is another way to think about it:

$$\frac{X^5}{X^3} = X^2 \quad \text{and} \quad \frac{X^3}{X^5} = X^{-2}$$

Since X^5/X^3 and X^3/X^5 are reciprocals of each other, X^2 and X^{-2} are also reciprocals of each other. Therefore, we can generalize:

$$X^{-n} = \frac{1}{X^n} \quad \text{and} \quad \frac{1}{X^{-n}} = X^n$$

Incidentally, it should be clear why any nonzero number raised to the power of 0 equals 1. By definition, any nonzero quantity divided by itself is 1:

$$\frac{X}{X} = 1 \quad \text{or, more generally,} \quad \frac{X^n}{X^n} = 1$$

So when we subtract the exponent of the denominator from that of the numerator:

$$\frac{X^n}{X^n} = X^{n-n} = X^0$$

$X^n/X^n = 1$; and also, $X^n/X^n = X^0$. Again, because things equal to the same thing are equal to each other, $X^0 = 1$, and this is true for *any* real number $X \neq 0$.

Very small quantities are expressed in scientific notation (Appendix B) in this manner, with a negative exponent. Suppose that a calculation gives an answer of 0.0000000368241756. If we have three significant digits, we round off the calculation to 0.0000000368 (three hundred sixty-eight ten-billionths):

$$0.0000000368 = \frac{368}{10,000,000,000} = \frac{368}{10^{10}}$$

Placing a decimal point after the first digit of the numerator (so that our answer will be in scientific notation form) means dividing the numerator by 100. Then to preserve the value of the quotient, we must also divide the denominator by 100:

$$\frac{368}{10^{10}} = \frac{3.68}{10^8}$$

And finally:

$$\frac{3.68}{10^8} = 3.68 \cdot \frac{1}{10^8} = 3.68 \cdot 10^{-8}$$

D. Dimensional Units with Negative Exponents

The concept that a negative exponent defines a reciprocal (Appendix C) applies to dimensional units as well as to pure numbers. For example, if a drug is to be given at a dosage of 500 milligrams per day, the mathematical meaning of "per day" is that the time unit is the denominator of a fraction—that is, 500 mg/day. Therefore, we can also express this dosage as the product of the numerator and the reciprocal of the denominator, as indicated by a negative exponent:

$$500 \, \frac{\text{mg}}{\text{day}} = 500 \, \text{mg} \cdot \frac{1}{\text{day}} = 500 \, \text{mg} \cdot \text{day}^{-1}$$

Dosages for children are most often based on body weight—a certain number of milligrams of drug per kilogram of weight, per day (often in divided doses). For example, the formula might be 5 milligrams per kilogram, per day (usually written confusingly as "5 mg/kg/day"). This expression is really a *complex fraction*—a fraction with a fractional numerator and/or denominator:

$$5 \text{ mg/kg/day} = \frac{5 \dfrac{\text{mg}}{\text{kg}}}{\text{day}}$$

The dosage-by-weight is 5 mg/kg (the fractional numerator), and this amount is to be given "per day" (the denominator). Alternate expressions of the same dosage all have the same meaning:

$$\frac{5 \dfrac{\text{mg}}{\text{kg}}}{\text{day}} = \frac{5 \text{ mg} \cdot \text{kg}^{-1}}{\text{day}} = \frac{5 \text{ mg}}{\text{kg} \cdot \text{day}} = 5 \text{ mg} \cdot \text{kg}^{-1} \cdot \text{day}^{-1}$$

For a child who weighs 27 kg, the dosage would be:

$$\frac{5 \text{ mg} \cdot \text{kg}^{-1}}{\text{day}} \cdot 27 \text{ kg} = 135 \text{ mg} \cdot \text{day}^{-1}$$

Similarly, chemotherapeutic agents for cancer patients are usually given in doses based on body surface area (BSA; see Chapter 1). For example, the formula might be 700 milligrams per square meter, per day:

$$\frac{700 \dfrac{\text{mg}}{\text{m}^2}}{\text{day}} = \frac{700 \text{ mg} \cdot \text{m}^{-2}}{\text{day}} = \frac{700 \text{ mg}}{\text{m}^2 \cdot \text{day}} = 700 \text{ mg} \cdot \text{m}^{-2} \cdot \text{day}^{-1}$$

For a patient with a standard BSA of 1.73 m², the dosage would be:

$$\frac{700 \text{ mg} \cdot \text{m}^{-2}}{\text{day}} \cdot 1.73 \text{ m}^2 \approx 1210 \text{ mg} \cdot \text{day}^{-1}$$

The major advantage of this use of negative exponents on dimensional units is that it lets us avoid complex fractions; in fact, we could avoid fractions altogether. Some scholarly scientific textbooks and journals favor this style, expressing all fractions and ratios as products with appropriate positive and negative exponents on the numeric and dimensional factors.

E. Unit Conversions

A logical way to convert measurements or values from one system of units to another is to employ *conversion factors*.

In Chapter 1, we see that 1 inch equals 2.54 centimeters. Since these measures are equal, we can place either of them into the numerator of a fraction and the other into the denominator, thus forming a pair of mutually reciprocal fractions, equal to each other and both equal to 1:

$$\frac{2.54 \text{ cm}}{\text{inch}} = \frac{\text{inch}}{2.54 \text{ cm}} = 1$$

Similarly, 1 foot equals 12 inches:

$$\frac{12 \text{ inch}}{\text{ft}} = \frac{\text{ft}}{12 \text{ inch}} = 1$$

These reciprocal fractions are conversion factors. *Because conversion factors are equal to 1, we can multiply a given measurement by a conversion factor—in either of its reciprocal forms—without altering the value of the measurement.*

Set up an equation with the given information on the left side and the desired dimensional units on the right. On the left, place the necessary conversion factors in their appropriate reciprocal forms so that the undesired units (the units we are converting *from*) each appear exactly once in a numerator (any numerator) and once in any denominator. These undesired units will then cancel out during multiplication, leaving only the desired units (the units we are converting *to*).

For example, in converting a height of exactly 6 feet to centimeters:

$$6 \text{ ft} \cdot \frac{12 \text{ inch}}{\text{ft}} \cdot \frac{2.54 \text{ cm}}{\text{inch}} = \underline{\hspace{2em}} \text{ cm}$$

The undesired dimensional unit *feet* is in the given information (which is the numerator of a fraction whose denominator is 1) and in the denominator of the first conversion factor; *inch* appears in the numerator of the first conversion factor and the denominator of the second. These units will therefore drop out. The desired unit, *centimeter,* appears only in a numerator (in the second conversion factor) and therefore remains in the answer as we perform the computation:

$$6 \text{ ft} \cdot \frac{12 \text{ inch}}{\text{ft}} \cdot \frac{2.54 \text{ cm}}{\text{inch}} = 6 \cdot 12 \cdot 2.54 \text{ cm} \approx 183 \text{ cm}$$

Note that the answer has three digits (Appendix A). The given information was "exactly 6 feet," which means exactly 72 inches; and the three-digit conversion factor, 2.54 cm/inch, is also exact. The $\approx$ sign ("is approximately equal to") is used because the numeric computation is rounded off to the allowed three digits.

A conversion factor may also be written as the product of the numerator and the reciprocal of the denominator, by using negative exponents (Appendix D).

$$\frac{2.54 \text{ cm}}{\text{inch}} = 2.54 \text{ cm} \cdot \text{inch}^{-1}$$

$$\frac{\text{inch}}{2.54 \text{ cm}} = \text{inch} \cdot (2.54 \text{ cm})^{-1} = \text{inch} \cdot 2.54^{-1} \cdot \text{cm}^{-1}$$

(Note that 2.54 cm = 2.54 · cm; thus, when the denominator has a numeric factor *and* a dimensional unit, as in the second case, *both* take a negative exponent.) Arrange the conversion factors so that each undesired unit *u* appears once with a positive exponent and once with a negative exponent of the same magnitude:

$$u^n \cdot u^{-n} = u^n \cdot \frac{1}{u^n} = \frac{u^n}{u^n} = 1$$

And so, the unwanted units drop out of the equation. As an example, we'll start with a height given as 178 cm and convert to inches:

$$178 \text{ cm} \cdot (\text{inch} \cdot 2.54^{-1} \cdot \text{cm}^{-1}) = \frac{178 \text{ inch}}{2.54} \approx 70.1 \text{ inch}$$

Conversion factors may be exact definitions (such as 12 inch/ft), or close approximations (such as 2.2 lb/kg). Both types are handled the same way. With approximations, use the ≈ sign for the answer, even if no rounding off is needed and every other element in the conversion is exact.

A routine clinical application is the conversion of a laboratory value from one system of units to another. Suppose that the concentration of some substance in the serum is 1.2 mg/L, and we want to see what this value would be in units of μg/dL. We use two conversion factors—one for the weight units (from mg to μg) and the other for the volume units (from L to dL):

$$1.2 \frac{\text{mg}}{\text{L}} \cdot \frac{1000 \ \mu\text{g}}{\text{mg}} \cdot \frac{\text{L}}{10 \ \text{dL}} = \frac{1.2 \cdot 1000 \ \mu\text{g}}{10 \ \text{dL}} = 120 \ \mu\text{g} \cdot \text{dL}^{-1}$$

As a final example, let's convert a serum glucose reading of 129 milligrams per deciliter (weight concentration) to the equivalent value in millimoles per liter (particle-count concentration). This conversion illustrates how we go from conventional units to SI units, as discussed in Chapter 1:

$$129 \frac{\text{mg}}{\text{dL}} \cdot \frac{\text{mmol}}{180 \ \text{mg}} \cdot \frac{10 \ \text{dL}}{\text{L}} = \frac{129 \cdot 10 \ \text{mmol}}{180 \ \text{L}} \approx 7.17 \ \text{mmol} \cdot \text{L}^{-1}$$

The first conversion factor is based on the formula weight for glucose: 1 mole of glucose = 180 grams, so 1 millimole = 180 milligrams. The second conversion factor is an exact definition: 10 deciliters = 1 liter. By choosing the appropriate reciprocal of each conversion factor, the undesired units cancel out.

F. Logarithms and Logarithmic Scales

A *logarithm* is an exponent—specifically, it is the power to which a certain base number must be raised in order to reach another number, called the *antilogarithm.*

In "common logarithms" (abbreviated log or $\log_{10}$), the base number is 10. By raising the number 10 to various powers, we can reach any given positive antilogarithm (antilog).[1] In other words, for any real number $x > 0$, there exists some real number n such that $10^n = x$. The exponent n is the logarithm of x (written "log x"). If $10^n = x$, log $x = n$.

For example, $10^2 = 100$; therefore, log $100 = 2$. Similarly:

$$\text{Log } 1000 = 3 \text{ } because \text{ } 10^3 = 1000$$

$$\text{Log } 10,000 = 4 \text{ } because \text{ } 10^4 = 10,000$$

$$\text{Log } 100,000 = 5 \text{ } because \text{ } 10^5 = 100,000, \text{ etc.}$$

Going in the other direction, we know that any number whose exponent is 1 equals itself, any number whose exponent is 0 equals 1, and any number whose exponent is negative is the reciprocal of the same number with a positive exponent of the same magnitude. Therefore:

$$\text{Log } 10 = 1 \text{ } because \text{ } 10^1 = 10$$

$$\text{Log } 1 = 0 \text{ } because \text{ } 10^0 = 1$$

$$\text{Log } 0.1 = -1 \text{ } because \text{ } 10^{-1} = 0.1$$

$$\text{Log } 0.01 = -2 \text{ } because \text{ } 10^{-2} = 0.01$$

$$\text{Log } 0.001 = -3 \text{ } because \text{ } 10^{-3} = 0.001, \text{ etc.}$$

For any antilogarithm that is *not* an exact integer power of 10, the logarithm will be an *irrational* number.[2] For example, 6841 is between 1000 and 10,000 (that is, between 10^3 and 10^4); therefore, its logarithm is an irrational number between 3 and 4. And indeed:

$$\text{Log } 6841 = 3.83511959\ldots$$

---

[1]Base-10 logarithms are also called Briggsian logarithms, for British mathematician Henry Briggs (1561–1630). "Natural logarithms" work the same way as common logarithms, but their base number is the transcendental constant e, whose value ($2.71828182\ldots$) is obtained from $(1 + x^{-1})^x$ as x increases without limit. Natural logarithms are also called Napierian logarithms, for Scottish mathematician John Napier (1550–1617), who invented the system and whose work was edited for publication by Briggs. Natural logarithms appear in equations relating to transmembrane voltage (see Chapter 2) and pharmacokinetics (see Chapter 7).

[2]A *rational* number represents the *ratio* of two integers, which means that it can be expressed as a fraction. The quotient is a decimal that either terminates in zeros ($1/4 = 0.25000\ldots = 0.25$) or continues as an endlessly repeating digit ($13/6 = 2.1666\ldots$) or sequence of digits ($7/11 = 0.636363\ldots$). An *irrational* number does not represent any ratio of integers and therefore cannot be expressed as a fraction; it appears as an endless decimal that never displays any repeating pattern. Examples: log $17 = 1.2304489\ldots$; $\pi = 3.1415926\ldots$; $\sqrt{2} = 1.4142135\ldots$.

Appendix: Basic Mathematical Concepts

Antilogarithms That Are *Not* Integer Powers of 10

If a Number Is Between:	Its Logarithm Is Between:
0.00001 and 0.0001	−5 and −4
0.0001 and 0.001	−4 and −3
0.001 and 0.01	−3 and −2
0.01 and 0.1	−2 and −1
0.1 and 1	−1 and 0
1 and 10	0 and 1
10 and 100	1 and 2
100 and 1000	2 and 3
1000 and 10,000	3 and 4
10,000 and 100,000	4 and 5

And so on, in both directions.

Which means (limiting the endless decimal to four places):

$$10^{3.8351} \approx 6841$$

Likewise, 0.0591 is between 0.01 and 0.1, so the logarithm is between −2 and −1:

$$\text{Log } 0.0591 = -1.2284125\ldots$$

$$10^{-1.2284} \approx 0.0591$$

If we need the quotient of 6841 divided by 0.0591, we could do the division:

$$\frac{6841}{0.0591} \approx 115{,}753$$

Or we could use the rounded-off logarithms to get a rougher approximation:

$$\frac{10^{3.8351}}{10^{-1.2284}} = 10^{3.8351-(-1.2284)} = 10^{3.8351+1.2284} = 10^{5.0635}$$

$$10^{5.0635} \approx 115{,}744$$

The more decimal places we use in the logarithms, the more accurate the result will be. With just one additional decimal place:

$$10^{3.83512+1.22841} = 10^{5.06353} \approx 115{,}752$$

Positive and Negative Logarithms

A positive logarithm indicates that the antilogarithm is a number greater than 1. A negative logarithm indicates that the antilogarithm is a number *between* 0 and 1; that is, greater than 0 but less than 1. Note that when the logarithm is negative, the antilogarithm is still a positive number. There is no logarithm for zero or for any negative number.

Summarizing, for any real number x:

- If $x > 1$, $\log x > 0$
- If $1 > x > 0$, $\log x < 0$
- If $x = 1$, $\log x = 0$
- If $x \leq 0$, there is no logarithm

If you use a calculator to find the logarithm of a number between 0 and 1, you will automatically get the correct negative logarithm. Printed tables give only the *mantissa*—the portion of the logarithm following the decimal point. The mantissa is the same for any given sequence of digits in the antilogarithm, regardless of the decimal point position. For example, the mantissa for 5670, 567.0, 56.7, 5.67, 0.567, 0.0567, and so on, is (rounded off to four digits) 7536.

However, the *characteristic*—the portion of the logarithm to the left of the decimal point—is determined by the decimal point position in the antilogarithm. The characteristic equals the number of places the decimal point would have to move to follow the first digit of the antilogarithm; the sign is positive if the decimal point moves leftward, negative if it moves rightward. For an antilog such as 567, the decimal point would move *two* places to the *left* (yielding 5.67), so the characteristic is +2. For 0.000567, the decimal point would move *four* places to the *right* (again yielding 5.67), so the characteristic is −4. And for 5.67, the decimal point does not move at all, so the characteristic is 0.

But we don't just stick the characteristic in front of the mantissa; rather, we *add* the signed-number value of the characteristic to the value of the mantissa:

$$\text{Log } 567 = 0.7536 + 2 = 2.7536$$

$$\text{Log } 56.7 = 0.7536 + 1 = 1.7536$$

$$\text{Log } 5.67 = 0.7536 + 0 = 0.7536$$

This is where it gets tricky. For the next item (log 0.567), the decimal point moves one place to the right, so the characteristic is −1; but log 0.567 is *not* −1.7536. Again, we take the *algebraic sum* of the characteristic and the mantissa:

$$\text{Log } 0.567 = 0.7536 + (-1) = 0.7536 - 1 = -0.2464$$

$$\text{Log } 0.0567 = 0.7536 + (-2) = 0.7536 - 2 = -1.2464$$

$$\text{Log } 0.00567 = 0.7536 + (-3) = 0.7536 - 3 = -2.2464, \text{ etc.}$$

Working with Logarithms

Multiplication becomes addition and division becomes subtraction when working with logarithms:

$$\text{If } A \cdot B = C, \text{ then } \log (A \cdot B) = \log A + \log B = \log C$$

$$\text{If } A/B = C, \text{ then } \log (A/B) = \log A - \log B = \log C$$

The negative logarithm of a number is the logarithm of its reciprocal (but do not confuse $-\log x$ with $\log x < 0$, which occurs when $1 > x > 0$):

$$-\log x = \log (1/x)$$

The natural logarithm (ln) of $10 \approx 2.3026$, and $\log 10 = 1.0000$; therefore, the common and natural logarithms for $x > 0$ are related as follows:

$$\ln x \approx 2.3026 \cdot \log x$$

$$\log x \approx 0.4343 \cdot \ln x$$

If you plot these points on the Number Line, you will see that they are equidistant, a unit apart, even as they cross the zero point between positive and negative:

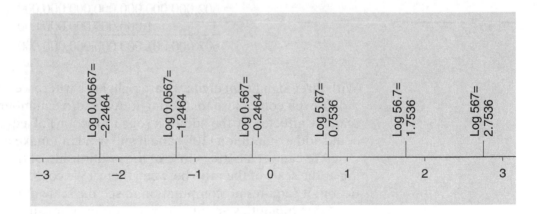

A logarithmic scale is one in which the intervals are calibrated not in unit quantities but in logarithmic powers. For certain clinical measurements, it is useful to employ a "semilog graph," with a logarithmic scale on the vertical axis and a normal numeric interval scale on the horizontal axis.

In Chapter 7, there are graphs showing the decline in plasma drug levels over time. Here are two graphs representing the same decline:

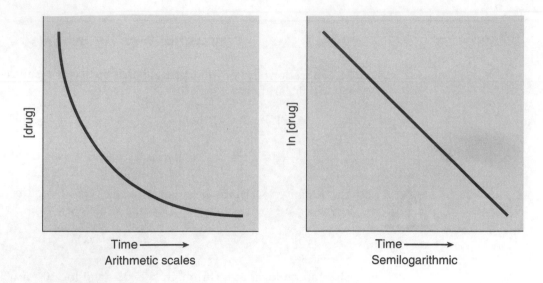

Time ——→
Arithmetic scales

Time ——→
Semilogarithmic

On the ordinary graph, the decline trails off as a gradually flattening curve; on the semilog graph (with the natural logarithm of drug concentration on the vertical scale), the decline appears as a straight line.

G. Order of Magnitude

When two quantities differ vastly in size, it may be convenient to compare them by *order of magnitude*. The comparison is expressed on a base-10 logarithmic scale (Appendix F) rather than as an arithmetic difference.

In Chapter 2, we look at Avogadro's Number, whose three-digit scientific notation is $6.02 \cdot 10^{23}$. The qualifying comment, " . . . plus or minus a few trillion," is not an exaggeration. Take Avogadro's Number and add a quadrillion (10^{15}):

$$
\begin{array}{r}
602{,}000{,}000{,}000{,}000{,}000{,}000{,}000 \\
+ \quad 1{,}000{,}000{,}000{,}000{,}000 \\
\hline
602{,}000{,}001{,}000{,}000{,}000{,}000{,}000
\end{array}
$$

With three significant digits, we are right back where we started, $6.02 \cdot 10^{23}$. We didn't even come close to altering it. Avogadro's number is so huge that it is scarcely affected by the addition (or subtraction) of a quadrillion. In fact, we could add a quintillion (10^{18}) and it still wouldn't make a notable difference.

We can get a better sense of the relationship of these numbers by looking at the scale of the ratio between them. (We could determine the exact ratio, but it requires no computation to see the scale of the ratio.)

Two quantities are of "the same order of magnitude" if the larger, L, is less than 10 times the smaller, S. At a scale of "one order of magnitude," L is at least 10 times but less than 100 times S. At "two orders of magnitude," L is at least 100 times but less than 1000 times S; and so on:

Same order of magnitude: $L < 10S$

One order of magnitude: $10S \leq L < 100S$

Two orders of magnitude: $100S \leq L < 1000S$, etc.

With both quantities expressed in proper scientific notation (Appendix B), create a fraction with the larger quantity in the numerator and the smaller quantity in the denominator. The larger quantity is that which has the larger exponential factor; if the exponential factors are the same, the one that has the larger numeric factor. The ratio of these quantities—that is, the quotient of the fraction we have just created—determines the scale. In the following examples, we'll compute the ratios; but you can see the order of magnitude just by looking at the quantities.

If the numerator and denominator have the *same exponential factor,* the quantities are of the same order of magnitude, because their ratio will always be more than 1 and less than 10. For example:

$$\frac{7.0 \cdot 10^8}{1.3 \cdot 10^8} \approx 5.4 \qquad \frac{8.59 \cdot 10^{14}}{6.68 \cdot 10^{14}} \approx 1.29$$

If the numerator has a *higher numeric factor* than the denominator, and the difference between the powers of the exponential factors is n, the ratio is between 10^n and 10^{n+1}, and the scale is n orders of magnitude. For example:

$$\frac{6.332 \cdot 10^{10}}{4.207 \cdot 10^6} \approx 1.505 \cdot 10^4 \approx 15,050$$

The difference between the powers is $10 - 6 = 4$, the ratio is between 10^4 and 10^5, and so the scale is four orders of magnitude. We can also say that the scale "exceeds" four orders of magnitude.

If the numerator has a *lower numeric factor* than the denominator, and the difference between the powers of the exponential factors is n, the ratio is between 10^{n-1} and 10^n, and the scale is $n - 1$ orders of magnitude:

$$\frac{4.04 \cdot 10^{11}}{7.38 \cdot 10^4} \approx 0.547 \cdot 10^7 = 5.47 \cdot 10^6$$

The difference between the powers is $11 - 4 = 7$, the ratio is between 10^6 and 10^7, and so the scale is (or exceeds) six orders of magnitude.

The numeric factors determine whether the scale is or is not an exact integer power of 10. However, the larger the ratio, the less we care about this distinction. For example, comparing Avogadro's number and 1 quadrillion:

$$\frac{6.02 \cdot 10^{23}}{10^{15}} = 6.02 \cdot 10^8$$

The numerator is more than 600 million times greater than the denominator. At such a huge ratio, it hardly matters whether we say, "The scale *is* eight orders of magnitude" or "The scale *exceeds* eight orders of magnitude."

H. Quadratic Equations

A *quadratic equation* (also called a second-degree equation) is one in which the unknown, x, is squared (raised to the second power); x may or may not also appear as a first-power (or *linear*) term.

The general format of a quadratic is:

$$ax^2 + bx + c = 0$$

The coefficients a and b, and the constant c are real numbers; $a \neq 0$. If $b = 0$, there is no first-power term, and the solutions are simply the positive and negative square roots of $(-c/a)$:

$$ax^2 + 0x + c = 0$$

$$ax^2 + c = 0$$

$$ax^2 = -c$$

$$x^2 = \frac{-c}{a}$$

$$x = \left(\frac{-c}{a}\right)^{1/2} = \pm\sqrt{\frac{-c}{a}}$$

However, if $b \neq 0$, then x also appears in a first-power term. In some cases, the general quadratic can be factored; for example:

$$x^2 + 4x - 21 = 0$$

$$(x + 7)(x - 3) = 0$$

$$x = -7 \quad \text{or} \quad x = 3$$

Otherwise, the solution requires the use of a formula. The first step is to see that the equation is in the standard format:

$$ax^2 + bx + c = 0$$

If the quadratic is in any different format (that is, $ax^2 = bx + c$; or $ax^2 + bx = c$; or $ax^2 + c = bx$), rearrange it as shown, with all the terms on one side set equal to zero on the other. Remember to reverse the sign of any term that must be moved to the opposite side of the equation in order to achieve this format.

Now plug the values for a, b, and c into the quadratic formula:

$$x = \frac{-b \pm \sqrt{b^2 - 4ac}}{2a}$$

Again, $a \neq 0$; however, the formula[3] works even if b and/or $c = 0$. If $b = 0$, there is no need to use the formula, because the solution is $(-c/a)^{1/2}$, as previously shown. There is also no need to use the formula if the quadratic can be factored, but we may not always see see the solution by factoring. Let's try the formula on the factorable equation shown above:

$$x^2 + 4x - 21 = 0$$

With $a = 1$, $b = 4$, and $c = -21$:

$$x = \frac{-b \pm \sqrt{b^2 - 4ac}}{2a}$$

$$x = \frac{-4 \pm \sqrt{4^2 - 4(1)(-21)}}{2(1)}$$

$$= \frac{-4 \pm \sqrt{16 + 84}}{2}$$

$$= \frac{-4 \pm \sqrt{100}}{2}$$

$$= \frac{-4 \pm 10}{2}$$

$$= \frac{6}{2} \quad or \quad \frac{-14}{2}$$

$$= 3 \quad or \quad -7$$

Now let's take an equation whose solution requires the use of the formula. In Chapter 6, we encounter this quadratic:

$$x^2 = 2.4 - 0.4x$$

Rearranging the terms to set the equation into standard format:

$$x^2 + 0.4x - 2.4 = 0$$

---

[3]This method was described in words by the ninth-century Arabian mathematician Mohammed al-Khowarizmi. The transliterated title of his major work contains the term *al-Jabr* ("reduction"), from which comes the word *algebra*. It was not until the late sixteenth century that anything resembling modern algebraic notation appeared.

Then $a = 1$, $b = 0.4$, and $c = -2.4$, and the solution is as follows:

$$x = \frac{-0.4 \pm \sqrt{0.4^2 - 4(1)(-2.4)}}{2(1)}$$

$$= \frac{-0.4 \pm \sqrt{0.16 + 9.6}}{2}$$

$$= \frac{-0.4 \pm \sqrt{9.76}}{2}$$

$$\approx \frac{-0.4 \pm 3.12}{2}$$

$$= \frac{2.72}{2} \quad or \quad \frac{-3.52}{2}$$

$$= 1.36 \quad or \quad -1.76$$

When x is an actual physical measurement, a negative root is meaningless.

The radicand in the numerator of the formula ($b^2 - 4ac$) is called the *discriminant*. If the discriminant is positive, there are two real-number roots, either or both of which may be positive or negative. If the discriminant is zero, there is one real-number root (given by $-b/2a$). If the discriminant is negative, there are no real-number solutions.

The quadratic formula is easy to use and useful to know—and fortunately, we do not encounter anything beyond the quadratic in entry-level clinical science. By comparison, the formulas for solving third- and fourth-degree equations are, respectively, harder and much harder. Equations of the fifth degree or higher are altogether beyond the limits of algebra; they can be approached by more advanced mathematical methods, but there are no algebraic formulas by which one can plug in coefficients and crank out roots.

I. Vectors

A *vector* is a mathematical entity possessing definite magnitude measured in a definite direction. It is symbolized by an arrow whose length is proportional to the magnitude, pointing in the direction of the measurement. (In contrast, a *scalar* is a pure magnitude, with no specific direction.)

For example, the vector *ten meters leftward* could be symbolized by an arrow 10 cm long (with 1 cm representing 1 meter), extending to the left from its point of origin. Then in the same system, the vector *five meters rightward* would be an arrow half the length and pointing in the opposite direction.

Any type of measurement that has a definite direction can be indicated by a vector arrow. Suppose that two *forces* are operating within a system—6 newtons applied along a line defined as 0°, and 4 newtons applied at 90°. If we let 1 cm represent a force of 1 newton, the two vectors are symbolized by arrows measuring 6 cm and 4 cm respectively, perpendicular to each other:

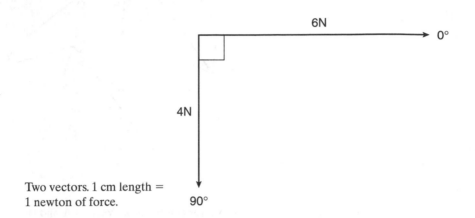

Two vectors. 1 cm length =
1 newton of force.

 Don't confuse *position* and *direction*. Because parallel lines extend infinitely in the same direction, a vector arrow can be moved along its own line or to any parallel line with no change in value—as long as there is no change in magnitude or direction, altering the position of a vector arrow is irrelevant.

Because the position is irrelevant, vectors can be combined. Two vectors operating in the same system can be shown as arrows originating from the same point. If we then consider these two vectors as adjacent sides of a parallelogram, their sum is the diagonal of the parallelogram. In other words, two vectors can be replaced by one that represents their combined magnitudes and directions:

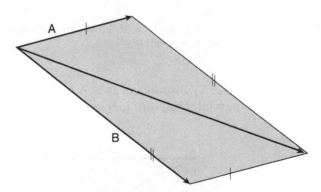

Vectors A and B can be replaced by the vector that forms
the diagonal of a parallelogram.

But if two vectors can be replaced by one, then *any* number of vectors can be replaced, pair by pair, until there is just one, representing the sum of all:

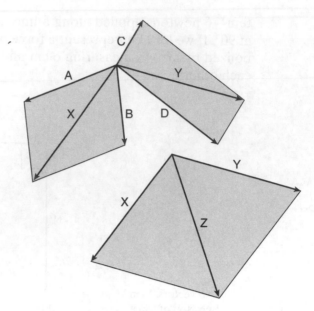

Vector combination. Vectors A and B are replaced by
X. Vectors C and D are replaced by Y. Then vectors X
and Y are replaced by Z.

Conversely, a given vector can be split in two. If a rectangle is constructed around a vector as the diagonal, the vector represents the sum of the adjacent sides of the rectangle. The sides of the constructed rectangle are vectors, because they have definite length and definite direction. In other words, the original vector can be replaced by a pair of perpendicular vectors whose lengths indicate the contribution of the original vector to each of the two new directions:

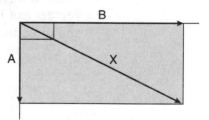

Vector X is split into perpendicular
vectors A and B.

Any number of different rectangles can be constructed around a given diagonal. Here are just two:

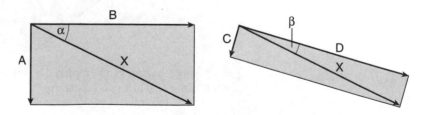

The same vector can be split up in limitless ways. On the left, the projection along B is equal to the magnitude of vector X times the cosine of angle α. On the right, the projection along D is equal to X times cosine β.

In the constructed rectangle, the diagonal vector and the length projected along an adjacent side form a certain angle and define a right (90°) triangle; the vector is the hypotenuse of the triangle and the projected length is the side adjacent to the angle. Therefore, we can use the trigonometric cosine function to determine the projected length:

$$\text{Cosine} = \frac{\text{adjacent side}}{\text{hypotenuse}} = \frac{\text{projected length}}{\text{vector}}$$

$$\text{Projected length} = \text{vector} \cdot \text{cosine}$$

Suppose the angle formed by the vector and an adjacent side is 60°; the cosine of 60° happens to be exactly 0.5, so the length projected along the adjacent side is just half the length or magnitude of the vector.

The closer the vector lies to the adjacent side of the constructed rectangle, the smaller the angle formed. As the angle decreases toward 0°, its cosine increases toward 1.0, and the projected length becomes a larger and larger fraction of the vector. When the vector is exactly parallel to the adjacent side, the angle = 0°, its cosine = 1.0, and 100% of the vector is projected. Conversely, as the angle increases toward 90°, its cosine decreases toward 0, and the projected length becomes a smaller and smaller fraction of the vector. When the vector is exactly perpendicular to the adjacent side, the angle = 90°, its cosine = 0, and no part of the vector is projected. As the angle increases past 90° and approaches 180°, its cosine moves into the negative range, approaching −1.0, and the projected length becomes a larger and larger fraction of the vector, but in the opposite direction:

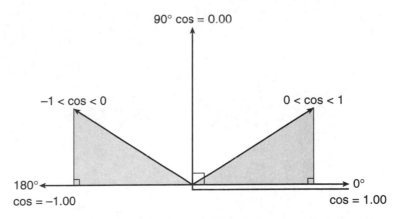

The reference line extends rightward. With the vector at 0° to the reference line, the full length is projected. At 90°, no part of the vector is projected. At 180° the full length is projected in the opposite direction.

Note that if most of the vector length is projected along one adjacent side of the constructed rectangle, very little is projected along the other adjacent side. In contrast, if the vector is equidistant from both adjacent sides (45° from each), the projections along the adjacent sides are of equal length, indicating equal magnitude in those perpendicular directions. Specifically:

$$\text{Projected length} = \text{vector} \cdot \text{cosine } 45° \approx \text{vector} \cdot 0.707$$

The main clinical application of vectors is in electrocardiography. The axis of electrical activity in the heart is represented by a single vector arrow whose direction denotes the overall course of the wave of electrical activity, and whose length is proportional to the voltage measured in that direction. (A totally unrelated clinical meaning of *vector* is a biological transmitter of infectious disease; for example, certain ticks are vectors for Lyme disease.)

Appendix: Basic Mathematical Concepts

Answers to Quiz Questions

Results of calculations are given to the correct number of significant digits.

Chapter 1: Clinical Metrics

1. To convert a height of 5 ft 7 in and a weight of 190 lb to metric units, we use conversion factors:

$$5 \text{ ft } 7 \text{ inch} = 67 \text{ inch}$$

$$67 \text{ inch} \cdot \frac{2.54 \text{ cm}}{\text{inch}} \approx 170 \text{ cm} = 1.7 \text{ m}$$

$$190 \text{ lb} \cdot \frac{\text{kg}}{2.2 \text{ lb}} \approx 86.4 \text{ kg}$$

Converting a height of 1.8 m and a weight of 73 kg to nonmetric units:

$$1.8 \text{ m} \cdot \frac{100 \text{ cm}}{\text{m}} = 180 \text{ cm}$$

$$180 \text{ cm} \cdot \frac{\text{inch}}{2.54 \text{ cm}} \approx 70.9 \text{ inch} \approx 5 \text{ ft } 11 \text{ inch}$$

$$73 \text{ kg} \cdot \frac{2.2 \text{ lb}}{\text{kg}} \approx 161 \text{ lb}$$

2. For the individual who stands 67 inches and weighs 190 pounds, the reading on the body surface area (BSA) nomogram is 1.98 m². Body mass index (BMI) by both formulas:

$$\text{BMI} = \frac{\text{weight-in-kilograms}}{(\text{height-in-meters})^2} = \frac{86.4}{1.70^2} \approx \frac{86.4}{2.89} \approx 29.9$$

$$BMI = \frac{(\text{weight-in-pounds}) \cdot 705}{(\text{height-in-inches})^2} = \frac{190 \cdot 705}{67^2} = \frac{133,950}{4489} \approx 29.8$$

For the individual at 1.8 m and 73 kg, the BSA is 1.92 m²; BMI:

$$BMI = \frac{73 \text{ kg}}{(1.8 \text{ m})^2} = \frac{73}{3.24} \approx 22.5$$

$$BMI = \frac{161 \text{ lb} \cdot 705}{(71 \text{ in})^2} = \frac{113,505}{5041} \approx 22.5$$

3. Given a height of 66 inches and BSA of 1.84 m², the corresponding weight on the nomogram is 75 kg or 165 lb. For the height in meters (reading directly from the BSA nomogram or using a conversion factor as in Question 1), 66 in ≈ 1.68 m. Therefore, BMI in metric and non-metric units is computed as follows:

$$BMI = \frac{75 \text{ kg}}{(1.68 \text{ m})^2} \approx \frac{75}{2.82} \approx 26.6$$

$$BMI = \frac{165 \text{ lb} \cdot 705}{(66 \text{ in})^2} = \frac{116,325}{4356} \approx 26.7$$

4. Converting a temperature of 38.6°C to Fahrenheit:

$$\frac{9}{5} \cdot 38.6 \approx 69.5$$

$$69.5 + 32 = 101.5°F$$

As a quick mental approximation, 38.6 is 1.6 more than 37°C, so add twice that difference to the corresponding Fahrenheit temperature:

$$98.6°F + 3.2 = 101.8°F$$

Converting a temperature of 103.2°F to Celsius:

$$103.2 - 32 = 71.2$$

$$71.2 \cdot \frac{5}{9} \approx 39.6°C$$

As a quick mental approximation, 103.2 is 0.8 less than 104°F, so subtract half that difference from the corresponding Celsius temperature:

$$40°F - 0.4 = 39.6°C$$

5. Since $1\text{ mg} = 10^{-3}$ g and $1\ \mu\text{g} = 10^{-6}$ g, it follows that $1\text{ mg} = 10^3\ \mu$g; similarly, $1\text{ mg} = 10^6\text{ ng} = 10^9$ pg. Using these equalities as conversion factors [in brackets]:

$$36\text{ mg} \cdot [\text{g} \cdot (1000\text{ mg})^{-1}] = 0.036\text{ g} = 3.6 \cdot 10^{-2}\text{ g}$$

$$36\text{ mg} \cdot [1000\ \mu\text{g} \cdot \text{mg}^{-1}] = 36{,}000\ \mu\text{g} = 3.6 \cdot 10^{4}\ \mu\text{g}$$

$$36\text{ mg} \cdot [10^6\text{ ng} \cdot \text{mg}^{-1}] = 3.6 \cdot 10^{7}\text{ ng}$$

$$36\text{ mg} \cdot [10^9\text{ pg} \cdot \text{mg}^{-1}] = 3.6 \cdot 10^{10}\text{ pg}$$

6. Since $1\text{ mL} = 10^{-3}$ L and $1\text{ dL} = 10^{-1}$ L, it follows that $1\text{ mL} = 10^{-2}$ dL; similarly, $1\text{ mL} = 10^3\ \mu\text{L} = 10^{12}$ fL. Using these equalities as conversion factors:

$$64\text{ mL} \cdot [\text{L} \cdot (1000\text{ mL})^{-1}] = 0.064\text{ L} = 6.4 \cdot 10^{-2}\text{ L}$$

$$64\text{ mL} \cdot [\text{dL} \cdot (100\text{ mL})^{-1}] = 0.64\text{ dL} = 6.4 \cdot 10^{-1}\text{ dL}$$

$$64\text{ mL} \cdot [1000\ \mu\text{L} \cdot \text{mL}^{-1}] = 64{,}000\ \mu\text{L} = 6.4 \cdot 10^{4}\ \mu\text{L}$$

$$64\text{ mL} \cdot [10^{12}\text{ fL} \cdot \text{mL}^{-1}] = 6.4 \cdot 10^{13}\text{ fL}$$

7. As in the last two problems:

$$48\text{ mg} \cdot \text{dL}^{-1} \cdot \ldots$$

$$\ldots [10\text{ dL} \cdot \text{L}^{-1}] = 480\text{ mg} \cdot \text{L}^{-1}$$

$$\ldots [\text{dL} \cdot (100\text{ mL})^{-1}] = 0.48\text{ mg} \cdot \text{mL}^{-1}$$

$$\ldots [10^3\ \mu\text{g} \cdot \text{mg}^{-1}] \cdot [10\text{ dL} \cdot \text{L}^{-1}] = 4.8 \cdot 10^{5}\ \mu\text{g} \cdot \text{L}^{-1}$$

$$\ldots [10^3\ \mu\text{g} \cdot \text{mg}^{-1}] = 4.8 \cdot 10^{4}\ \mu\text{g} \cdot \text{dL}^{-1}$$

$$\ldots [10^3\ \mu\text{g} \cdot \text{mg}^{-1}] \cdot [\text{dL} \cdot (100\text{ mL})^{-1}] = 480\ \mu\text{g} \cdot \text{mL}^{-1}$$

$$\ldots [10^3\ \mu\text{g} \cdot \text{mg}^{-1}] \cdot [\text{dL} \cdot (10^5\ \mu\text{L})^{-1}] = 0.48\ \mu\text{g} \cdot \mu\text{L}^{-1}$$

$$\ldots [10^6\text{ ng} \cdot \text{mg}^{-1}] \cdot [10\text{ dL} \cdot \text{L}^{-1}] = 4.8 \cdot 10^{8}\text{ ng} \cdot \text{L}^{-1}$$

$$\ldots [10^6\text{ ng} \cdot \text{mg}^{-1}] = 4.8 \cdot 10^{7}\text{ ng} \cdot \text{dL}^{-1}$$

$$\ldots [10^6\text{ ng} \cdot \text{mg}^{-1}] \cdot [\text{dL} \cdot (100\text{ mL})^{-1}] = 4.8 \cdot 10^{5}\text{ ng} \cdot \text{mL}^{-1}$$

$$\dots [10^6 \, \text{ng} \cdot \text{mg}^{-1}] \cdot [\text{dL} \cdot (10^5 \, \mu\text{L})^{-1}] = 480 \, \text{ng} \cdot \mu\text{L}^{-1}$$

$$\dots [10^9 \, \text{pg} \cdot \text{mg}^{-1}] \cdot [10 \, \text{dL} \cdot \text{L}^{-1}] = 4.8 \cdot 10^{11} \, \text{pg} \cdot \text{L}^{-1}$$

$$\dots [10^9 \, \text{pg} \cdot \text{mg}^{-1}] = 4.8 \cdot 10^{10} \, \text{pg} \cdot \text{dL}^{-1}$$

$$\dots [10^9 \, \text{pg} \cdot \text{mg}^{-1}] \cdot [\text{dL} \cdot (100 \, \text{mL})^{-1}] = 4.8 \cdot 10^8 \, \text{pg} \cdot \text{mL}^{-1}$$

$$\dots [10^9 \, \text{pg} \cdot \text{mg}^{-1}] \cdot [\text{dL} \cdot (10^5 \, \mu\text{L})^{-1}] = 4.8 \cdot 10^5 \, \text{pg} \cdot \mu\text{L}^{-1}$$

8. Converting the units to g/L, the concentration of Solution A is:

$$\frac{35 \, \text{mg}}{80 \, \text{mL}} = \frac{0.035 \, \text{g}}{0.08 \, \text{L}} \approx 0.44 \, \text{g} \cdot \text{L}^{-1}$$

The concentration of Solution B is:

$$\frac{0.84 \, \text{g}}{2.1 \, \text{L}} = 0.4 \, \text{g} \cdot \text{L}^{-1}$$

Since $0.44/0.4 = 1.1$, the concentration of A is about 10% greater than that of B. (Work it out the other way, by converting the units in Solution B to mg/mL.)

9. If the erythrocyte (red blood cell) count is $4.8 \cdot 10^6 \cdot \mu\text{L}^{-1}$ and the total blood volume is 5.2 L, the total number of RBCs in circulation is a simple calculation:

$$\frac{4.8 \cdot 10^6 \, \text{RBCs}}{\mu\text{L}} \cdot \frac{10^6 \, \mu\text{L}}{\text{L}} \cdot 5.2 \, \text{L} \approx 2.5 \cdot 10^{13} \, \text{RBCs}$$

10. If the dosage is $8.5 \, \text{mg} \cdot \text{kg}^{-1} \cdot \text{day}^{-1}$, to be given in four equally divided doses to a child who weighs 68 lb, the size of each dose is calculated as follows:

$$68 \, \text{lb} \cdot \frac{\text{kg}}{2.2 \, \text{lb}} \approx 31 \, \text{kg}$$

$$31 \, \text{kg} \cdot \frac{8.5 \, \text{mg} \cdot \text{kg}^{-1}}{\text{day}} \approx 264 \, \text{mg} \cdot \text{day}^{-1}$$

$$\frac{264 \, \text{mg}}{\text{day}} \cdot \frac{\text{day}}{4 \, \text{doses}} = 66 \, \text{mg} \cdot \text{dose}^{-1}$$

Chapter 2: Solutions and Body Fluids

1. The patient's total body water is 42 L (25.6 L ICF, 16.4 L ECF), at a concentration of about 293 mosmol/L. The solute load in each compartment is:

$$ECF: \quad 6.4\,L \cdot \frac{293\ \text{mosmol}}{L} \approx 4800\ \text{mosmol}$$

$$ICF: \quad 25.6\,L \cdot \frac{293\ \text{mosmol}}{L} \approx 7500\ \text{mosmol}$$

The combined load for the whole body is:

$$4800 + 7500 = 12{,}300\ \text{mosmol}$$

Administration of 1.5 liters of twice-normal saline contributes additional solute:

$$1.5\,L \cdot \frac{600\ \text{mosmol}}{L} \approx 900\ \text{mosmol}$$

This solute enters the ECF but it also adds to the total load for the whole body:

$$ECF: \quad 4800 + 900 = 5700\ \text{mosmol}$$

$$Body: \quad 12{,}300 + 900 = 13{,}200\ \text{mosmol}$$

Total body water increases by the volume of fluid administered:

$$42\,L + 1.5\,L = 43.5\,L$$

The resulting new osmolality for the whole body is:

$$\frac{13{,}200\ \text{mosmol}}{43.5\,L} \approx 303\ \text{mosmol} \cdot L^{-1}$$

Since this osmolality is the same in both compartments, the final volumes are:

$$ECF: \quad 5700\ \text{mosmol} \cdot \frac{L}{303\ \text{mosmol}} \approx 18.8\,L$$

$$ICF: \quad 7500\ \text{mosmol} \cdot \frac{L}{303\ \text{mosmol}} \approx 24.8\,L$$

The patient's original osmolality has essentially been restored, but total body water is now about 3.5 liters larger than it was at the start.

2. The molecular weight of magnesium chloride ($MgCl_2$) is approximately:

$$24 + (2 \cdot 35) = 24 + 70 = 94$$

A 1.2% solution contains 1.2 grams per deciliter of fluid. The millimolar concentration of this solution is:

$$\frac{1.2 \text{ g}}{\text{dL}} \cdot \frac{\text{mol}}{94 \text{ g}} \approx 0.013 \text{ mol} \cdot \text{dL}^{-1}$$

$$\frac{0.013 \text{ mol}}{\text{dL}} \cdot \frac{1000 \text{ mmol}}{\text{mol}} \cdot \frac{10 \text{ dL}}{\text{L}} \approx 130 \text{ mmol} \cdot \text{L}^{-1}$$

With complete ionic dissociation of 1 liter of this 1.2% solution:

$$130 \text{ mmol}_{MgCl_2} \rightarrow 130 \text{ mmol}_{Mg^{+2}} + (2 \cdot 130) \text{ mmol}_{Cl^-} = 390 \text{ mmol}_{total}$$

And because each particle is osmotically active, the osmolality is 390 mosmol/L.

3. The molecular weight for glucose ($C_6H_{12}O_6$) is 180; that is, one mole of glucose weighs 180 grams. Since a 5% solution contains 5 grams per deciliter, the molar concentration is:

$$\frac{5 \text{ g}}{\text{dL}} \cdot \frac{\text{mol}}{180 \text{ g}} \approx \frac{0.028 \text{ mol}}{\text{dL}} = \frac{0.28 \text{ mol}}{\text{L}} = 280 \text{ mmol} \cdot \text{L}^{-1}$$

And because glucose is a nonionic, nondissociating compound, each molecule is a separate particle in solution, and the number of millimoles equals the number of milliosmoles. So the osmotic concentration is 280 mosmol/L. Note that this value is in the range of the normal osmolality of body fluids—which is why 5% glucose, like 0.9% saline, is sometimes given intravenously for fluid replacement.

4. The molecular weight of sodium bicarbonate ($NaHCO_3$) is approximately:

$$23 + 1 + 12 + (3 \cdot 16) = 84$$

We have 12.6 grams of this compound dissolved in 4 liters of water. The number of moles in this weight of solute is:

$$12.6 \text{ g} \cdot \frac{\text{mol}}{84 \text{ g}} \approx 0.15 \text{ mol}$$

The molar concentration of this solution is:

$$\frac{0.15 \text{ mol}}{4 \text{ L}} = \frac{0.0375 \text{ mol}}{\text{L}} \approx 38 \text{ mmol} \cdot \text{L}^{-1}$$

After dissociation into ions of sodium (Na^+) and bicarbonate (HCO_3^-), 1 liter of the solution contains:

$$38 \text{ mmol}_{NaHCO_3} \rightarrow 38 \text{ mmol}_{Na^+} + 38 \text{ mmol}_{HCO_3^-} = 76 \text{ mmol}_{total}$$

The osmolality is 76 mosmol/L.

5. The number of equivalents in 100 grams of calcium is calculated as follows:

$$100 \text{ g} \cdot \frac{\text{mol}}{40 \text{ g}} = 2.5 \text{ mol}$$

$$2.5 \text{ mol} \cdot \frac{2 \text{ eq}}{\text{mol}} = 5 \text{ eq}$$

Chlorine and calcium combine in a 2-to-1 ratio. If we have 2 moles of chloride ions, we need 1 mole of calcium ions to form 1 mole of calcium chloride:

$$2Cl^- + Ca^{+2} \rightarrow CaCl_2$$

If the serum concentration of nonionized calcium is 9.0 mg/dL:

$$9.0 \frac{\text{mg}}{\text{dL}} \cdot \frac{\text{mmol}}{40 \text{ mg}} \cdot \frac{10 \text{ dL}}{\text{L}} \approx 2.3 \text{ mmol} \cdot \text{L}^{-1}$$

Note that in the first part of the problem, we used the conversion factor mol/40 g, and in the last part we used mmol/40 mg. In general, for a substance of atomic or molecular weight W, W g = 1 mol, W mg = 1 mmol, W μg = 1 μmol, etc.

6. To find the hematocrit in a patient whose plasma volume and total blood volume are given, subtract the plasma volume from the total blood volume to determine the volume of the red blood cells; then calculate the percentage of the total blood volume that is represented by the RBC volume:

$$\text{RBC volume} = 5.2 \text{ L} - 3.3 \text{ L} = 1.9 \text{ L}$$

$$\text{Hematocrit} = \frac{1.9 \text{ L}}{5.2 \text{ L}} \approx 0.37 = 37\%$$

7. The difference in concentration between a 2-molar formulation and a 2-molal formulation of the same substance is so small as to be clinically irrelevant. However, we can compare them theoretically if we make the following two assumptions: first, that all measurements are *exact;* and second, that introducing solute into a measured volume of water creates a solution whose volume is infinitessimally greater than the original volume of water. In a 2-molar formulation, *exactly* 2 moles of solute are in a solution whose volume is *exactly* 1 liter. In a 2-molal formulation, 2 moles of solute are *added* to 1 kilogram of pure water—that is, added to a volume that was already *exactly* 1 liter.

   Then the volume of the resulting 2-molal solution is some minuscule quantity greater than 1 liter. We must remove that same minuscule quantity of the solution to get the total volume back down to *exactly* 1 liter. As a result, 1 liter of the 2-molal solution contains very slightly less than 2 moles of solute, whereas 1 liter of the 2-molar formulation contains exactly 2 moles of solute. Therefore, the 2-molar formulation is very slightly more concentrated.

8. Converting a plasma copper level of 20 μmol/L to conventional weight-units, we use a conversion factor based on the atomic weight of copper:

$$20\,\frac{\mu\text{mol}}{\text{L}} \cdot \frac{63.5\,\mu\text{g}}{\mu\text{mol}} \cdot \frac{\text{L}}{10\,\text{dL}} \approx 127\,\mu\text{g} \cdot \text{dL}^{-1}$$

In the other direction, if N is the numeric of a concentration by weight ($N\,\mu$g/dL), the conversion factor to go directly to SI units (μmol/L) is computed as follows:

$$N\,\frac{\mu\text{g}}{\text{dL}} \cdot \frac{\mu\text{mol}}{63.5\,\mu\text{g}} \cdot \frac{10\,\text{dL}}{\text{L}} \approx 0.157N\,\mu\text{mol} \cdot \text{L}^{-1}$$

Dividing by 63.5 and multiplying by 10 is the same as dividing by 6.35, which is the same as multiplying by its reciprocal ($1/6.35 \approx 0.157$). In other words, multiply N by 0.157 to go directly from the weight concentration (in μg/dL) to the equivalent value in SI units (μmol/L).

9. In the formula for estimating serum osmolality, the weight concentrations of glucose and BUN must be converted to SI units of particle-count per liter. But in the first laboratory report, all the values are given in SI units, so we can just add:

$$\text{Osmolality} \approx (2 \cdot \text{sodium}) + \text{glucose} + \text{BUN}$$

$$\approx (2 \cdot 136) + 6.4 + 9.5 = 288\,\text{mosmol} \cdot \text{L}^{-1}$$

This glucose level is slightly high (normal, 3.9–6.1 mmol/L = 70–110 mg/dL):

$$\frac{6.4\,\text{mmol}}{\text{L}} \cdot \frac{180\,\text{mg}}{\text{mmol}} \cdot \frac{\text{L}}{10\,\text{dL}} \approx 115\,\text{mg} \cdot \text{dL}^{-1}$$

The second laboratory report gives glucose and BUN in units of weight concentration. Therefore, to get the particle-count concentration, we divide the weight concentration of glucose by 18, and of BUN by 2.8 (approximated as 3). For sodium, a monovalent ion, the number of milliequivalents equals the number of millimoles. Now we can estimate osmolality by the formula:

$$(2 \cdot \text{sodium}) + \frac{\text{glucose}}{18} + \frac{\text{BUN}}{3} = (2 \cdot 142) + \frac{108}{18} + \frac{22}{3}$$

$$\approx 284 + 6 + 7 = 297\,\text{mosmol} \cdot \text{L}^{-1}$$

10. To find the equilibrium potential for an ion—the transmembrane voltage at which the electrical gradient would exactly offset the effects of the chemical concentration gradient—we employ the Nernst equation. For the divalent cation magnesium, at certain extracellular and intracellular concentrations:

$$E = \frac{61.5}{valence} \cdot \log\left(\frac{[ion_{ECF}]}{[ion_{ICF}]}\right)$$

$$E = \frac{61.5}{+2} \cdot \log\left(\frac{2.5\ meq \cdot L^{-1}}{25\ meq \cdot L^{-1}}\right)$$

$$\approx 30.8 \cdot \log 0.1 = 30.8 \cdot (-1) = -30.8\ mV$$

Note that the magnitude of the equilibrium potential for potassium is much greater (about −95 mV). Both potassium and magnesium are cations with higher concentrations inside cells than outside, so the concentration gradient pushes outward for both; hence, the equilibrium voltage is negative (pulling a cation inward) for both. However, the concentration gradient is smaller for magnesium than for potassium. So *aside* from the fact that we must divide by 2 because magnesium is divalent, it takes a smaller charge gradient to offset the effect of the concentration gradient for magnesium than for potassium.

Chapter 3: Circulatory Function

1. The mean arterial pressure (MAP) for a hypertensive patient whose blood pressure is measured at 180/110 mm Hg can be computed in two ways that are really just mathematical rearrangements of each other:

$$MAP = \frac{systolic + (2 \cdot diastolic)}{3} = \frac{180 + (2 \cdot 110)}{3} \approx 133\ mm\ Hg$$

$$MAP = diastolic + \frac{pulse\ pressure}{3} = 110 + \frac{(180 - 110)}{3}$$
$$\approx 133\ mm\ Hg$$

Comparing the effects of pressure reduction to 165/110 mm Hg versus 180/95 mm Hg:

$$MAP = \frac{165 + (2 \cdot 110)}{3} \approx 128\ mm\ Hg$$

$$MAP = 95 + \frac{(180 - 95)}{3} \approx 123\ mm\ Hg$$

Note that when the systolic pressure is reduced by 15 mm Hg, MAP is reduced by 5 mm Hg; but when diastolic pressure is reduced by 15 mm Hg, MAP is reduced by 10 mm Hg. If the other pressure is unchanged, a given change in diastolic pressure has twice the impact on MAP as the same change in systolic pressure.

2. The pulse pressure is the difference between the systolic and diastolic pressures. Given MAP and the systolic pressure, we find the diastolic pressure as follows:

$$MAP = 95 = \frac{135 + (2 \cdot \text{diastolic})}{3}$$

$$285 = 135 + (2 \cdot \text{diastolic})$$

$$\text{Diastolic} = \frac{285 - 135}{2} = \frac{150}{2} = 75 \text{ mm Hg}$$

Pulse pressure = systolic − diastolic = 135 − 75 = 60 mm Hg.

3. In going from an artery of luminal radius r to an artery half as wide ($0.5r$), resistance increases and flow decreases as the fourth power of the radius:

$$\text{Resistance in smaller artery} \propto 0.5^{-4} = 16 \cdot \text{original resistance}$$

$$\text{Flow rate in smaller artery} \propto 0.5^4 = 0.0625 \cdot \text{original flow}$$

That is, resistance increases 16-fold, and flow is decreased by almost 94%.

4. Conversions between pressure units:

$$25 \text{ cm H}_2\text{O} \cdot \frac{0.735 \text{ mm Hg}}{\text{cm H}_2\text{O}} \approx 18 \text{ mm Hg}$$

$$25 \text{ cm H}_2\text{O} \cdot \frac{0.735 \text{ mm Hg}}{\text{cm H}_2\text{O}} \cdot \frac{133.3 \text{ Pa}}{\text{mm Hg}} \approx 2.5 \cdot 10^3 \text{Pa} = 2.5 \text{ kPa}$$

$$25 \text{ mm Hg} \cdot \frac{1.36 \text{ cm H}_2\text{O}}{\text{mm Hg}} = 34 \text{ cm H}_2\text{O}$$

$$25 \text{ mm Hg} \cdot \frac{133.3 \text{ Pa}}{\text{mm Hg}} \approx 3.3 \cdot 10^3 \text{ Pa} = 3.3 \text{ kPa}$$

$$3000 \text{ Pa} \cdot \frac{0.0075 \text{ mm Hg}}{\text{Pa}} \approx 23 \text{ mm Hg}$$

$$3000 \text{ Pa} \cdot \frac{0.0075 \text{ mm Hg}}{\text{Pa}} \cdot \frac{1.36 \text{ cm H}_2\text{O}}{\text{mm Hg}} \approx 31 \text{ cm H}_2\text{O}$$

5. In this problem, stroke volume will be obtained as the quotient of cardiac output divided by heart rate, a rearrangement of the definition of output as heart rate times stroke volume. We know the heart rate, but we must calculate output from the other items of given information—the oxygen consumption rate and the oxygen content in arterial blood and in venous blood. Using the Fick method:

$$\text{Output} = \frac{O_2 \text{ consumption}}{\text{A-V difference}} = \frac{240 \text{ mL} \cdot \text{min}^{-1}}{(18 - 14) \text{ mL} \cdot \text{dL}^{-1}} = 60 \text{ dL} \cdot \text{min}^{-1}$$

$$\text{Stroke volume} = \frac{\text{output}}{\text{heart rate}} = \frac{60 \text{ dL} \cdot \text{min}^{-1}}{68 \cdot \text{min}^{-1}} \approx 0.88 \text{ dL} = 88 \text{ mL}$$

6. Here, we are looking at cardiac output as the quotient of MAP divided by peripheral resistance. With blood pressure given as 130/85 mm Hg and resistance given as 1600 dyne $\cdot$ sec $\cdot$ cm^{-5}, the first step is to determine MAP and convert that value to units of force per unit-area:

$$\text{MAP} = \frac{\text{systolic} + (2 \cdot \text{diastolic})}{3} = \frac{130 + (2 \cdot 85)}{3} = 100 \text{ mm Hg}$$

$$100 \text{ mm Hg} \cdot \frac{133.3 \text{ Pa}}{\text{mm Hg}} = 1.33 \cdot 10^4 \text{ Pa} = 1.33 \cdot 10^4 \text{ N} \cdot \text{m}^{-2}$$

Recall that the SI unit for pressure, the pascal, is defined as 1 newton (N) of force per square meter. However, resistance is given in dynes and centimeters, so we must place MAP and resistance in like units, using the following conversion factors for force and area:

$$\frac{10^5 \text{ dynes}}{N} = \frac{10^{-5} \text{ N}}{\text{dyne}} = 1$$

$$\frac{(100 \text{ cm})^2}{\text{m}^2} = \frac{10^4 \text{ cm}^2}{\text{m}^2} = \frac{10^{-4} \text{ m}^2}{\text{cm}^2} = 1$$

We'll convert MAP to units of dynes per square centimeter:

$$\frac{1.33 \cdot 10^4 \text{ N}}{\text{m}^2} \cdot \frac{10^5 \text{ dynes}}{N} = \frac{1.33 \cdot 10^9 \text{ dynes}}{\text{m}^2}$$

$$\frac{1.33 \cdot 10^9 \text{ dynes}}{\text{m}^2} \cdot \frac{10^{-4} \text{ m}^2}{\text{cm}^2} = \frac{1.33 \cdot 10^5 \text{ dynes}}{\text{cm}^2}$$

Now we can calculate cardiac output:

$$\text{Output} = \frac{\text{MAP}}{\text{resistance}} = \frac{1.33 \cdot 10^5 \text{ dyne} \cdot \text{cm}^{-2}}{1.6 \cdot 10^3 \text{ dyne} \cdot \text{sec} \cdot \text{cm}^{-5}}$$

$$\approx \frac{0.83 \cdot 10^2 \text{ cm}^3}{\text{sec}} = 83 \text{ mL} \cdot \text{sec}^{-1} \approx 5 \text{ L} \cdot \text{min}^{-1}$$

7. Given the stroke volume, heart rate, and cardiac index, the body surface area (BSA) can be determined by computing cardiac output and rearranging the formula for cardiac index:

$$\text{Output} = \text{stroke volume} \cdot \text{heart rate} = 55\,\frac{\text{mL}}{\text{beat}} \cdot 80\,\frac{\text{beats}}{\text{min}}$$

$$= 4400\,\frac{\text{mL}}{\text{min}} = 4.4\,\text{L} \cdot \text{min}^{-1}$$

$$\text{BSA} = \frac{\text{cardiac output}}{\text{cardiac index}} = \frac{4.4\,\text{L} \cdot \text{min}^{-1}}{2.8\,\text{L} \cdot \text{min}^{-1} \cdot \text{m}^{-2}} \approx 1.57\,\text{m}^2$$

8. Cardiac output and heart rate are given in two patients, and we are told that end-diastolic volumes are the same. To compare the filling pressures, we must first determine the stroke volumes by rearranging the formula for cardiac output:

Patient A: $\text{Stroke volume} = \dfrac{\text{output}}{\text{heart rate}} = \dfrac{5000\,\text{mL} \cdot \text{min}^{-1}}{76 \cdot \text{min}^{-1}} \approx 66\,\text{mL}$

Patient B: $\text{Stroke volume} = \dfrac{4500\,\text{mL} \cdot \text{min}^{-1}}{78 \cdot \text{min}^{-1}} \approx 58\,\text{mL}$

Since end-diastolic volumes are the same, the ejection fraction (stroke volume divided by end-diastolic volume) is larger for Patient A than for Patient B. Then Patient B has a greater volume of blood left in the ventricle after each contraction, which implies a higher filling pressure.

9. If the width of the QRS complex is 0.6 cm = 6 mm on an ECG tracing at standard speed, the time duration is:

$$6\,\text{mm} \cdot \frac{0.04\,\text{sec}}{\text{mm}} = 0.24\,\text{sec}$$

If the QRS complexes occur at intervals of 0.75 cm = 7.5 mm, the heart rate is:

$$\frac{1500\,\text{mm} \cdot \text{min}^{-1}}{7.5\,\text{mm}} = 200 \cdot \text{min}^{-1}$$

Note that the interval 7.5 mm is halfway between 1 and 2 heavy-line boxes—yet we cannot take the average of the per-minute rates associated with occurrence at every box (300/min) and every other box (150/min). The average of 300 and 150 is 225, which is definitely *not* the correct answer.

10. If the axis is 15°, the angle to Lead II $= 60 - 15 = 45°$. The vector voltage is obtained as the quotient of the voltage on Lead II divided by the cosine of 45°:

$$\text{Vector voltage} = \frac{1.4 \text{ mV}}{0.707} \approx 2 \text{ mV}$$

The voltage on Lead I is then obtained as the product of this vector voltage and the cosine of the angle between the vector and Lead I (15°):

$$2 \text{ mV} \cdot 0.966 \approx 1.9 \text{ mV}$$

The voltage on Lead III can be obtained in the same way, using the cosine of the angle between Lead III and the vector ($120° - 15° = 105°$):

$$2 \text{ mV} \cdot (-0.259) \approx -0.5 \text{ mV}$$

Or we could employ Einthoven's law. The voltage in Lead III equals the difference between the voltages in Leads II and I: $1.4 - 1.9 = -0.5$ mV.

Chapter 4: Air, Blood, and Respiratory Function

1. The functional residual capacity, given as 2.0 L, is the volume of air left in the lungs at the end of normal expiration. At a tidal volume of 500 mL, the total volume at the end of normal inspiration is $2.0 + 0.5 = 2.5$ L. If the total lung capacity is 5.4 L, then the inspiratory reserve (the additional volume that can be forcibly inhaled after normal inspiration) is the difference between total lung capacity and the volume at the end of normal inspiration: $5.4 - 2.5 = 2.9$ L. The expiratory reserve (the additional volume that can be forcibly exhaled after normal expiration) is the difference between the volume at the end of normal expiration (that is, the functional residual capacity), and the residual volume (given as 0.9 L): $2.0 - 0.9 = 1.1$ L. Vital capacity is the difference between total lung capacity and the residual volume: $5.4 - 0.9 = 4.5$ L.

2. Given, the forced vital capacity (FVC) $= 4.4$ L and the forced expiratory volume in 1 second (FEV_1) $= 2.8$ L, the FEV_1/FVC ratio is $2.8/4.4 \approx 0.64$. For an adult male, the total volume of forced exhalation (FVC) is rather small, suggesting some form of restrictive lung disease; at the same time, the flow rate (FEV_1) is quite slow, and the ratio is low, suggesting an obstructive condition as well. Obviously, restrictive and obstructive conditions are not mutually exclusive; this man has combined disorders.

3. If the mid-expiratory flow rate from the 25th to the 75th percentile (MEF_{25-75}) is 4 L/sec and the actual time for this interval is 0.6 sec, the volume between the 25th and 75th percentile is:

$$4 \frac{\text{L}}{\text{sec}} \cdot 0.6 \text{ sec} = 2.4 \text{ L}$$

This volume represents half of the FVC—the middle two quartiles. Therefore, FVC is just twice this volume, 4.8 L.

4. Given, $P_aO_2 = 96$ mm Hg, $P_aCO_2 = 42$ mm Hg. At normal sea-level atmospheric conditions (total pressure = 760 mm Hg, $FiO_2 = 21\%$), the alveolar partial pressure (P_AO_2) is calculated by correcting for the increased amount of water vapor and subtracting the estimated carbon dioxide concentration (1.2 times the P_aCO_2):

$$P_AO_2 = [(760 - 47) \cdot 0.21] - (42 \cdot 1.2)$$

$$\approx 150 - 50 = 100 \text{ mm Hg}$$

Then the alveolar-arterial (A-a) oxygen gradient is the difference between this calculated value for the alveolar oxygen tension and the measured oxygen tension in the arterial blood:

$$\text{A-a oxygen gradient} = 100 - 96 = 4 \text{ mm Hg}$$

5. In air containing 50% oxygen, $FiO_2 = 0.5$; at normal atmospheric pressure, the partial pressure of oxygen is $760 \cdot 0.5 = 380$ mm Hg. If P_aO_2 is 78 mm Hg and P_aCO_2 is 54 mm Hg, we compute the A-a oxygen gradient as follows:

$$P_AO_2 = [(760 - 47) \cdot 0.5] - (54 \cdot 1.2)$$

$$\approx 357 - 65 = 292 \text{ mm Hg}$$

$$\text{A-a oxygen gradient} = 292 - 78 = 214 \text{ mm Hg}$$

Such a large value is certainly abnormal. However, even a healthy individual would exhibit a significant gradient while breathing 50% oxygen, because the amount of oxygen that can be carried in the blood is limited. The relevant point here is not that there is a large gradient, but that arterial oxygen tension is on the low side *despite* the very high alveolar oxygen tension—something is interfering with blood aeration. For example, chronic bronchitis represents a ventilatory defect while emphysema represents a loss of alveolar surface area; both of these components of chronic obstructive lung disease can interfere with gas exchange. *Hypercapnia* (the high carbon dioxide tension) is consistent with this scenario. In healthy individuals, the main stimulus for respiration is a rising carbon dioxide level and its effect on acid-base balance. However, patients with chronic obstructive lung disease can become physiologically accustomed to hypercapnia, and their respiratory stimulus becomes the need for oxygen. Giving high concentrations of oxygen in this situation carries a certain degree of risk, because it may leave the patient with no stimulus to spontaneous respiration at all.

6. To use the oxygen-hemoglobin dissociation curve (Figure 4.9), read up from the given oxygen tension and then across to read the value on the saturation scale. The arterial oxygen saturation is approximately 88% at $P_{aO_2} = 60$ mm Hg; 92% at 70 mm Hg; and 95% at 80 mm Hg.

7. Retention of carbon dioxide in the blood creates a condition of acidity. Acidity shifts the oxygen-hemoglobin curve rightward. With a rightward shift, oxygen saturation would be lower than normal at any given oxygen tension; that is, oxygen is more easily delivered in this chemical environment.

8. An arterial oxygen tension of 98 mm Hg is normal. At this level of aeration, each gram of hemoglobin carries around 1.3 mL of oxygen. Then if the hemoglobin concentration is 12 g/dL, the *volume* of oxygen in the arterial blood is measured as follows:

$$12 \, \frac{g}{dL} \cdot 1.3 \, \frac{mL}{g} = 15.6 \, mL \cdot dL^{-1}$$

If oxygen saturation in venous blood is 75%, each gram of hemoglobin carries about 1.0 mL of oxygen:

$$12 \, \frac{g}{dL} \cdot 1.0 \, \frac{mL}{g} = 12.0 \, mL \cdot dL^{-1}$$

Then the A-V difference (the difference between arterial and venous oxygen content) is $15.6 - 12.0 = 3.6$ mL for each deciliter of blood passing through the systemic capillaries. If we take a typical blood flow rate of 6 L/min = 60 dL/min, the actual rate of oxygen delivery would be:

$$3.6 \, \frac{mL}{dL} \cdot 60 \, \frac{dL}{min} = 216 \, mL \cdot min^{-1}$$

9. The most direct effect of removal of a lung would be a reduction in FVC—one lung cannot move as much air as two lungs. For the same reason, FEV_1 will also be reduced. If no other pulmonary problem is present, the reductions in volume and airflow rate will be in proportion to each other, and the FEV_1/FVC ratio may still be normal. Compliance reflects the elasticity of the lung tissue, and there is no reason to assume that it would be abnormal, unless the compliance of the removed lung was altered by a disease that is also present in the remaining lung. Effects on oxygen tension and carbon dioxide tension are hard to predict; the opportunity for gas exchange is only half of what it had been, but whether that reduced exchange is sufficient to meet the body's needs depends on such variables as the level of physical activity and the metabolic rate. Since carbon dioxide diffuses across the alveolar-capillary interface so much more easily than oxygen, a drop in oxygen tension is more likely than a buildup in carbon dioxide tension.

10. From the given information, the mean corpuscular volume (MCV), mean corpuscular hemoglobin (MCH), and mean corpuscular hemoglobin concentration (MCHC) are obtained as follows:

$$\text{MCV} = \frac{\text{percent hematocrit} \cdot 10}{\text{erythrocytes (millions per } \mu\text{L)}} = \frac{38 \cdot 10}{4.6} \approx 83 \text{ fL}$$

$$\text{MCH} = \frac{\text{hemoglobin concentration} \cdot 10}{\text{erythrocytes (millions per } \mu\text{L)}} = \frac{12 \cdot 10}{4.6} \approx 26 \text{ pg}$$

$$\text{MCHC} = \frac{\text{hemoglobin concentration} \cdot 100}{\text{percent hematocrit}} = \frac{12 \cdot 100}{38} \approx 32 \text{ g} \cdot \text{dL}^{-1}$$

Chapter 5: Renal Function

1. From the given information, the rate of entry of glucose into the tubular fluid is well below the Tm:

$$\text{Rate of entry} = [\text{glucose}]_{\text{serum}} \cdot \text{GFR}$$

$$= 1.35 \, \frac{\text{mg}}{\text{mL}} \cdot 96 \, \frac{\text{mL}}{\text{min}} \approx 130 \, \text{mg} \cdot \text{min}^{-1}$$

This value is much lower than the Tm, and we may assume that there is no glycosuria; in any case, the serum glucose itself is well below the usual renal threshold (the serum level at which glycosuria appears), which is about 200 mg/dL.

2. The likelihood that substance S will build up in the blood is greater than the likelihood that substance R will spill over into the urine. The rate at which any substance in serum or plasma enters the peritubular capillaries is usually about four times greater than the rate at which it enters the glomerular filtrate. Given the same serum levels, substance S in the nonfiltered blood enters the peritubular vessels that much faster than substance R in the glomerular filtrate enters the nephron tubule; and because the Tm for reabsorption of R is the same as the Tm for secretion of S, the entry of S into the peritubular capillaries would surpass its secretion Tm sooner than the entry of R into the nephron would surpass its reabsorption Tm.

3. In Patient A, creatinine clearance is calculated as follows:

$$\text{Clearance}_{\text{creatinine}} = \frac{[\text{creatinine}]_{\text{urine}} \cdot \text{volume}_{\text{urine}} \cdot 0.07}{[\text{creatinine}]_{\text{serum}}}$$

$$= \frac{0.8 \, \dfrac{\text{mg}}{\text{mL}} \cdot 900 \, \dfrac{\text{mL}}{\text{day}} \cdot 0.07}{1.3 \, \dfrac{\text{mg}}{\text{dL}}} \approx 39 \, \text{mL} \cdot \text{min}^{-1}$$

In Patient B, we are given the total amount of creatinine present in a 24-hour collection, rather than the concentration and volume of that collection. But the product of the creatinine concentration and the volume of the collection *is* the total amount collected, so the numerator is simply that amount multiplied by the conversion factor 0.07:

$$\text{Clearance}_{\text{creatinine}} = \frac{1400\,\dfrac{\text{mg}}{\text{day}} \cdot 0.07}{1.2\,\dfrac{\text{mg}}{\text{dL}}} \approx 82\ \text{mL} \cdot \text{min}^{-1}$$

In both patients, the estimated GFR is considerably below normal; but the estimate is worse in Patient A, because the value is lower than in Patient B. In other words, the true GFR in Patient B is below 82 mL/min by some margin; but the true GFR in Patient A is below 39 mL/min by an even larger margin, which means that the true GFR in Patient A is desperately low.

4. From the given information, total renal plasma flow is calculated as follows:

$$\frac{594\,\dfrac{\text{mL}}{\text{min}}}{0.9} = 660\ \text{mL} \cdot \text{min}^{-1}$$

At a GFR of 125 mL/min, the filtration fraction is:

$$\frac{125\ \text{mL} \cdot \text{min}^{-1}}{660\ \text{mL} \cdot \text{min}^{-1}} \approx 0.19 = 19\%$$

The rate of entry of nonfiltered glucose into the peritubular capillaries is the product of the serum concentration and the nonfiltration rate:

$$\text{Rate of entry} = [\text{glucose}]_{\text{serum}} \cdot (\text{renal plasma flow} - \text{GFR})$$

$$= 0.9\,\frac{\text{mg}}{\text{mL}} \cdot (660 - 125)\,\frac{\text{mL}}{\text{min}} \approx 482\ \text{mg} \cdot \text{min}^{-1}$$

If the hematocrit is 37%, then the plasma represents $100 - 37 = 63\%$ of the blood volume. Using the calculated renal plasma flow, the total renal blood flow is:

$$\frac{660\,\dfrac{\text{mL}}{\text{min}}}{0.63} \approx 1048\,\frac{\text{mL}}{\text{min}} \approx 1.05\ \text{L} \cdot \text{min}^{-1}$$

And finally, if the total blood volume is 5.40 liters, the percentage of blood cycling through the kidneys each minute is:

$$\frac{1.05\ \text{L}}{5.40\ \text{L}} \approx 0.194 = 19.4\%$$

5. Calculating creatinine clearance from the given information:

$$\text{Clearance}_{\text{creatinine}} = \frac{[\text{creatinine}]_{\text{urine}} \cdot \text{volume}_{\text{urine}} \cdot 0.07}{[\text{creatinine}]_{\text{serum}}}$$

$$= \frac{1.1 \, \frac{\text{mg}}{\text{mL}} \cdot 1300 \, \frac{\text{mL}}{\text{day}} \cdot 0.07}{1.6 \, \frac{\text{mg}}{\text{dL}}} \approx 63 \, \text{mL} \cdot \text{min}^{-1}$$

To estimate creatinine clearance using the Cockcroft-Gault formula, we need the patient's ideal weight in kilograms. If he weighs 200 pounds and is 15% overweight, his ideal weight is calculated as follows:

$$200 \, \text{lb} \cdot \frac{\text{kg}}{2.2 \, \text{lb}} \approx 91 \, \text{kg}$$

$$\frac{91 \, \text{kg}}{1.15} \approx 79 \, \text{kg}$$

Now, plugging this weight, his age, and the serum creatinine level into the formula:

$$\text{Clearance}_{\text{creatinine}} \approx \frac{(140 - \text{age}) \cdot \text{ideal weight}}{72 \cdot [\text{creatinine}]_{\text{serum}}}$$

$$= \frac{(140 - 46) \cdot 79}{72 \cdot 1.6} \approx 64 \, \text{mL} \cdot \text{min}^{-1}$$

In this case, the calculated creatinine clearance is virtually the same as the estimate obtained with the Cockroft-Gault formula. Keep in mind, however, that creatinine clearance may be significantly higher than the true GFR when the true GFR is low.

6. At a serum glucose concentration of 220 mg/dL and a GFR of 105 mL/min:

$$\text{Rate of delivery} = 2.2 \, \frac{\text{mg}}{\text{mL}} \cdot 105 \, \frac{\text{mL}}{\text{min}} = 231 \, \text{mg} \cdot \text{min}^{-1}$$

As a general rule, the renal threshold is lower than the Tm—that is, glycosuria appears at a serum concentration lower than that at which the rate of delivery into the tubular fluid actually surpasses the Tm. We do not know this patient's individual Tm for glucose reabsorption, but we can assume that the tubular load at which glycosuria appears is somewhat lower.

7. Ammonia is one of the urinary buffers; it combines with secreted hydrogen ion to form ammonium ion, which is excreted in the urine.

Sodium and bicarbonate are reabsorbed and recovered into the general circulation.

8.   Removal of one kidney places a greater burden on the remaining organ. The arterial pressure to the remaining kidney will be somewhat higher, and some part of that higher pressure will be transmitted through to the glomeruli. Because a higher intraglomerular pressure enhances the rate of filtration, GFR will probably stay within a normal range even though the patient now has only half as many nephrons as before.

9.   The normal response to dehydration is to conserve body fluid. Therefore, ADH activity will increase; as a result, a larger-than-normal volume of water will be extracted from the collecting ducts and recovered into circulation, yielding a smaller-than-normal volume of highly concentrated urine. As the urinary solute concentration rises, so does the urine specific gravity.

10.   At a serum glucose level of 550 mg/dL and a GFR of 110 mL/min:

$$\text{Rate of entry} = 5.5 \, \frac{\text{mg}}{\text{mL}} \cdot 110 \, \frac{\text{mL}}{\text{min}} = 605 \, \text{mg} \cdot \text{min}^{-1}$$

The spillover rate is the difference between this rate of entry and the Tm for reabsorption:

$$\text{Rate of spillover} = 605 \, \frac{\text{mg}}{\text{min}} - 360 \, \frac{\text{mg}}{\text{min}} = 245 \, \text{mg} \cdot \text{min}^{-1}$$

The presence of this amount of glucose in the urine acts as an *osmotic diuretic;* that is, water is pulled along, producing a large volume of urine.

Chapter 6: Acid-Base Balance

1.   Given the bicarbonate level and the carbon dioxide tension, the predicted pH is obtained by plugging the given values into the Henderson-Hasselbalch equation:

$$\text{pH} = \text{pK} + \log\left(\frac{[\text{HCO}_3^-]}{\text{PaCO}_2 \cdot 0.03}\right)$$

$$= 6.10 + \log\left(\frac{28}{48 \cdot 0.03}\right)$$

$$= 6.10 + \log\left(\frac{28}{1.44}\right)$$

$$\approx 6.10 + \log 19.4$$

$$\approx 6.10 + 1.29 = 7.39$$

The carbon dioxide tension is slightly high; however, the bicarbonate level is also slightly high, so the bicarbonate/CO_2 ratio is almost normal. As a result, the pH is normal. This condition could represent respiratory acidosis due to chronic hypercapnia, with metabolic compensation; or less likely, metabolic alkalosis with respiratory compensation. In either case, the compensatory mechanism has kept the pH well within the normal range.

2. This time, we are given the pH and the bicarbonate level, and asked to compute the carbon dioxide tension:

$$7.52 = 6.10 + \log\left(\frac{25}{PaCO_2 \cdot 0.03}\right)$$

$$1.42 = \log\left(\frac{25}{PaCO_2 \cdot 0.03}\right)$$

$$\text{Antilog } 1.42 = \frac{25}{PaCO_2 \cdot 0.03}$$

$$26.3 \approx \frac{25}{PaCO_2 \cdot 0.03}$$

$$PaCO_2 = \frac{25}{26.3 \cdot 0.03} \approx \frac{25}{0.789} \approx 31.7 \text{ mm Hg}$$

The condition is respiratory alkalosis—a high pH due to hypocapnia. No metabolic compensation is noted, which means that the condition is acute rather than chronic.

3. Similarly, if we are given the pH and the carbon dioxide tension, we can compute the bicarbonate level as follows:

$$7.37 = 6.10 + \log\left(\frac{[HCO_3^-]}{33 \cdot 0.03}\right)$$

$$1.27 = \log\left(\frac{[HCO_3^-]}{0.99}\right)$$

$$\text{Antilog } 1.27 = \frac{[HCO_3^-]}{0.99}$$

$$18.6 \approx \frac{[HCO_3^-]}{0.99}$$

$$18.6 \cdot 0.99 = [HCO_3^-] \approx 18.4 \text{ meq} \cdot L^{-1}$$

The condition is a metabolic acidosis due to the low bicarbonate level. The carbon dioxide tension is also slightly low, indicating partial respiratory compensation (the acidemia would have been worse were it not for the low carbon dioxide tension).

4. The combination of low bicarbonate and low carbon dioxide tension in a patient with acidemia (low pH) is exactly what we saw in Question 3—metabolic acidosis with partial respiratory compensation. If bicarbonate and carbon dioxide tension are both low but the pH is normal, it would mean that the respiratory response is providing full compensation for the primary metabolic disturbance.

5. The combination of a normal bicarbonate level and a high carbon dioxide tension in a patient with a low pH represents acute respiratory acidosis—acidemia due to hypercapnia, with no metabolic compensation.

6. From the given data, the bicarbonate/CO_2 ratio is:

$$\frac{20}{46 \cdot 0.03} \approx \frac{20}{1.38} \approx 14.5$$

The ratio is much lower than the normal value of about 20, so the logarithm of the ratio will be low; when that low value is added to the constant 6.10, the pH will be low, as well. In this case, the low bicarbonate level and the high carbon dioxide tension both contribute to the acidemia, and there is no compensatory mechanism. With combined respiratory and metabolic acidosis, the pH can be very low.

7. Introducing 24 mmol of weak acid HA into three liters of water creates a formulation of 8 mmol/L. At equilibrium, the measured concentration of intact compound is 7 mmol/L, which means that 1 mmol dissociated, releasing 1 mmol each of hydrogen ions and conjugate-base anions:

$$HA \rightleftharpoons H^+ + A^-$$
$$7 1 1$$

The dissociation constant is computed as follows:

$$K = \frac{[H^+] \cdot [A^-]}{[HA]} = \frac{1 \cdot 1}{7} = \frac{1}{7} \approx 0.143$$

The pK is the negative logarithm of the dissociation constant:

$$\log 0.143 \approx -0.845$$

$$-\log 0.143 \approx 0.845$$

8. With the same compound as in Question 7, introducing 24 mmol into two liters of water creates a formulation of 12 mmol/L. Using the known dissociation constant, we set up an equilibrium equation with x representing the number of millimoles of hydrogen ion and conjugate-base

anion released by the dissociation of x mmol of HA, and $12 - x$ representing the number of moles of intact compound:

$$0.143 = \frac{x \cdot x}{12 - x}$$

$$1.72 - 0.143x = x^2$$

From this quadratic equation, $x \approx 1.24$. So at 12 mmol/L, the predicted concentration for hydrogen ion and conjugate-base anion is 1.24 mmol/L; and for the intact compound, $12 - 1.24 = 10.76$ mmol/L. Note that at 8 mmol/L (as in Question 7), 1 mmol dissociates; 1 of $8 = 12.5\%$. At 12 mmol/L (as we see here), 1.24 mmol dissociates; 1.24 of $12 \approx 10.3\%$.

9. Nitric acid dissociates virtually completely, releasing equal quantities of hydrogen ion and nitrate:

$$HNO_3 \rightarrow H^+ + NO_3^-$$

Introducing 0.5 mmol in two liters of water, the formulation is 0.25 mmol/L, with a hydrogen ion concentration of 0.25 mmol/L. Therefore:

$$[H^+] = 0.25 \, \frac{mmol}{L} = 0.00025 \, \frac{mol}{L}$$

$$\log 0.00025 \approx -3.6$$

$$pH = 3.6$$

10. Carbonic acid and its salt, sodium bicarbonate, form a buffer pair. The action of a buffer is to minimize the pH change caused by the introduction of a strong acid or a strong base. If this buffer had been in the water prior to the introduction of nitric acid (as in Question 9), the new pH would not have come down so far from the pH of the water—that is, the pH would have remained somewhat closer to 7.0.

Chapter 7: Pharmacokinetics

1. For drugs with first-order elimination, half-life ($t\frac{1}{2}$) and elimination constant (K_e) are related to each other via the natural logarithm of 2:

$$t\frac{1}{2} = 0.693/K_e \quad \text{and} \quad K_e = 0.693/t\frac{1}{2}$$

If $K_e = 0.063/hr$:

$$t\frac{1}{2} = \frac{0.693}{0.063 \, hr^{-1}} = 11 \, hr$$

If $t\frac{1}{2} = 15$ hr:

$$K_e = \frac{0.693}{15 \text{ hr}} \approx 0.046 \text{ hr}^{-1}$$

2. This problem asks for the half-life, using the given information on clearance and apparent volume of distribution (V_d). Recall that the basic relationship is:

$$Cl = V_d \cdot K_e$$

Therefore, the strategy is to find K_e and then use that to calculate the half-life:

$$K_e = \frac{Cl}{V_d} = \frac{0.04 \text{ L} \cdot \text{hr}^{-1} \cdot \text{kg}^{-1}}{0.48 \text{ L} \cdot \text{kg}^{-1}} = \frac{0.04 \text{ hr}^{-1}}{0.48} \approx 0.083 \text{ hr}^{-1}$$

And, as in Problem 1:

$$t\frac{1}{2} = \frac{0.693}{0.083 \text{ hr}^{-1}} = 8.35 \text{ hr}$$

3. In this problem, we are given plasma concentrations at two different times and asked to find the elimination constant. Easy! K_e is the slope of the straight-line terminal segment of the declining plasma concentration. The slope is the change in y (natural logarithm of the plasma concentration) divided by the change in x (time):

$$K_e = \frac{\ln[\text{drug}]_2 - \ln[\text{drug}]_1}{\text{time}_2 - \text{time}_1} = \frac{\ln 10 - \ln 15}{9 \text{ hr} - 1 \text{ hr}}$$

$$\approx \frac{2.3 - 2.7}{8 \text{ hr}} = \frac{-0.4}{8 \text{ hr}} = -0.05 \text{ hr}^{-1}$$

In other words, about 5% of the amount present is removed each hour.

4. The Henderson-Hasselbalch equation gives the ionization in plasma (pH $= 7.4$):

$$\log\left(\frac{[\text{nonionized}]}{[\text{ionized}]}\right) = pK - pH = 6.7 - 7.4 = -0.7$$

Antilog $-0.7 \approx 0.20$. The ratio of nonionized drug to ionized drug is 1 to 5; then 5/6 or about 83% of the amount present in plasma is ionized. (Don't mix up the ratio with the percentage of the amount. If the ratio is 1/5, it means that for every nonionized molecule, there are 5 ionized molecules. So we are dealing with 6 molecules at a time, of which 1 is nonionized and 5 are ionized; that is, 5/6 or about 83% is ionized.) For the same drug in urine (pH = 6.2):

$$\log\left(\frac{[\text{nonionized}]}{[\text{ionized}]}\right) = pK - pH = 6.7 - 6.2 = 0.5$$

Antilog $0.5 \approx 3.2$. The ratio of nonionized drug to ionized drug is 3.2 to 1 or 16 to 5; then 5/21 or about 24% of the amount present in urine is ionized. If a drug is 33.3% ionized in plasma, the ratio of nonionized drug to ionized drug is 2 to 1. Since $pK - pH$ is the logarithm of this ratio, we can easily find the drug's pK:

$$pK - pH = \log 2$$

$$pK - 7.4 \approx 0.3$$

$$pK = 0.3 + 7.4 = 7.7$$

5. A drug's apparent volume of distribution (V_d) is affected by the extent of protein-binding in plasma and the extent of ionization that occurs. Other things being equal, V_d is larger when there is less protein-binding, because more of the drug can leave the intravascular space and enter the interstitial fluids, thus expanding V_d to the entire extracellular fluid compartment. Likewise, V_d is larger when there is less ionization, because more of the drug can cross cell membranes and enter the intracellular compartment, thus expanding V_d to the total body water. With respect to protein-binding, Drug B (50% bound) should have a larger V_d than Drug A (90% bound). With respect to ionization:

Drug A: $\log\left(\dfrac{[\text{nonionized}]}{[\text{ionized}]}\right) = pK - pH = 5.0 - 7.4 = -2.4$

Drug B: $\log\left(\dfrac{[\text{nonionized}]}{[\text{ionized}]}\right) = 8.2 - 7.4 = 0.8$

For Drug A, antilog $-2.4 \approx 0.004$; over 99% is ionized, so very little will cross cell membranes. For Drug B, antilog $0.8 \approx 6.3$; only about 14% is ionized, so a lot can cross cell membranes. By virtue of less protein-binding and less ionization, Drug B should have the larger V_d.

6. The straightforward approach to determining a drug's V_d is to divide the amount administered by the resulting plasma concentration:

$$V_d = \frac{500 \text{ mg}}{12.5 \text{ mg} \cdot L^{-1}} = 40 \text{ L}$$

7. If a patient is showing an appropriate clinical response to a drug, the free concentration is probably normal. Then if the total concentration in plasma is high, the implication is that there is an excess in carrier protein.

In this circumstance, the high total concentration is due to the higher-than-normal bound concentration. (In the opposite situation—an appropriate clinical response in the face of a low plasma concentration—the implication would be a deficit in carrier protein.)

8. A drug's AUC is determined mainly by its bioavailability, which in turn is determined by the extent (but not the rate) of absorption after an oral dose. The *shape* or contour of the curve of plasma concentration over time varies with the speed of absorption, but the *area under* that curve does not. In short, the drug with the higher bioavailability has the larger AUC.

9. For a drug with $t\frac{1}{2} = 9$ hr:

$$K_e = \frac{0.693}{9 \text{ hr}} \approx 0.077 \text{ hr}^{-1}$$

That is, about 7.7% of the amount present is removed each hour. Then if we start with 1 gram = 1000 mg, the amounts removed over the first three hours are as follows:

1st hour: $1000 - (0.077 \cdot 1000) = 1000 - 77 = 923$ mg

2nd hour: $923 - (0.077 \cdot 923) = 923 - 71 = 852$ mg

3rd hour: $852 - (0.077 \cdot 852) = 852 - 66 = 786$ mg

After three hours, more than three-fourths of the original amount is left.

10. Drug elimination involves metabolism and excretion, and the liver contributes to both of these processes. The liver is the body's main chemical processing plant, utilizing a variety of enzyme systems to metabolize drugs. In addition, the liver extracts substances out of the blood passing through it, as we saw in the "first-pass effect"—orally administered drugs that have reached the liver via the portal system are partially removed before any of the drug enters the systemic circulation.

Chapter 8: Energy and Metabolism

1. With water accounting for half the 116-gram weight of the burger, the nutritive portion is 58 grams; with protein and fat present in equal measure, there are 29 grams of each. The bun adds a gram of protein and of fat, so the total is 30 grams each. The carbohydrate total is 72 grams (27 from the bun, 45 from the soda). The meal provides a total of 690 kilocalories:

Carbohydrate: $72 \text{ g} \cdot 4.0 \text{ kcal/g} = 288 \text{ kcal}$

Protein: $30 \text{ g} \cdot 4.0 \text{ kcal/g} = 120 \text{ kcal}$

Fat: $30 \text{ g} \cdot 9.4 \text{ kcal/g} = 282 \text{ kcal}$

2. Calculating the percentage of kilocalories derived from each nutrient:

Carbohydrate: $288/690 \approx 0.42 = 42\%$

Protein: $120/690 \approx 0.17 = 17\%$

Fat: $282/690 \approx 0.41 = 41\%$

In fact, fast-food burgers are often considerably higher in fat content.

3. The oxygen consumption rate is measured as the product of the ventilatory volume per minute and the difference in percentage of oxygen between inspired and expired air. Assuming 21% oxygen in inspired air, the oxygen difference is $21\% - 16\% = 5\%$. Oxygen consumption at the given ventilation rate of 45 L/min is calculated as follows:

$$\dot{V}_{O_2} = \frac{45\ L}{min} \cdot 0.05 = 2.25\ L \cdot min^{-1}$$

4. In a 35-year-old sedentary male, expected $\dot{V}_{O_2}max$ is computed as:

$$57.8 - (0.445 \cdot 35) \approx 57.8 - 15.6 = 42.2\ mL \cdot kg^{-1} \cdot min^{-1}$$

Since the actual $\dot{V}_{O_2}max$ is given as $37.5\ mL \cdot kg^{-1} \cdot min^{-1}$, the degree of functional impairment is calculated as follows:

$$Impairment = \frac{42.2 - 37.5}{42.2} = \frac{4.7}{42.2} = 0.11 = 11\%$$

5. Basal metabolic rate based on oxygen consumption measured at 0.2 L/min:

$$BMR = 0.2\ \frac{L}{min} \cdot \frac{1440\ min}{day} \cdot \frac{4.825\ kcal}{L} \approx 1390\ kcal \cdot day^{-1}$$

6. If basal $\dot{V}_{O_2}$ is 240 mL/min and ventilatory volume is $6\ L = 6000\ mL$ per minute:

$$240\ \frac{mL}{min} = \frac{6000\ mL}{min} \cdot (0.21 - expired\ O_2\ content)$$

$$\frac{240\ mL \cdot min^{-1}}{6000\ mL \cdot min^{-1}} = 0.21 - expired\ O_2\ content$$

$$0.04 = 0.21 - expired\ O_2\ content$$

$$Expired\ O_2\ content = 0.21 - 0.04 = 0.17 = 17\%$$

7. Expected maximum heart rate in a 38-year-old man = $220 - 38 = 182$ beats per minute. At 75% of this rate (137 beats per minute), his oxygen consumption per minute is given as 2 L = 2000 mL. Since he weighs 154 lb:

$$154 \text{ lb} \cdot \frac{\text{kg}}{2.2 \text{ lb}} = 70 \text{ kg}$$

Determining $\dot{V}_{O_2}$ per kilogram of body weight:

$$\dot{V}_{O_2} = \frac{2000 \text{ mL} \cdot \text{min}^{-1}}{70 \text{ kg}} \approx 28.6 \text{ mL} \cdot \text{kg}^{-1} \cdot \text{min}^{-1}$$

In METS:

$$\frac{28.6 \text{ mL} \cdot \text{kg}^{-1} \cdot \text{min}^{-1}}{3.5 \text{ mL} \cdot \text{kg}^{-1} \cdot \text{min}^{-1} \cdot \text{MET}^{-1}} \approx 8.2 \text{ METS}$$

8. First, we determine the expected $\dot{V}_{O_2}$max in a sedentary 35-year-old woman:

$$42.3 - (0.356 \cdot 35) \approx 42.3 - 12.5 = 29.8 \text{ mL} \cdot \text{kg}^{-1} \cdot \text{min}^{-1}$$

If her $\dot{V}_{O_2}$max is 6% above this expected value:

$$29.8 \cdot 1.06 \approx 31.6 \text{ mL} \cdot \text{kg}^{-1} \cdot \text{min}^{-1}$$

In METS:

$$\frac{31.6 \text{ mL} \cdot \text{kg}^{-1} \cdot \text{min}^{-1}}{3.5 \text{ mL} \cdot \text{kg}^{-1} \cdot \text{min}^{-1} \cdot \text{MET}^{-1}} \approx 9.0 \text{ METS}$$

9. If basal $\dot{V}_{O_2} = 206$ mL/min and the woman weighs 52 kg, BMR in METS is calculated as follows:

$$\dot{V}_{O_2} = \frac{206 \text{ mL} \cdot \text{min}^{-1}}{52 \text{ kg}} \approx 3.96 \text{ mL} \cdot \text{kg}^{-1} \cdot \text{min}^{-1}$$

$$\frac{3.96 \text{ mL} \cdot \text{kg}^{-1} \cdot \text{min}^{-1}}{3.5 \text{ mL} \cdot \text{kg}^{-1} \cdot \text{min}^{-1} \cdot \text{MET}^{-1}} \approx 1.13 \text{ METS}$$

10. There is no way to predict who has the higher basal metabolic rate. The man might have the higher rate on the basis of gender and living in a cold environment; or the lower rate on the basis of being short, heavy, and sedentary.

1. Since the patient's weight on Scale A is given in pounds, we must convert to grams for comparison with the other data. Using the converting factor 2.2 lb/kg:

$$149.8 \text{ lb} \cdot \frac{\text{kg}}{2.2 \text{ lb}} \approx 68.1 \text{ kg}$$

$$150.1 \text{ lb} \cdot \frac{\text{kg}}{2.2 \text{ lb}} \approx 68.2 \text{ kg}$$

$$150.3 \text{ lb} \cdot \frac{\text{kg}}{2.2 \text{ lb}} \approx 68.3 \text{ kg}$$

Compared to the confirmed weight (68,236 g = 68.236 kg), these readings are all correct to within 0.2 kg, and they encompass a range of 0.2 kg. The readings on Scale B (68.841, 68.841, and 68.842 kg) are correct only to within 0.7 kg, but the range they encompass is smaller, 0.001 kg. In quality-control terms, Test A is more accurate (results are closer to the true result) but Test B is more precise (results are more consistent).

2. It is *not* possible for a nonspecific test to yield an accurate result, because data representing entities other than the intended object of measurement will be included. It *is* possible for an imprecise test to yield an accurate result once but not repeatedly; in that case, the accuracy is by chance. It is *not* possible for an imprecise test to be considered valid, because validity means that a test is precise, as well as specific and accurate.

3. The diagnostic test is positive in 398 of 500 people who are known to have the condition, so there are 102 false negatives. *Sensitivity* is the number of true positives divided by the sum of true positives and false negatives:

$$\text{Sensitivity} = \frac{\text{true positive}}{\text{true positive} + \text{false negative}}$$

$$= \frac{398}{398 + 102} = \frac{398}{500} = 79.6\%$$

The same test is negative in 428 of 500 people who are known to be free of the condition in question, so there are 72 false positives. *Specificity* is the number of true negatives divided by the sum of true negatives and false positives:

$$\text{Specificity} = \frac{\text{true negative}}{\text{true negative} + \text{false positive}}$$

$$= \frac{428}{428 + 72} = \frac{428}{500} = 85.6\%$$

4. We have not been given any information about the prevalence of the condition, so we know nothing about the test's predictive values.

5. The disease prevalence in the general population is given as $6.6 \cdot 10^{-3}$. Therefore, if the test is conducted in 100,000 people:

$$100,000 \cdot \frac{6.6}{1000} = 660$$

That is, there should be approximately 660 people who have the disease, leaving 99,340 people who do not. If the test sensitivity is 85%, then we expect the following results among the 660 people with the disease:

$$660 \cdot 0.85 = 561 \text{ true positives}$$

$$660 - 561 = 99 \text{ false negatives}$$

Similarly, if the test specificity is 72%, then among the 99,340 people who do not have the disease:

$$99,340 \cdot 0.72 \approx 71,525 \text{ true negatives}$$

$$99,340 - 71,525 = 27,815 \text{ false positives}$$

The predictive values can now be computed:

$$\text{Positive predictive value} = \frac{\text{true positive}}{\text{true positive} + \text{false positive}}$$

$$= \frac{561}{561 + 27,815} = \frac{561}{28,376} \approx 0.0198 = 1.98\%$$

$$\text{Negative predictive value} = \frac{\text{true negative}}{\text{true negative} + \text{false negative}}$$

$$= \frac{71,525}{71,525 + 99} = \frac{71,525}{71,624} \approx 0.999 = 99.9\%$$

6. Predictive values are different when the test is used in patients with signs and symptoms of the disease, because the prevalence is much higher in this group. If the test is conducted in 1000 people suspected of having the condition and the prevalence is $4.0 \cdot 10^{-1}$ in this population:

$$1000 \cdot \frac{4.0}{10} = 400$$

There should be approximately 400 people who have the disease, leaving 600 people who do not. At 85% sensitivity:

$$400 \cdot 0.85 = 340 \text{ true positives}$$

$$400 - 340 = 60 \text{ false negatives}$$

And at 72% specificity:

$$600 \cdot 0.72 = 432 \text{ true negatives}$$

$$600 - 432 = 168 \text{ false positives}$$

Therefore:

$$\text{Positive predictive value} = \frac{340}{340 + 168} = \frac{340}{508} \approx 0.669 = 66.9\%$$

$$\text{Negative predictive value} = \frac{432}{432 + 60} = \frac{432}{492} \approx 0.878 = 87.8\%$$

These results dramatically demonstrate a basic principle about the use of diagnostic tests. When the prevalence is low (as in the general population), a positive result may very well turn out to be false; but when the prevalence is high, a positive result is more likely to be true. For this reason, tests should be used mainly in situations in which there is good clinical reason to suspect that the condition in question is present.

7. A disease marker that is present at a titer of 1:32 exceeds the minimum diagnostic criterion of 1:16, so this result is positive. The question is whether this positive result is a true positive, and we do not know for sure. Aside from a gold standard test, we assume that all tests sometimes yield false positive results (that is, specificity < 100%). Therefore, this positive test result by itself is not conclusive proof that the patient has the disease in question.

8. The atomic weight of iron is given as 55.85; therefore, 55.85 μg = 1 μmol. If the serum iron level is 94 μg/dL:

$$94 \, \frac{\mu\text{g}}{\text{dL}} \cdot \frac{\mu\text{mol}}{55.85 \, \mu\text{g}} \cdot \frac{10 \, \text{dL}}{\text{L}} \approx 16.83 \, \mu\text{mol} \cdot \text{L}^{-1}$$

We multiplied the weight-in-micrograms per deciliter by 10 to get the weight per liter, and divided by 55.85 to get the number of micromoles per liter. In one step, we could divide by 5.585 or multiply by the reciprocal 5.585:

$$94 \, \frac{\mu\text{g}}{\text{dL}} \cdot \frac{1}{5.585} \approx 94 \cdot 0.1791 \approx 16.84 \, \mu\text{mol} \cdot \text{L}^{-1}$$

For direct conversion of an iron concentration given as a weight concentration in $\mu g/dL$ to SI units of $\mu mol/L$, multiply the weight concentration by 0.1791.

9. To compare the creatinine concentrations, they must be in the same units. Converting weight concentration to SI units, 1.7 mg/dL turns out to be higher than 95 $\mu mol/L$:

$$1.7 \frac{mg}{dL} \cdot \frac{mmol}{113\ mg} \cdot \frac{1000\ \mu mol}{mmol} \cdot \frac{10\ dL}{L} \approx 150\ \mu mol \cdot L^{-1}$$

By the way, the direct multiplication factor for this conversion is the product of the three conversion factors used: $1{,}000 \cdot 10 \cdot 113^{-1} \approx 88.5$; thus, $1.7 \cdot 88.5 \approx 150$. Going in the opposite direction, 95 $\mu mol/L$ is lower than 1.7 mg/dL:

$$95 \frac{\mu mol}{L} \cdot \frac{113\ \mu g}{\mu mol} \cdot \frac{mg}{1000\ \mu g} \cdot \frac{L}{10\ dL} \approx 1.07\ mg \cdot dL^{-1}$$

10. Screening requires a highly sensitive test, since the goal is detection. Highly specific tests are typically less sensitive, resulting in many false negatives. Also, if the prevalence of the condition in question is low in the general population, the positive predictive value is poor. However, the prevalence is higher in the select group whose screening tests are positive, and using a highly specific test to confirm these screening positives yields a much better positive predictive value; that is, a positive result is much more likely to be a true positive.

Chapter 10: Biostatistics

1. We are given two sets of data on age at menarche, but they are not in numeric order. First, we'll arrange the data in ascending sequence (as needed for the median):

    Set #1:   11.7, 11.9, 12.3, 12.8, 12.8, 12.9, 13.3, 13.5, 14.0

    The mean is the sum of these data divided by the number of data:

    $$\text{Mean} = \frac{115.2}{9} \approx 12.8 \text{ years}$$

    The median is that datum whose position is given by $(N+1)/2$. In this case, $(9+1)/2 = 10/2 = 5$. The median is the fifth datum, 12.8 years. There is a mode at 12.8 years (reported twice; however, this may be an artifact due to small sample size). The fact that all three measures of central tendency are the same in this set is coincidental.

    Set #2:   10.8, 11.1, 11.7, 11.9, 12.2, 12.3, 13.0, 13.1, 14.4, 14.6

    $$\text{Mean} = \frac{125.1}{10} \approx 12.5 \text{ years}$$

For the median, $(10+1)/2 = 11/2 = 5.5$. Taking the mean of the fifth and sixth data, $(12.2+12.3)/2 = 24.5/2 = 12.25$ years. This data set has no mode.

2. We start by computing the variance, for which we square the difference between each datum and the mean of the set. For Set #1:

$$(11.7 - 12.8)^2 = (-1.1)^2 = 1.21$$

$$(11.9 - 12.8)^2 = (-0.9)^2 = 0.81$$

$$(12.3 - 12.8)^2 = (-0.5)^2 = 0.25$$

$$(12.8 - 12.8)^2 = (0)^2 = 0$$

$$(12.8 - 12.8)^2 = (0)^2 = 0$$

$$(12.9 - 12.8)^2 = (0.1)^2 = 0.01$$

$$(13.3 - 12.8)^2 = (0.5)^2 = 0.25$$

$$(13.5 - 12.8)^2 = (0.7)^2 = 0.49$$

$$(14.0 - 12.8)^2 = (1.2)^2 = 1.44$$

The sum of these squared differences is 4.46, and $N = 9$. Now we can compute the variance:

$$s^2 = \frac{\sum (X_i - \overline{X})^2}{N - 1} \approx \frac{4.46}{8} \approx 0.558$$

The standard deviation is the positive square root of the variance:

$$s = \sqrt{0.558} \approx 0.747$$

The standard error of the mean is the standard deviation divided by the square root of the number of data:

$$\text{SEM} = \frac{s}{\sqrt{N}} = \frac{0.747}{\sqrt{9}} = \frac{0.747}{3} = 0.249$$

The coefficient of variation is the standard deviation divided by the mean:

$$\text{Coefficient of variation} = \frac{s}{\overline{X}} = \frac{0.747}{12.8} \approx 0.0584 = 5.84\%$$

The z-score for the youngest age in the set is found by subtracting the mean from that score and dividing the difference by the standard deviation:

$$z_{11.7} = \frac{X - \overline{X}}{s} = \frac{11.7 - 12.8}{0.747} = \frac{-1.1}{0.747} \approx -1.47$$

Similarly, for Set #2, the sum of the squared differences is approximately 14.6 and $N = 10$. Therefore:

$$s^2 = \frac{\sum (X_i - \overline{X})^2}{N - 1} \approx \frac{14.6}{9} \approx 1.62$$

$$s = \sqrt{1.62} \approx 1.27$$

$$\text{SEM} = \frac{s}{\sqrt{N}} = \frac{1.27}{\sqrt{10}} \approx \frac{1.27}{3.16} \approx 0.402$$

$$\text{Coefficient of variation} = \frac{s}{\overline{X}} = \frac{1.27}{12.5} \approx 0.102 = 10.2\%$$

$$z_{10.8} = \frac{X - \overline{X}}{s} = \frac{10.8 - 12.5}{1.27} = \frac{-1.7}{1.27} \approx -1.34$$

3. Age at menarche is an example of parametric data (normal distribution in a larger set); therefore, we can employ the t-test. The first step is to determine the combined standard deviation for the two sets of data:

$$s_c = \sqrt{\frac{\sum_1 (X_i - \overline{X}_1)^2 + \sum_2 (X_i - \overline{X}_2)^2}{N_1 + N_2 - 2}} = \sqrt{\frac{4.46 + 14.6}{9 + 10 - 2}}$$

$$\approx \sqrt{\frac{19.1}{17}} \approx \sqrt{1.12} \approx 1.06$$

Now this value is used in the t-test formula:

$$t = \frac{\overline{X}_1 - \overline{X}_2}{s_c \sqrt{\dfrac{1}{N_1} + \dfrac{1}{N_2}}} = \frac{12.8 - 12.5}{1.06 \sqrt{\dfrac{1}{9} + \dfrac{1}{10}}} = \frac{0.3}{1.06 \sqrt{\dfrac{19}{90}}}$$

$$\approx \frac{0.3}{1.06 \sqrt{0.21}} \approx \frac{0.3}{1.06 \cdot 0.46} \approx \frac{0.3}{0.49} \approx 0.616$$

In the question, the probability table excerpt is part of the t-table, for the appropriate df ($N_1 + N_2 - 2 = 17$):

df	Value of p:			
	0.50	0.10	0.05	0.01
17	0.689	1.74	2.11	2.90

For $t = 0.616$, $p > 0.5$ (not even close to significant). The Null Hypothesis stands; there is no meaningful difference in age at menarche between the two groups.

4. The pre- and posttreatment data are shown as mean $\pm$ standard error (SEM) instead of mean $\pm$ standard deviation (s). Since $SEM = s/N^{1/2}$, $s = SEM \cdot N^{1/2}$. In this case, $s = SEM \cdot 25^{1/2} = SEM \cdot 5$. The pretreatment mean is not 243 ± 6, but 243 ± 30 mg/dL; the posttreatment mean is not 190 ± 9, but 190 ± 45. Reporting the data as mean $\pm$ SEM falsely suggests tighter clustering than actually exists.

5. If 3200 new cases of a condition are diagnosed over a five-year period, the average annual number of new cases is $3200/5 = 640$. Then the annual incidence in a population of 1,000,000 is:

$$\text{Incidence} = \frac{640}{10^6} = \frac{6.4}{10^4} = 6.4 \cdot 10^{-4}$$

That is, 6.4 new cases per 10,000 people in the defined population, annually. From the given data, we cannot say anything about the prevalence halfway through the five-year period; we would need to know how many people who had developed the condition at *any* time earlier still had it at that point in time.

6. In this question, we have to work backward from a given p value to find the corresponding data difference. In the coin-toss model, df $= 1$, and $p = 0.05$ corresponds to a χ^2 value of 3.841. Then for a trial of 64 tosses:

$$3.841 = \frac{(\text{greater} - \text{lesser})^2}{64}$$

$$3.841 \cdot 64 = (\text{greater} - \text{lesser})^2$$

$$245.8 = (\text{greater} - \text{lesser})^2$$

$$\sqrt{245.8} \approx 16 = \text{greater} - \text{lesser}$$

The expected numbers are 32 heads and 32 tails, so a difference of 16 means either 24 heads and 40 tails or 40 heads and 24 tails as the smallest statistically significant heads-tails difference in a trial of 64 coin tosses.

7. The 95% confidence interval is defined by the mean ± 1.96 SEM. We can compute SEM from the data given (mean $= 76 \pm 9.4$, $N = 53$):

$$\text{SEM} = \frac{s}{\sqrt{N}} = \frac{9.4}{\sqrt{53}} \approx \frac{9.4}{7.3} \approx 1.3$$

Therefore:

$$\text{Cl}_{95} = 76 \pm (1.96 \cdot 1.3) \approx 76 \pm 2.5$$

(That is, a 95% chance that the true population mean lies somewhere within the range of values from 73.5 to 78.5.)

8. In a set of normally distributed data from 625 subjects, mean $= 100 \pm 16$. Then a z-score of -3.00 refers to a value three standard deviations below the mean:

$$100 - (3 \cdot 16) = 100 - 48 = 52$$

Scores of 84 and 132 are, respectively, $1s$ below the mean ($100 - 16 = 84$) and $2s$ above the mean ($100 + 32 = 132$). Referring to Figure 10.15, the interval between $-1s$ and the mean includes approximately 34% of the data in the set; the interval between the mean and $+2s$ includes approximately 47.5% of the data. Therefore, the combined interval from the score of 84 (at $-1s$) to 132 (at $+2s$) includes approximately 81.5% of the data. Of the 625 subjects, the number with scores within this range is:

$$625 \cdot 0.815 \approx 509$$

9. From the information provided, the design of this study is so flawed that an observed difference in divorce rates in patients with depression versus other psychiatric diagnoses cannot be used to draw any inferences or conclusions, even if the difference is statistically significant. Apart from uncertainties about the random sampling procedure, there is no justification for assuming that data obtained from patients seen in private psychiatric practices in one community are representative of patients seen in clinic settings or in other communities; much less that these data are applicable to the general population. Given these design flaws, the conclusion of a meaningful correlation between depression and divorce may well represent a Type I error (inappropriately rejecting the Null Hypothesis). The fact that this study was flawed and therefore inconclusive does not disprove any such correlation; but it will take a well-designed study to demonstrate a meaningful correlation.

10. By artificially arranging the data on test scores (variable X) and heights (variable Y) in ascending numeric order, we fabricated a spurious positive correlation as shown by the Spearman rank test. Now we are using these arranged data in the Pearson product-moment test:

Subject	X	$X - \overline{X}$	Y	$Y - \overline{Y}$	$(X - \overline{X}) \cdot (Y - \overline{Y})$
AB	12	−50.3	155	−16	+804.8
CD	29	−33.3	156	−15	+499.5
EF	64	+1.7	160	−11	−18.7
GH	66	+3.7	168	−3	−11.1
IJ	66	+3.7	169	−2	−7.4
KL	67	+4.7	173	+2	+9.4
MN	68	+5.7	175	+4	+22.8
OP	70	+7.7	176	+5	+38.5
QR	70	+7.7	178	+7	+53.9
ST	74	+11.7	181	+10	+117.0
UV	99	+36.7	186	+15	+550.5

The sum of the difference products is approximately 2059. As previously shown, the standard deviations for the data on test scores and heights are 23.1 and 10.1, respectively; $N = 11$. Therefore:

$$r = \frac{\sum [(X_i - \overline{X}) \cdot (Y_i - \overline{Y})]}{s_x \cdot s_y \cdot (N - 1)} = \frac{2059}{23.1 \cdot 10.1 \cdot 10} \approx 0.88$$

Recall that $r = 1$ indicates perfect positive correlation; so we successfully fabricated a strong positive correlation by arranging both data sets in ascending order. Now arrange the data in opposite order (test scores in ascending order, heights in descending order) to fabricate a strong negative correlation. When you apply the Pearson formula, the value of r should be close to −1.

Index

Pascal (Pa), 16, 66, 97, 239
Torr, 97, 239

METRIC-NONMETRIC CONVERSIONS

MATHEMATICAL, SCIENTIFIC, AND CLINICAL CONCEPTS